Bluff-Body Wakes, Dynamics and Instabilities

International Union of Theoretical and Applied Mechanics

H. Eckelmann, J. M. R. Graham,
P. Huerre, P. A. Monkewitz (Eds.)

Bluff-Body Wakes, Dynamics and Instabilities

IUTAM Symposium, Göttingen, Germany
September 7-11, 1992

Springer-Verlag
Berlin Heidelberg New York
London Paris Tokyo
Hong Kong Barcelona Budapest

Prof. Dr. Helmut Eckelmann
Institut für Angewandte Mechanik und
Strömungsphysik der Universität Göttingen
und Max-Planck-Institut für Strömungsforschung
Bunsenstraße 10
37073 Göttingen
Germany

Prof. Dr. J. Michael R. Graham
Department of Aeronautics
Imperial College of Science, Technology and Medicine
South Kensington, London SW7 2BY
United Kingdom

Prof. Dr. Patrick Huerre
Laboratoire d'Hydrodynamique (LADHYX)
Ecole Polytechnique
91128 Palaiseau Cedex
France

Prof. Dr. Peter A. Monkewitz
Mechanical, Aerospace and Nuclear Engineering Dept.
University of California, Los Angeles
405 Hilgard Ave.
Los Angeles, CA 90024-1597
USA

ISBN 978-3-662-00416-6 **ISBN 978-3-662-00414-2 (eBook)**
DOI 10.1007/978-3-662-00414-2

Softcover reprint of the hardcover 1st edition 1993

Typesetting: Camera ready by authors

61/3020-5 4 3 2 1 0 – Printed on acid-free paper.

Scientific Committee

P.W. Bearman (UK)
E. Berger (Germany)
H. Eckelmann (Germany), Chairman
J.H. Gerrard (UK)
J.M.R. Graham (UK), Chairman
P. Huerre (France)
M. Kiya (Japan)
V.Ya. Levchenko (Russia)
P.A. Monkewitz (U.S.A.)
G.V. Parkinson (Canada)
A.E. Perry (Australia)
A. Roshko (U.S.A.)
K.R. Sreenivasan (U.S.A.)
L. van Wijngaarden (Netherlands)

Local Organizing Committee

H. Eckelmann
U. Eberhardt
M. König
B.R. Noack
F. Ohle (Conference Secretary)

Symposium Sponsors

International Union of Theoretical and Applied Mechanics (IUTAM)
DANTEC Elektronische Meßgeräte GmbH, Ettlingen, Germany
Druckerei und Verlag Martin Saß & Co., Göttingen, Germany
Sparkasse Göttingen, Germany
TSI Optische und elektron. Meßtechnik für Forschung und Industrie, Aachen, Germany
Volkswagen AG, Wolfsburg, Germany
Prof. John R. Weske, Sandy Spring, MD, USA

Preface

The idea for the IUTAM Symposium on *Bluff-Body Wakes, Dynamics and Instabilities* originated at a workshop organized at Lehigh University in May 1989 in the same subject area. At this workshop, which was arranged with the support of the United States Office of Naval Research, it was felt that the topic was of considerable general interest and deserved discussion at an international meeting with a much wider participation. As a result the two co-chairmen of the present meeting applied to IUTAM with a proposal, which was granted, for a Symposium under IUTAM auspices. The planning for the Symposium which was held in Göttingen from 7-11 September 1992 at the Max-Planck-Institut für Strömungsforschung began in October 1990.

The earlier view about the wide interest in bluff-body wakes and their instabilities has been fully justified by the number of papers submitted to the committee and the number of participants who attended the Symposium. These numbers were larger than had been expected when the meeting was planned and as a result about half of the good quality papers which were selected for the Symposium had to be presented in two poster sessions, introduced by short oral presentations to the meeting beforehand. In addition to the invited lectures, oral paper presentations and poster presentations, a session was given over to the assessment of the performance of computational fluid dynamics codes against well documented experimental results for the wakes of circular cylinders. This was achieved by requests to appropriate participants to bring results of computations or experiments for a specified range of conditions and in a common format to the meeting.

It was apparent at the Symposium that significant progress had been made in the understanding of bluff-body wake phenomena. The first sessions were concerned with the structure of vortex street wakes, numerical simulation of these wakes at both low and high Reynolds numbers and the effect of gross perturbations on vortex wakes. Numerical simulations up to Reynolds numbers of 10^6 were shown but it was clear from the discussion about these at the time and in the later session dealing with comparisons between numerical simulation and experiment that important questions still remain. Two dimensional simulations cannot be regarded yet as valid representations above the specific low Reynolds number two–dimensional shedding regime and three–dimensional simulations cannot be taken much beyond this regime with accuracy because of limitations of grid size/computer power. However, considerable progress has been made in numerical simulation. This, together with the greater understanding of the low Reynolds number modes of vortex shedding from cylinders and the ability to control two–dimensionality has led to a good convergence of numerical and experimental results in this regime. Later sessions dealt with stability theory. The understanding of the roles of absolute and convective instability in wakes has developed significantly in the last few years. As a result of this it has become apparent that the prediction of instability is more complicated than was believed at the earlier meeting and many challenges still remain for theoreticians in this area. Solutions of the Ginsberg–Landau equation have been used to reveal many of the features of oblique shedding and cellular patterns obtained by experimenters under controlled conditions. The development of large scales and three–dimensional structures in the far wake is also better understood. Presentations were also made on wake control at both low and higher Reynolds numbers through the action of acoustic input, body motion and spanwise perturbations among others. The phenomenon of lock-in was not, however, a main theme of the meeting since this has been extensively studied in many other conferences over the last two or three decades. Many experimental techniques were presented but it is clear that a technique with great promise for quantitative elucidation of vortex wake phenomena is particle image velocimetry, and some impressive results taken in vortex wakes using this technique were presented at the meeting.

The Symposium was also felt to be successful in terms of communication between groups working in North America, Europe, Asia and Australia. The larger number of people from former Eastern bloc countries, in particular Russia, was welcome. This has not occurred in many previous meetings on vortex wakes. The success of the meeting would have been impossible without the members of the Scientific Committee who assisted by word and deed. Their help is highly acknowledged.

The organizing and execution of the Symposium was performed by the Max-Planck-Institut für Strömungsforschung and by the Institut für Angewandte Mechanik und Strömungsphysik der Georg-August-Universität in Göttingen. The detailed organizing up to and during the Symposium was in large part due to Ursula Eberhardt, Michael König, Bernd R. Noack and Frank Ohle (Symposium Secretary). Many other persons from both Institutes contributed to the success of the Symposium. In particular, we are indebted to Martin Brede, Uwe Fey, Benoît Gossieaux, Frank Jacobitz, Christian Klein, Jürgen Knorr, Karl-Heinz Nörtemann, Martin Schäfer, Burkhard Thiele and Matthias Weiland. Last but not least we want to thank Edith Rohrmoser who took care of the social programme and of the ladies programme.

There was a reception by the mayor in the ancient city hall of Göttingen and at "half time" of the Symposium an excursion to Wernigerode, an old town on the eastern side of the Harz mountains with a castle and many half-timbered houses. On the way back to

Göttingen the remains of the former "iron curtain" were inspected. This frontier had not only divided Germany but also the world into an eastern and western hemisphere. The Symposium was terminated by visits of the facilities of the Max-Planck-Institut and of the Deutsche Forschungsanstalt für Luft- und Raumfahrt (DLR, Göttingen).

The Chairmen and Scientific Committee of this Symposium are very grateful to IUTAM for making this meeting possible. Very helpful financial support for the Symposium was provided by IUTAM, by the above listed companies and institutions and by Prof. John R. Weske. We would like to thank them very much.

Finally, we appreciate the help and efficiency of the Springer Verlag for the publication of the Symposium proceedings.

Göttingen, November 1992

Helmut Eckelmann, J. Michael R. Graham

List of Participants

B.K. Ahlborn	*CDN*	Department of Physics University of British Columbia Vancouver, British Columbia V6Z 1Z1
R.H. Arkell	*UK*	Department of Aeronautics Imperial College of Science, Technology and Medicine Prince Consort Road London SW7 2BY
G.K.F. Bärwolff	*D*	Strömungstechnisches Labor Berlin-Adlershof (WIP/KAI e.V.) Rudower Chaussee 6 D-1199 Berlin
S.P. Bardakhanov	*Russia*	Inst. of Theor. and Applied Mechanics Sibirian Division of Russian Academy of Sciences 630090 Novosibirsk Akademgorodok
P.W. Bearman	*UK*	Department of Aeronautics Imperial College of Science, Technology and Medicine London SW7 2BY
H. Bestek	*D*	Institut für Aero- und Gasdynamik Universität Stuttgart Pfaffenwaldring 21 D-7000 Stuttgart 80
G. Bosch	*D*	Institut für Hydromechanik Universität Karlsruhe Kaiserstr. 12 D-7500 Karlsruhe
M. Braza	*F*	Institut de Mécanique des Fluides de Toulouse Avenue du Professeur Camille Soula F-31400 Toulouse Cedex
M. Brede	*D*	Max-Planck-Institut für Strömungsforschung Bunsenstr. 10 D-3400 Göttingen

G.L. Brown	*USA*	Department of Mechanical and Aerospace Engineering Princeton University Princeton, NJ 08540
M.K. Bull	*AUS*	Mechanical Engineering Dept. University of Adelaide G.P.O. Box 498 Adelaide, S.A. 5000
I.P. Castro	*UK*	Dept. of Mechanical Engineering Surrey University Guildford, Surrey GU2 5XH
J.-L. Champion	*F*	Laboratoire de Mécanique des Fluides Université de Poitiers 40, Avenue du Recteur Pineau F-86022 Poitiers Cedex
J.R. Chaplin	*UK*	Department of Civil Engineering City Univerisity London Northampton Square London EC1V 0HB
T.C. Corke	*USA*	Mechanical and Aerospace Engineering Department Illinois Institute of Technology 3110 S. State St. Chicago, IL 60616
U.C. Dallmann	*D*	Institut für Theoretische Strömungsmechanik der DLR Bunsenstr. 10 D-3400 Göttingen
A. Diese	*D*	GKSS Forschungszentrum Geesthacht Institut für Anlagentechnik Abt. AM Max-Planck-Str. D-2054 Geesthacht
E.G. Duell	*USA*	Department of Mechanical and Aerospace Engineering Cornell University Upson Hall, Box 28 Ithaca, NY 14853
H. Eisenlohr	*D*	BFI Automation Dipl.-Ing. K.-H. Mindermann GmbH. Eggerscheidter Str. 112 D-4030 Ratingen 6 (Hösel)

H. Eckelmann	*D*	Institut für Angewandte Mechanik und Strömungsphysik Georg-August-Universität Göttingen Bunsenstr. 10 D-3400 Göttingen
U. Fey	*D*	Max-Planck-Institut für Strömungsforschung Bunsenstr. 10 D-3400 Göttingen
J.F. Foss	*USA*	Department of Mechanical Engineering Michigan State University B100 Research Complex - Engineering East Lansing, MI 48824
G. Gerbeth	*D*	Institut für Sicherheitsforschung Forschungszentrum Rossendorf e.V. Postfach 19 D-8051 Dresden
J.H. Gerrard	*UK*	Department of Engineering Simon Engineering Laboratory University of Manchester Oxford Rd. Manchester M13 9PL
K. Gersten	*D*	Institut für Thermo- und Fluiddynamik Lehrstuhl für Strömungsmechanik Ruhr-Universität Bochum Postfach 10 21 48 D-4630 Bochum 1
M. Gharib	*USA*	Department of Applied Mathematics and Engineering Sciences University of California at San Diego 9500 Gilman Drive La Jolla, CA 92093-0411
T.J. Gieseke	*USA*	Mechanical Engineering Department Ohio State University 206 W. 18th Avenue Columbus, OH 43210
J.M.R. Graham	*UK*	Department of Aeronautics Imperial College of Science, Technology and Medicine London SW7 2BY
Y.G. Guezennec	*USA*	Mechanical Engineering Department Ohio State University 206 W. 18th Avenue Columbus, OH 43210

V.A. Gushchin	*Russia*	Institute of Computer Aided Design Russian Academy of Sciences 19/18, 2nd Brestskaya Moscow, 123 865
F.R. Hama	*D*	Wilramstr. 17/0 D-8000 München 80
M. Hammache	*USA*	Institute for Nonlinear Science University of California at San Diego La Jolla, CA 92093-0402
T. Hayashi	*J*	Department of Mechanical Engineering Tottori University Minami 4-101, Koyama, Tottori 680
F. Hayot	*USA*	Department of Physics Ohio State University 174 W. 18th Avenue Columbus, OH 43210
H. Higuchi	*USA*	Department of Mechanical and Aerospace Engineering Syracuse University 141 Link Hall Syracuse, NY 13244-1240
W.J. Hiller	*D*	Max-Planck-Institut für Strömungsforschung Bunsenstr. 10 D-3400 Göttingen
C.-M. Ho	*USA*	Department of Mechanical, Aerospace and Nuclear Engineering University of California, Los Angeles 405 Hilgard Avenue Los Angeles, CA 90024–1597
P. Huerre	*F*	Laboratoire d'Hydrodynamique Ecole Polytechnique Route de Saclay F-91128 Palaiseau Cedex
R.E. Hunt	*UK*	Department of Applied Mathematics and Theoretical Physics University of Cambridge Silver Street Cambridge CB2 9EW

F. Jacobitz	*D*	Max-Planck-Institut für Strömungsforschung Bunsenstr. 10 D-3400 Göttingen
P. Jonáš	*CS*	Institute of Thermomechanics Czechoslovak Academy of Sciences Dolejškova 5 CS-182 00 Praha 8
G.E. Karniadakis	*USA*	Department of Mechanical and Aerospace Engineering Princeton University Princeton, NJ 08540
C.R. Kaykayoğlu	*TR*	Faculty of Aeronautics and Astronautics Istanbul Technical University Maslak Istanbul 80626
M. Kiya	*J*	Department of Mechanical Engineering Hokkaido University Sapporo, 060
Ch. Klein	*D*	Max-Planck-Institut für Strömungsforschung Bunsenstr. 10 D-3400 Göttingen
W. Koch	*D*	Institut für Theoretische Strömungsmechanik der DLR Bunsenstr. 10 D-3400 Göttingen
V.V. Kozlov	*Russia*	Inst. of Theor. and Applied Mechanics Siberian Division of Russian Academy of Sciences 630090 Novosibirsk Akademgorodok
M. König	*D*	Max-Planck-Institut für Strömungsforschung Bunsenstr. 10 D-3400 Göttingen
W. Kühn	*D*	Deutsche Airbus GmbH. Hünefeldstr. 1-5 Postfach 10 78 45 D-2800 Bremen 1
K. Kuwahara	*J*	The Institute of Space and Astronautical Science 3-1-1 Yoshinodai Sagamihara-shi Kanagawa 229

S. Le Dizès	*F*	Laboratoire d'Hydrodynamique Ecole Polytechnique Route de Saclay F-91128 Palaiseau Cedex
P. Leehey	*USA*	Massachusetts Institute of Technology Department of Mechanical Engineering Building 3, Room 262 Cambridge, MA 02139
V.Ya. Levchenko	*Russia*	Inst. of Theor. and Applied Mechanics Sibirian Division of Russian Academy of Sciences 630090 Novosibirsk Akademgorodok
T. Leweke	*F*	Laboratoire de Recherche en Combustion Université de Provence Centre de Saint Jérôme Service 252 F-13397 Marseille Cedex 13
C.G. Lewis	*USA*	Department of Applied Mathematics and Engineering Sciences University of California, San Diego La Jolla, CA 92093-0411
S.C. Luo	*Singapore*	Mechanical and Production Engineering Department National University of Singapore 10 Kent Ridge Cresent Singapore 0511
E. Meiburg	*USA*	Department of Aerospace Engineering University of Southern California Los Angeles, CA 90089-1191
G.E.A. Meier	*D*	Institut für Experimentelle Strömungsmechanik der DLR Bunsenstr. 10 D-3400 Göttingen
J.R. Meneghini	*UK*	Department of Aeronautics Imperial College of Science Technology and Medicine London SW7 2BY
R.W. Miksad	*USA*	Department of Aerospace Engineering and Engineering Mechanics University of Texas at Austin Austin, TX 78712-1085

S. Minevich	*Russia*	Moscow Aviation Institute Kirov Strasse 20-12 140 005 Luberzy
Y. Mochimaru	*J*	Department of Mechanical Engineering Tokio Institute of Technology 2-12-1, Ookayama, Meguro-ku Tokyo 152
P.A. Monkewitz	*USA*	Department of Mechanical, Aerospace and Nuclear Engineering University of California, Los Angeles 405 Hilgard Ave. Los Angeles, CA 90024-1597
M. Morzyński	*D*	Hermann-Föttinger-Institut für Thermo- und Fluiddynamik Technische Universität Berlin Straße des 17. Juni 135 D-1000 Berlin 12
T. Nakagawa	*J*	Department of Mechanical System Engineering Kanazawa Institute of Technology Kanazawa, Ishikawa 921
Y. Nakamura	*J*	Research Institute for Applied Mechanics Kyushu University Kasuga, 816
P.N. Nguyen	*USA*	Propulson Technology Branch Code 1544 David Taylor Model Basin Bethesda, MD 20084-5000
B.R. Noack	*D*	Max-Planck-Institut für Strömungsforschung Bunsenstr. 10 D-3400 Göttingen
C.K.H. Norberg	*S*	Department of Applied Thermodynamics and Fluid Mechanics Chalmers University of Technology Hörsalsvägen 7 S-412 96 Göteborg
F. Ohle	*D*	Institut für Angewandte Mechanik und Strömungsphysik Georg-August-Universität Göttingen Bunsenstr. 10 D-3400 Göttingen

K. Ohmi	*J*	Faculty of Engineering Osaka Sangyo University 3-1-1 Nakagaito Daito 574
A. Papangelou	*UK*	Department of Engineering University of Cambridge Peterhouse Trumpington Street Cambridge CB2 1PZ
H.-D. Papenfuß	*D*	Institut für Thermo- und Fluiddynamik Ruhr-Universität Bochum Postfach 10 21 48 D-4630 Bochum
P. Paranthoën	*F*	Faculté des Sciences et des Techniques Université de Rouen Place Emile Blondel - BP 118 F-76134 Mont Saint Aignan
D. Park	*USA*	Department of Aerospace Engineering University of Southern California Los Angeles, CA 90089-1191
G. Parkinson	*CDN*	Dept. of Mechanical Engineering University of British Columbia 2324 Main Mall Vancouver, British Columbia V6T 1Z4
P.S. Piccirillo	*USA*	AMES Department R-011 University of California, San Diego La Jolla, CA 92093
R. Piva	*I*	Dipartimento di Meccanica e Aeronautica Università degli Studi di Roma "La Sapienza" Via Eudossiana, 18 I-00184 Roma
G. Polidori	*F*	Laboratoire de Mécanique des Fluides Composante de l'URA CNRS 191 Université de Poitiers 40, Avenue du Recteur Pineau F-86022 Poitiers Cedex
M. Provansal	*F*	Laboratoire de Recherche en Combustion Université de Provence Centre de Saint Jérôme Service 252 F-13397 Marseille Cedex 13

E. Quandt	*USA*	David W. Taylor Naval Ship R&D Center Code 2704 Annapolis, MD 21402
S.E. Ramberg	*USA*	Office of Naval Research Code 1121 Ocean Engineering 800 North Quincy Street Arlington, VA 22217-5000
G. Riccardi	*I*	Dipartimento di Meccanica e Aeronautica Università degli Studi di Roma "La Sapienza" Via Eudossiana, 18 I-00184 Roma
D.O. Rockwell	*USA*	Department of Mechanical Engineering and Mechanics 354 Packard Laboratory # 19 Lehigh University Bethlehem, PA 18015
A. Roshko	*USA*	California Institute of Technology Graduate Aeronautical Laboratories Mail Code 105-50 Pasadena, CA 91125
K. Roussopoulos	*UK*	Engineering Department Cambridge University Trumpington Street Cambridge CB2 3HU
M. Schäfer	*D*	Max-Planck-Institut für Strömungsforschung Bunsenstr. 10 D-3400 Göttingen
G. Schewe	*D*	Institut für Aeroelastik der DLR Bunsenstr. 10 D-3400 Göttingen
V. Schwarz	*D*	Institut für Aero- und Gasdynamik Universität Stuttgart Pfaffenwaldring 21 D-7000 Stuttgart 80
G. Seifert	*D*	Strömungstechnisches Labor Berlin-Adlershof (KAJ e.V./WIP) Rudower Chaussee 6 D-1199 Berlin

T.K. Sengupta	*IND*	Dept. of Aerospace Engineering Indian Institute of Technology Kanpur 208016
J. Sheridan	*AUS*	Mechanical Engineering Department Monash University Clayton, Vic. 3168
H. Stapountzis	*GR*	Department of Mechanicsl Engineering University of Thessaloniki Box 443 GR-540 06 Thessaloniki
Y. Sugiyama	*J*	Department of Aeronautical Engineering Nagoya University Furo-Cho, Chikusa-Ku Nagoya 464-01
G.K. Suryanarayana	*IND*	Experimental Aerodynamics Division National Aeronautical Laboratory Post Box 1779 Kodihally Bangalore 560 017
S. Szepessy	*D*	Institut für Hydromechanik Universität Karlsruhe Kaiserstr. 12 Postfach 6980 D-7500 Karlsruhe 1
A.A. Szewczyk	*USA*	Department of Aerospace and Mechanical Engineering University of Notre Dame Notre Dame, IN 46556
T. Tamura	*J*	Ohsaki Research Institute Shimiyu Corporation Fukoku Seimei Bldg. 2-2-2 Uchisaiwai-cho, Chiyoda-ku Tokyo 100
B. Thiele	*D*	Max-Planck-Institut für Strömungsforschung Bunsenstr. 10 D-3400 Göttingen
F. Thiele	*D*	Hermann Föttinger-Institut für Thermo- und Fluiddynamik Technische Universität Berlin Straße des 17. Juni 135 D-1000 Berlin 12

N. Tombazis	*UK*	Department of Aeronautics Imperial College of Science, Technology and Medicine Prince Consort Road London SW7 2BY
G.S. Triantafyllou	*USA*	Dept. of Mechanical Engineering Steinman 202 The Levich Inst. Convent Avenue at 140th St. The City College of New York New York, NY 10031
C. Tropea	*D*	Lehrstuhl für Strömungsmechanik Universität Erlangen-Nürnberg Cauerstr. 4 D-8520 Erlangen
M.F. Unal	*TR*	Faculty of Aeronautics and Astronautics Istanbul Technical University Maslak Istanbul 80626
V. Uruba	*CS*	Institute of Thermomechanics Czechoslovak Academy of Sciences Dolejškova 5 CS-182 00 Praha 8
R. Verzicco	*I*	Dipartimento di Meccanica e Aeronautica Università degli Studi di Roma "La Sapienza" Via Eudossiana, 18 I-00184 Roma
H. Vollmers	*D*	Institut für Theoretische Strömungsmechanik der DLR Bunsenstr. 10 D-3400 Göttingen
J.D.A. Walker	*USA*	Department of Mechanical Engineering and Mechanics Center for the Application of Mathematics Packard Lab. # 19 Lehigh University Bethlehem, PA 18015
B. Walter	*USA*	Department of Mechanical Engineering Michigan State University B100 Research Complex - Engineering East Lansing, MI 48824-1226
M. Weiland	*D*	Max-Planck-Institut für Strömungsforschung Bunsenstr. 10 D-3400 Göttingen

J.E. Wesfreid	*F*	Laboratoire d'Hydrodynamique et Mécanique Physique E.S.P.C.I. 10 Rue Vauquelin F-75231 Paris Cedex 05
D.R. Williams	*USA*	Illinois Institute of Technology Armour College of Engineering Fluid Dynamics Research Center Dept. of Mech. and Aerospace Eng. Chicago, IL 60616
C.H.K. Williamson	*USA*	Mechanical and Aerospace Engineering Cornell University 252 Upson Hall Ithaca, NY 14853-7501
N.I. Yavorsky	*Russia*	Institute of Thermophysics Sibirian Division of Russian Academy of Sciences 630090 Novosibirsk Akademgorodok
B.A. Younis	*UK*	Department of Civil Engineering City University London Northampton Square London EC1V 0HB
M.M. Zdravkovich	*UK*	Department of Aeronautical and Mechanical Engineering University of Salford Salford M5 4WT
V.L. Zhdanov	*Byelorus*	Institute of Heat and Mass Transfer of the Academy of Science of Byelorussian Republic P. Brovka Str. 15 Minsk 220728

Contents

Session 4
Poster discussion and presentations
Chairman: U. Dallmann

*) paper not available at time of printing

Photograph by U. Böcher

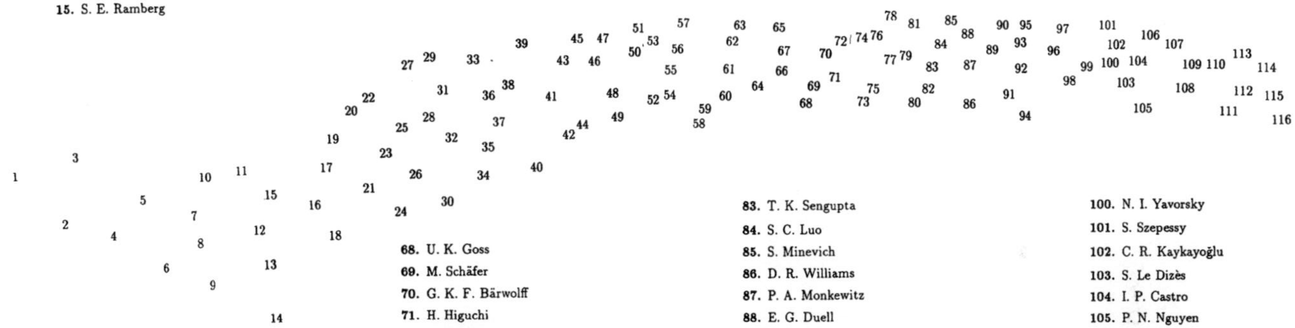

1. B. Gossieaux
2. B. Thiele
3. A. Diese
4. K. Roussopoulos
5. B. Walter
6. C. Klein
7.
8. U. Fey
9. M. Weiland
10. C. K. H. Norberg
11. H. Kronewetter
12. V. Uruba
13. P. Jonáš
14. Y. Mochimaru
15. S. E. Ramberg
16. E. Ahlborn
17. B. R. Ahlborn
18. P. W. Bearman
19. M. Kiya
20. V. Schwarz
21. G. V. Parkinson
22. G. Bosch
23. D. O. Rockwell
24. A. A. Szewczyk
25. S. P. Bardakhanov
26. C. Tropea
27. F. Ohle
28. Y. Sugiyama
29. A. Weismann
30. J. F. Foss
31. A. Papangelou
32. G. Riccardi
33. J. D. A. Walker
34. M. K. Bull
35. V. A. Gushchin
36. T. Leweke
37. R. Verzicco
38. D. Park
39. H. Bestek
40. H. Eckelmann
41. J.-L. Champion
42. H.-D. Papenfuß
43. E. Meiburg
44. P. Leehey
45. M. Brede
46. B. R. Noack
47. F. Thiele
48.
49. V. V. Kozlov
50. H. Eisenlohr
51. M. Morzyński
52. T. C. Corke
53. F. Jacobitz
54. T. Tamura
55. T. Hayashi
56. M. König
57. G. Gerbeth
58. V. Ya. Levchenko
59. H. Stapountzis
60. K. Kuwahara
61. M. M. Zdravkovich
62. C. H. K. Williamson
63. B. A. Younis
64. G. Glauben
65. J. R. Chaplin
66. A. Roshko
67. W. Kühn
68. U. K. Goss
69. M. Schäfer
70. G. K. F. Bärwolff
71. H. Higuchi
72. H. Vollmers
73. J. E. Wesfreid
74. J. H. Gerrard
75. G. Seifert
76. Y. Nakamura
77. G. L. Brown
78.
79. G. K. Suryanarayana
80. F. R. Hama
81. J. Sheridan
82. P. Paranthoën
83. T. K. Sengupta
84. S. C. Luo
85. S. Minevich
86. D. R. Williams
87. P. A. Monkewitz
88. E. G. Duell
89.
90. R. H. Arkell
91. C.-M. Ho
92. G. E. Karniadakis
93. J. M. R. Graham
94. T. Nakagawa
95. J. R. Meneghini
96. M. Gharib
97. M. F. Unal
98. R. E. Hunt
99. P. Huerre
100. N. I. Yavorsky
101. S. Szepessy
102. C. R. Kaykayoğlu
103. S. Le Dizès
104. I. P. Castro
105. P. N. Nguyen
106. E. Quandt
107. A. Quandt
108. F. Hayot
109. R. W. Miksad
110. E. Rohrmoser
111. M. Provansal
112. Y. G. Guezennec
113. U. Eberhardt
114. P. S. Piccirillo
115. C. G. Lewis
116. M. Hammache

Challenging Problems in Bluff Body Wakes

P.W.Bearman
Department of Aeronautics,
Imperial College of Science, Technology and Medicine,
London SW7 2BY, UK.

Abstract

A number of outstanding questions related to the near and middle wake structure of two-dimensional, or mildly three-dimensional, bluff bodies are presented. Experiments on a bluff body with a controlled three-dimensional geometric disturbance are described and it is shown that vortex splitting is a dominant feature of the wake flow. The use of CFD to study low Reynolds number bluff body wakes is discussed. Also similarities between vortex shedding in steady flows and in planar oscillatory flows are pointed out.

1 Introduction

Bluff body flows present a wide range of challenging problems and it is quite beyond the scope of this paper to attempt to describe all of them. Although it is by no means fully proved, it may be that very far downstream the wakes of all bluff bodies asymptote to a similar form. Discovering this structure is in itself a major challenge; however, this paper is restricted to discussing outstanding questions related to the near and middle wake structures and on how these depend on the details of the body geometry. The wakes generated by bluff bodies that have a two-dimensional, or mildly three-dimensional, geometry will be considered. Strictly, two-dimensional geometries and two-dimensional flows can only exist in computer simulations since all bodies have ends and, above some critical Reynolds number, all bluff body flows generate instabilities with some spanwise wavelength. Although great care may have been taken by both the experimentalist and the computationalist there can still be discrepancies between experimental and computational results for two-dimensional bluff body flows due to the presence of three-dimensional instabilities in physical flows. Two important developments have taken place to help resolve this problem. By carefully manipulating end conditions(1)(2)(3) it has been shown that vortex shedding behind bluff bodies in the low Reynolds number laminar unstable regime can be forced to be parallel to the body axis (the unproven assumption being that in an experiment parallel shedding is synonymous with two-dimensional flow). Secondly, computations are beginning to be carried out to simulate three-dimensional flow around two-dimensional bluff bodies. Hence there is now the opportunity to make very detailed comparisons between experiment and computation.

It has long been known that, given two shear layers with opposite sign vorticity, viscosity plays a minor role in the development of vortex shedding. In the case of a circular cylinder, feedback between vortex generation and flow separation position undoubtedly takes place but it is not the primary mechanism causing shedding. A descriptive model of the fluid dynamic processes at work in the vortex formation region, was introduced by Gerrard[4]. Huerre and Monkewitz[5] describe wake dynamics in terms of local and global modes related to absolute and convective instabilities and the near wake of a bluff body is identified as a region of absolute instability. It is interesting to note how robust this instability is because even a bluff body oscillated back and forth in still fluid exhibits the same type of vortex shedding behaviour. Strong vortex shedding occurs for oscillation amplitudes as small as two cylinder diameters and it is clear that the middle and far wake can play little part in this process.

Attractions of the hydrodynamic stability approach are that it provides a theoretical background and can give valuable insight into possible mechanisms for controlling bluff body flow. Control is achieved through affecting the near wake region by some passive or active means. Passive control, say by a splitter plate or by introducing bleed flow, has a direct effect on the near wake absolute instability region. Interesting possibilities exist for flow control using active means such as body oscillation or pulsed jets. However, some distinction should be made between open and closed loop active control. Most investigators have studied open loop systems where there is no feedback between the wake and the controller. Challenges abound in this area, particularly in the application of closed loop control to vortex suppression. However, research to date suggests that suppression may only be possible in a small range of Reynolds numbers slightly above the critical Reynolds number for the onset of shedding. At higher Reynolds numbers it is possible that while one mode can be suppressed another is excited. Also the three-dimensionality of vortex shedding introduces a further complication because while shedding might be attenuated at one spanwise position it could be amplified at another. The investigation of effective control strategies is an interesting challenge for computational fluid dynamicists because, compared to the difficulties facing the experimentalist, it is relatively easy to introduce active control into computer simulations.

In this paper three challenging problems will be discussed which it is hoped will throw some light on our understanding of bluff body wakes: three-dimensionality in nominally two-dimensional flows, including the effect of three-dimensional disturbances, the use of CFD to study low Reynolds number wakes and vortex shedding in oscillatory flows.

2 The Three-Dimensional Structure of Bluff Body Wakes

Gerrard[6] was one of the first to address in depth the question of the three-dimensionality of cylinder wakes. He worked across a wide range of Reynolds number and identified key three-dimensional features such as oblique shedding and looping of vortices across to the other side of the wake. In his paper he also presents a diagram which shows the phenomenon of vortex splitting or vortex dislocation. More recently, Eisenlohr and Eckelmann[2] showed that vortex splitting occurs when there are spanwise variations in the vortex shedding frequency. A vortex dislocation or split occurs at the boundary between adjacent cells and is a means by which vortices can connect with each other when their frequencies are different. One way to quantify the degree of three-dimensionality in a vortex wake is to measure the spanwise correlation length of some fluctuating quantity related to vortex shedding. At Reynolds numbers high enough to generate a wake of turbulent vortices, measurements indicate a typical value for the correlation length of a few body diameters. However, this length cannot be predicted and the mechanisms responsible for limiting spanwise correlation are not clearly understood.

The sensitivity of bluff body flow to end conditions and the influence of aspect ratio illustrate the important role of three-dimensional effects. Figure 1 shows measurements reported by Szepessy and Bearman(7) of the variation of the root mean square value of the fluctuating lift coefficient, measured at the centre span of a circular cylinder, as a function of aspect ratio. The ends of the cylinder were terminated by end plates and the Reynolds number of the experiment was 5.1×10^4. It can be seen that as aspect ratio is reduced below about 8, C'_L increases by as much as 50%. Even with end plates there is still some three-dimensional disturbance introduced into the flow but any strong mean cross flow near the ends is eliminated. At high Reynolds numbers there are unsteady pressure gradients imposed along the span, as evidenced by vortex shedding not being in phase over large spanwise distances. The unsteady cross flows induced by these pressure gradients are suppressed near an end plate and this seems to lead to an enhancement in the mean strength of the shed vortices in this region.

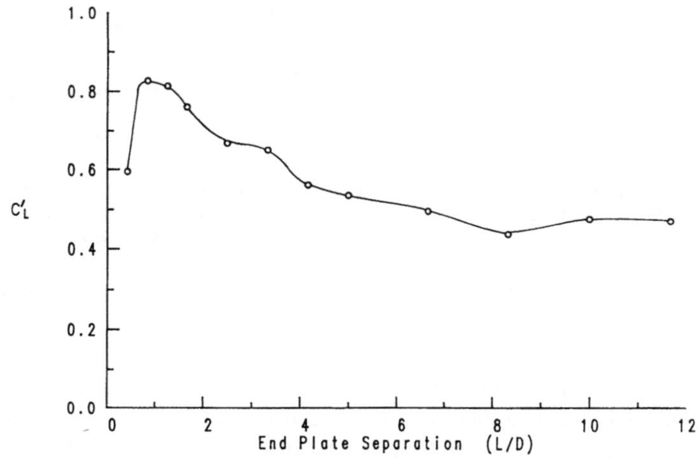

Fig 1 Variation of coefficient of fluctuating lift with aspect ratio, Re = 5.1×10^4

On large aspect ratio bluff bodies events leading to a departure from two-dimensional vortex shedding, such as a vortex dislocation, appear to occur at random points in time and space. In order to try and control these events, so that they can be studied in detail, a series of experiments has been undertaken on a bluff body with imposed three-dimensional geometric disturbances. Details of the blunt-trailing-edge body used are given in figure 2 where it can be seen that the separation points are fixed and the trailing edge is wavy. A number of different wavelengths were studied, all with a wave height equal to 25% of the base height, but most of the research was carried out on the configuration shown in figure 2. Compared to a straight edge, the

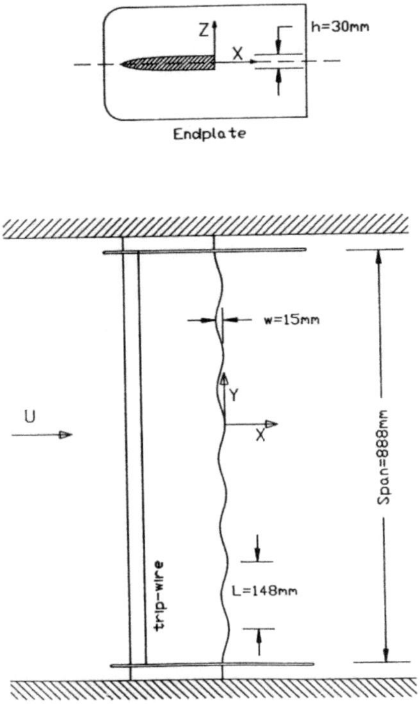

Fig.2 Geometry of three-dimensional blunt trailing edge model.

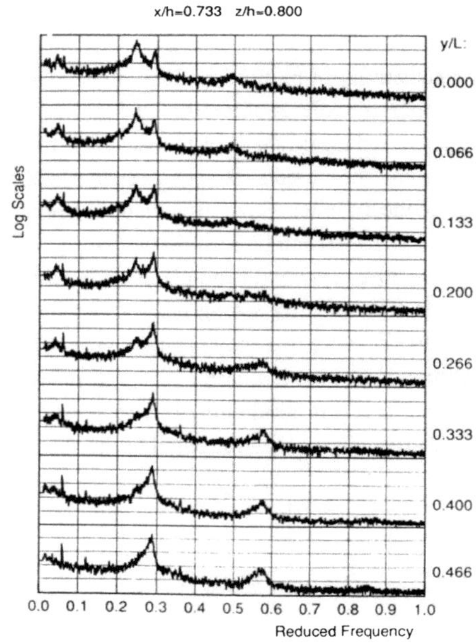

Fig.3 Variation of power spectrum of velocity, measured outside the near wake of the wavy model,with spanwise position.

introduction of a wavy trailing edge reduces base drag and the highest base suction occurs at a valley; i.e. where the chord of the model is a minimum.

Spectra of velocity signals recorded in the wake of the wavy model shown in figure 2 are plotted in figure 3. y/L=0 is a position corresponding to a peak in the waviness across the base and y/L=0.5 is a valley. It can be seen that at a valley the spectrum is centred around a single vortex shedding frequency whereas at a peak two frequencies are evident. The lower of the two frequencies is close to the shedding frequency for a straight edge and the low frequency hump rising above the background (the adjacent sharp peak is spurious) is equal to the difference between the two shedding frequencies. Using flow visualisation, two basic modes of vortex shedding have been observed around a peak in the spanwise waviness: a symmetric and an antisymmetric mode. The spectra show that the flow has to accommodate changes in the shedding frequency across the span and this is achieved via vortex splitting. A visualisation of the flow is shown in figure 4.

During the course of carrying out the above experiments an interesting additional three-dimensional effect was observed which was present on the straight-edged and the wavy models.

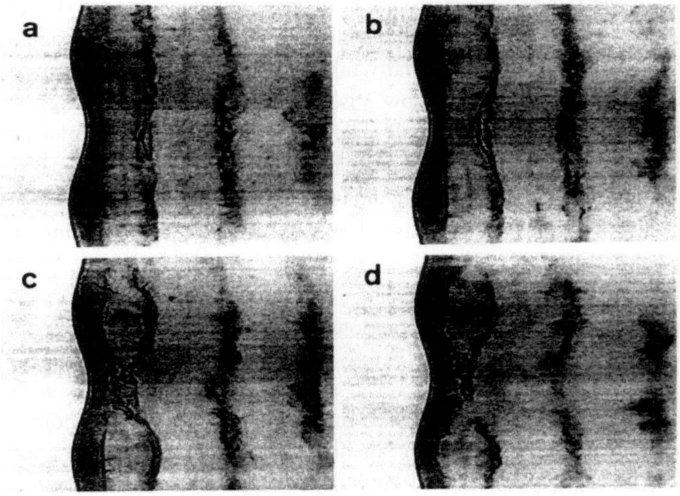

Fig.4 Visualisation showing the symmetric mode and the development of a double dislocation.

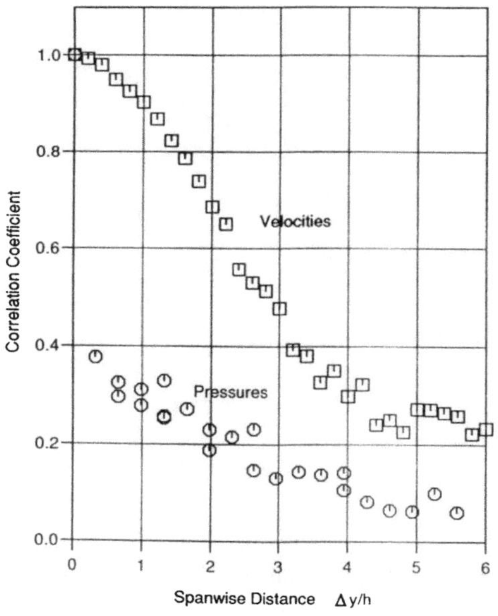

Fig.5 Measurements of correlation coefficient against spanwise separation for a straight trailing edge model: ◻, velocity outside the near wake; 0, pressure at the centre of base.

Figure 5 shows two sets of spanwise correlation measurements: one for fluctuating velocity measured just outside the near wake and the other for fluctuating base pressure. It can be seen that the base pressure correlation initially falls off very rapidly with spanwise separation, indicating strongly three-dimensional flow. Visualisation using a spanwise light sheet along the wake centre line shows a relatively small scale and apparently random vortex structure which may be turbulence entrained into the near wake or the remnants of a longitudinal vortex structure originating in the shear layers shed from the body. The remainder of the base pressure signal measured along the centre of the base is mostly at twice the shedding frequency and substantially better correlated.

3 Numerical Simulation of Bluff Body Flow

One of the most exciting aspects of numerical simulation of bluff body flow is the possibility it offers for studying in detail the mechanics of vortex formation and shedding. However, in attempting to compare computation with experiment two obvious difficulties arise. At Reynolds numbers up to about 150, where it is appropriate to carry out two-dimensional simulations, there is for good practical reasons a lack of detailed experimental measurements of mean and unsteady forces. At higher Reynolds numbers three-dimensional instabilities, both small scale and large scale, should be modelled but presently the very considerable computer resources required restrict activity in this area to just a few research groups.

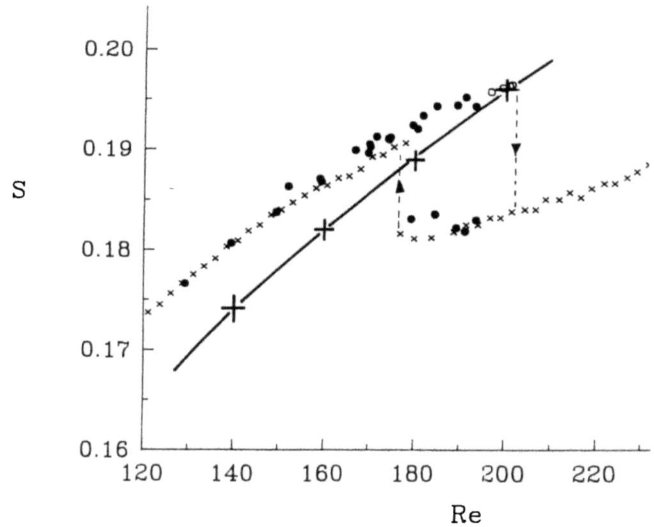

Fig.6 Strouhal number versus Reynolds number. x, Williamson[1], 0 and ●, suction end control, Williamson[9]; +, computation[8].

A paper by Meneghini and Bearman[8], to be presented at this meeting, describes the application of a two-dimensional discrete vortex code to study control of the flow around a circular cylinder. The control is applied through body oscillation and the numerical method incorporates an Eulerian vorticity diffusion scheme and a Lagrangian convection algorithm. In order to calculate the diffusion of vorticity correctly, great care has to be taken to ensure that the mesh used is sufficiently fine, particularly near the body surface. The Reynolds number of much of this work is 200 and this means that the results cannot be compared directly with experiment. However, Williamson[9] has shown that laminar flow with parallel vortex shedding can be extended to a Reynolds number of 200 if suction is applied at the edges of the wake. Measurements by Williamson of the Strouhal number for a circular cylinder are shown in figure 6 together with values obtained from the numerical method. It is clear that there is a systematic difference between the two sets of results. An obvious possibility is that there is an error in the simulation method although the present results agree well with those of Braza et al[10] who used a different numerical scheme. Alternatively is there some hidden three-dimensional effect in the experiment which is responsible for the differences?

4 Vortex Shedding in Oscillatory Flow

The flow around a cylinder undergoing relative sinusoidal motion along a diameter is a function of the Keulegan Carpenter KC and the Reynolds number Re. KC = UT/D, where U is the maximum velocity during a cycle, T is the period and D the cylinder diameter. Re is formed using U and D. An alternative viscous parameter is β, where $\beta = Re/KC = D^2/\upsilon T$. The configuration of vortex shedding is primarily a function of KC and the various patterns have been described by Bearman[11], Williamson[12], Obasaju et al[13] and Tatsuno and Bearman[14]. If the oscillatory flow is from side to side then, above a KC value of about 6 or 7, vortices form alternately from the upper and lower sides of the cylinder. As KC is increased by an increment of about 8, another fully formed vortex is shed in each half cycle. The mechanism by which vortices are convected from the cylinder is through mutual induction whereby a pair of vortices of opposite sign induce a velocity on each other. By the end of a cycle a vortex may not be fully formed and it will remain to influence shedding in the next half cycle. If it is sufficiently strong it will pass back over the same side of the cylinder as the one on which it was formed. This will help to accelerate the formation of an opposite sign vortex in the shear layer that forms from that side at the beginning of the next half cycle. If the vortex is weak it will pass around the cylinder and find its way into the similar sign vortex forming in the diametrically opposite shear layer in the next half cycle.

At high KC values, say in excess of 40 or 50, the cylinder might be expected to shed part of a vortex street in each half cycle. Bearman et al[15] have devised a simple quasi-steady model of vortex shedding to predict the fluctuating transverse or lift force on a cylinder in oscillatory flow. Since the magnitude of this force is expected to depend on the square of velocity, the oscillating flow will induce an amplitude modulation on the lift. Assuming a constant Strouhal number then in oscillating flow the fluctuating lift will also be frequency modulated following the variations in velocity. A fit of the model equation to experimental data for a circular cylinder is shown in figure 7. The interesting point about this approach is that the best fit is obtained when the Strouhal number takes its steady flow value of 0.2.

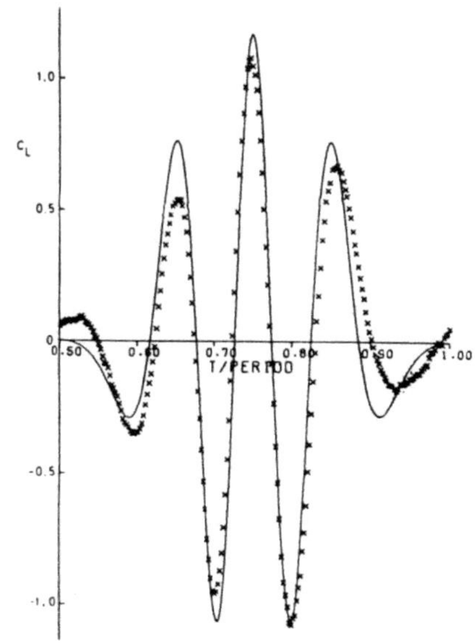

Fig.7 Coefficient of transverse force on a circular cylinder in oscillatory flow. x, experimental data; ——— quasi-steady model with S=0.2[15].

5 Outstanding Questions

Instead of attempting to present a series of conclusions to the work outlined above a number of questions are listed. These are offered in no particular order of importance.

(a) What determines the degree of spanwise correlation of bluff body flow?

(b) Are vortex dislocations a fundamental feature of high Reynolds number bluff body wakes?

(c) What new insights into bluff body wakes can be obtained from new experimental techniques that provide instantaneous spatial information, such as particle image velocimetry?

(d) What can be learnt about bluff body flows by applying passive or active control and is it possible to use active control to suppress vortex shedding at high Reynolds numbers?

(e) In the lock-in region what determines the phase between body or incident flow oscillations and vortex shedding?

(f) What can further study of low order dynamical systems offer towards the understanding of bluff body wakes?

(g) What is the future role for hydrodynamic stability theory in the study of bluff body wakes?

(h) Is there any need for more experiments at low Reynolds numbers with parallel shedding or can CFD simulate these flows adequately?

(i) What problems should researchers working on three-dimensional simulation be trying to solve?

(j) Can what has been learnt about vortex shedding in steady incident flows be applied in oscillatory flows?

6 Acknowledgements

In preparing this paper I have drawn upon the results and ideas of many people but I would particularly like to acknowledge the valuable contributions made by Julio Meneghini, Stefan Szepessy and Nicholas Tombazis. The research into the effect of controlled 3D disturbances on near wake flow is funded by the Office of Naval Research.

7 References

1 Williamson C.H.K. "Oblique and Parallel Modes of Vortex Shedding in the Wake of a Circular Cylinder at Low Reynolds Numbers". J.Fluid Mech. 206, p579, 1989.

2 Eisenlohr H. and Eckelmann H. "Vortex Splitting and its Consequences in the Vortex Street Wake of Cylinders at Low Reynolds Number". Phys. Fluid A1, p189, 1989

3 Hammache M. and Gharib M. "An Experimental Study of the Parallel and Oblique Vortex Shedding from Circular Cylinders". J.Fluid Mech. 232, p567, 1991.

4 Gerrard J.H. "The Mechanics of the Formation Region of Vortices behind Bluff Bodies". J.Fluid Mech. 25, p401, 1966.

5 Huerre P. and Monkewitz A. "Local and Global Instabilities in Spatially Developing Flows". Ann. Rev, Fluid Mech. 22, p473, 1990.

6 Gerrard J.H. "The Three-Dimensional Structure of the Wake of a Circular Cylinder". J.Fluid Mech. 25, p143, 1966.

7 Szepessy S. and Bearman P.W. "Aspect Ratio and End Plate Effects on Vortex Shedding from a Circular Cylinder". J.Fluid Mech. 234, p191, 1992.

8 Meneghini J.R. and Bearman P.W. "Numerical Simulation of Control Of Bluff Body Flow Using a Discrete Vortex Method Incorporating Viscous Diffusion" IUTAM Sym. Bluff-Body Wakes, Dynamics and Instabilities, Gottingen, 1992.

9 Williamson C.H.K. Private Communication, 1992.

10 Braza M., Chassaing P. and Ha Minh H. "Numerical Study and Physical Analysis of the Pressure and Velocity Fields in the Near Wake of a Circular Cylinder". J. Fluid Mech. 165, p79, 1986.

11 Bearman P.W. "Vortex Trajectories in Oscillatory Flow". Proc. Intl. Symp. on Separated Flow around Marine Structures, Trondheim, Norway, p133, 1985.
12 Williamson C.H.K. "Sinusoidal Flow Relative to Circular Cylinders". J.Fluid Mech. 155, p141, 1985
13 Obasaju E.D., Bearman P.W. and Graham J.M.R. "A Study of Forces, Circulation and Vortex Patterns around a Circular Cylinder in Oscillating Flow". J.Fluid Mech. 196, p467, 1988.
14 Tatsuno M. and Bearman P.W. "A Visual Study of the Flow around an Oscillating Circular Cylinder at Low Keulegan-Carpenter Numbers and Stokes Numbers". J.Fluid Mech. 211, p157, 1990.
15 Bearman P.W., Graham J.M.R. and Obasaju E.D. "A Model Equation for the Transverse Forces on Cylinders in Oscillatory Flows". Appl. Ocean Res. 6, p166, 1984.

Reverse Kármán Streets in the Wake of Flapping Foils for Application in Fish Propulsion

GEORGE S. TRIANTAFYLLOU and MICHAEL S. TRIANTAFYLLOU

The Levich Institute & Department of Mechanical Engineering, The City College of New York; and Department of Ocean Engineering, Massachusetts Institute of Technology

Summary

The time-averaged flow behind thrust-producing, oscillating foils has the shape of a jet, while over a wide parametric range a staggered array of vortices is formed, similar to the Karman street behind bluff bodies but with opposite rotational direction (*reverse Karman street*). It is found that the jet-like average flow is convectively unstable, having a narrow range of frequencies of maximum amplification. A stable co-existence of the jet profile and the reverse Karman street is ensured only at the frequency of maximum amplification, hence at this frequency optimal efficiency is obtained, i.e. maximum thrust per unit input energy. The non-dimensional frequency of maximum amplification (Strouhal number) is in the range of 0.25 to 0.35. Experiments confirm this result, while the analysis of a large number of data from observations on fish and cetaceans confirm that optimal fish propulsion is achieved within this range of Strouhal number.

Introduction

An oscillating foil (heaving and/or pitching) moving with forward speed produces thrust under certain conditions. The generation of steady thrust is associated with the formation of an average velocity profile behind the foil which has the form of a jet. This jet flow is convectively unstable, hence, under the influence of the harmonic excitation is in dynamic equilibrium with a staggered array of vortices, moving downstream and closely resembling a Karman vortex street behind bluff objects, but with reverse rotational direction. This mechanism was first proposed by Karman and Burgess [1], and then proven conclusively in a number of experimental studies through visualization (Ohashi and Ishikawa [2], Koochesfahani [3]).

It is found here, based on concepts and ideas from similar developments in bluff wake dynamics, that the average flow in the wake behind the foil is linearly unstable, acting as an amplifier with a relatively small band of frequencies of maximum disturbance amplification. It is proposed that the efficient development of thrust is associated with the formation of a reverse Karman street. Then the frequencies of maximum spatial growth must also be the frequencies of maximum foil efficiency, since they result in the most expedient formation of this array of vortices.

Theoretical Developments

Recent developments in bluff wake dynamics (Triantafyllou et al. [4]) have demonstrated the importance of the shape of the time-averaged flow on the frequency and wavelength of the large scale patterns that develop as a result of the inherent instability of the flow at high Reynolds numbers.

The flow behind a thrust-developing foil must have a time-averaged shape of a jet; this is indeed shown in the measurements by Koochesfahani [3]. Also, several visualization studies have demonstrated that a thrust-producing foil generates a reverse Karman street, as also expected from purely kinematical considerations (Ohashi and Ishikawa [2], Koochesfahani [3]). We analyzed velocity profiles measured behind a flapping foil by Koochesfahani [3], using the methods proposed in Triantafyllou et al. ([4]). The jet-like flow was found to be convectively unstable with a maximum spatial amplification Strouhal number in the range of 0.25 to 0.35, with a most likely value of about 0.30 (Triantafyllou et al. [5]; Triantafyllou et al. [6]). The Strouhal number is defined for simplicity as:

$$St=\frac{fA}{V}\ , \tag{1}$$

where f is the frequency of oscillation (Hz); A is the width of the wake, taken to be equal to the maximum excursion of the foil's trailing edge (double amplitude); and V is the average forward velocity.

The average, jet-like velocity profile is in dynamic equilibrium with the large scale flow patterns. An experimental investigation by Koochesfahani [3] has shown that thrust production is closely associated with a staggered array of vortices having only one pair of counter-rotating vortices shed per cycle (reverse Karman street), while any other configuration (for example, two or more vortices shed per half cycle) results in poor thrust production. As a result, we conjectured that optimal thrust production occurs at the frequency of maximum spatial amplification of the average velocity profile.

Experimental Data

We measured the thrust and lift of a foil capable of pitching and heaving while towed at constant speed. The tests were conducted in the MIT Ocean Engineering Testing Tank Facility and are described in Triantafyllou et al [5]. Figure 2 shows the thrust coefficient achieved by moving the foil in a combination of heave (amplitude $h=2.5$ inches) and pitch motion ϕ, 90 degrees out of phase with respect to heave, as function of the Strouhal number $St=fA/V$. The amplitude of pitch oscillation was adjusted to be a constant portion of the heave induced angle, i.e., $tan(\phi)=\theta h\omega/V$, where $\theta=0.7$. The forward speed was set at $V=0.33$ knots ($0.17\,m/s$) to avoid free surface effects (which nonetheless appear above $St=0.35$). Superimposed on the figure are data from two other sources, Koochesfahani [3] for a pitching foil and Isshiki and Murakami [7] for a foil subjected to a flow, which is equivalent to a heaving and pitching motion. Figure 3 shows the estimated efficiency as function of the Strouhal number, showing a peak at $St=0.25$, thus confirming that optimal efficiency is achieved in the range of frequencies of maximum amplification.

Summary of Data from Fish Propulsion

An analysis of observed fish and cetacean swimming characteristics confirms our prediction for a variety of fish, near their maximum free-swimming speeds (Triantafyllou et al. [6]):

1. dace, trout, goldfish St in the range of 0.28 to 0.31
2. dolphin: $St=0.30$
3. bream: $St=0.32$
4. bonito, striped mullet, buefish: St in the range of 0.28 to 0.33.
5. trout: St in the range of 0.30 to 0.33
6. jack mackerel: St in the range of 0.30 to 0.31
7. sockeye salmon: St in the range of 0.19 to 0.32.
8. bluefish: $St=0.345$
9. cod: St in the range 0.25 to 0.35

10. blacktip shark: average $St = 0.254$
11. mackerel: St in the range of 0.27 to 0.33
12. rainbow trout: St in the range of 0.25 to 0.38

Aknowledgments

The authors wish to aknowledge support from the M.I.T. Sea Grant Program under grant numbers NA86AA-D-SG089 and NA90AA-D-SG424; and support from the Office of Naval Research under contract number N00014-90-J-1312.

References

1. KARMAN VON T. AND BURGESS J. M. 1935 General aerodynamic theory: perfect fluids. In *Aerodynamic Theory* vol. II (ed. W. F. Durand), p. 308. Leipzig: Springer Verlag.
2. OHASHI H. AND ISHIKAWA N. 1972 Visualization study of a flow near the trailing edge of an oscillating airfoil. *Bulletin of the Japanese Society of Mechanical Engineers* **15**, 840-845.
3. KOOCHESFAHANI M. 1989 Vortical patterns in the wake of an oscillating airfoil. *AIAA Journal* **27**, 1200-1205.
4. TRIANTAFYLLOU G. S., KUPFER K. AND BERS A. 1987 Absolute instabilities and self-sustained oscillations in the wakes of circular cylinders. *Physical Review Letters* **59**, 1914-1917.
5. TRIANTAFYLLOU M. S., TRIANTAFYLLOU G. S. AND GOPALKRISHNAN R. 1991 Wake Mechanics for Thrust Generation in Oscillating Foils, *Physics of Fluids A*, **3** (12), pp. 2835-2837.
6. TRIANTAFYLLOU G. S., TRIANTAFYLLOU M. S. AND GROSENBAUGH M. A. 1992 Optimal Thrust Development in Oscillating Foils with Application to Fish Propulsion, *Journal of Fluids and Structures* (Accepted for Publication)
7. ISSHIKI H. AND MURAKAMI M. 1984 A Theory of wave devouring propulsion. *Society of Naval Architects of Japan* **156**, 102-114.

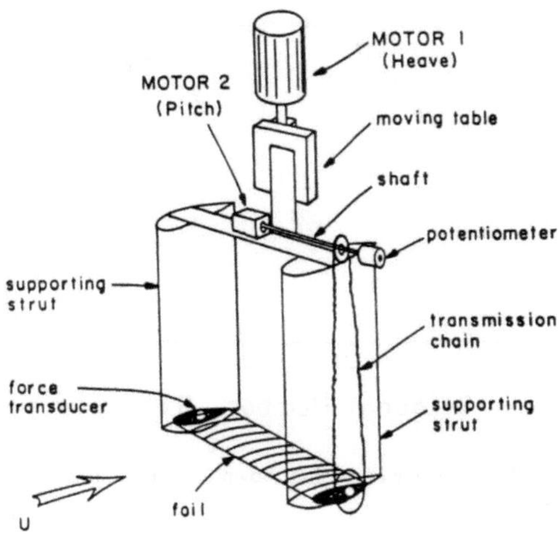

Figure 1: Sketch of experimental apparatus used.

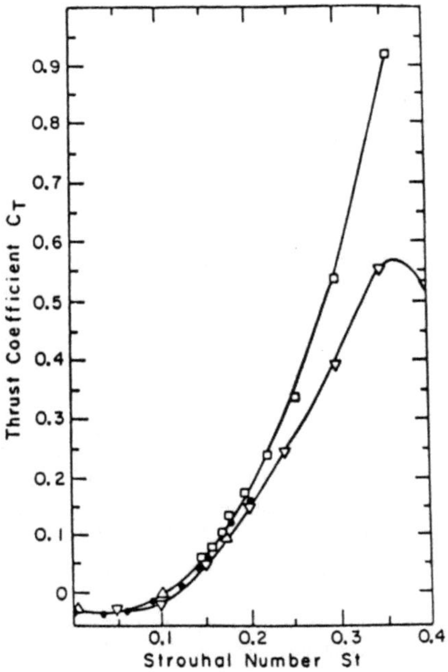

Figure 2: Experimental data for the average thrust coefficient of an airfoil: (o) 2 degrees pitch and (Δ) 4 degrees pitch about the quarter point (Koochesfahani [3]); (o) heaving and pitching foil (Isshiki & Murakami [7]); (∇) present data.

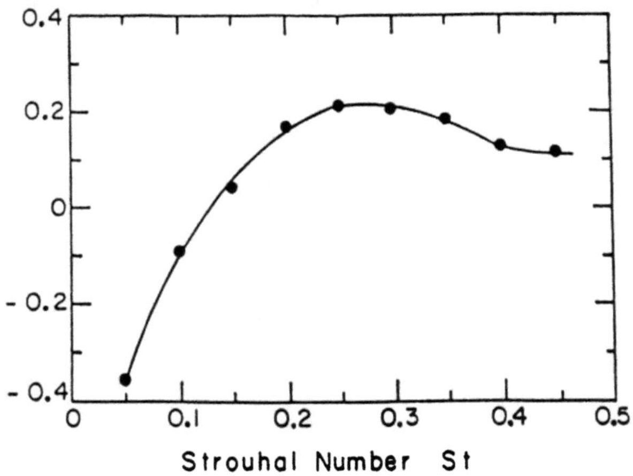

Figure 3: Experimental data for the efficiency of the foil shown in figure 1, heaving with amplitude 2.5 inches and for the feathering parameter equal to 0.7.

Flip-Flopping Flow Around Two Bluff Bodies in Tandem Arrangement

M. KIYA, O. MOCHIZUKI, Y. IDO, T. SUZUKI and T. ARAI

Department of Mechanical Engineering
Hokkaido University, Sapporo, 060 Japan

Summary

At sufficiently high Reynolds numbers flow around two tandem circular cylinders of the same diameter has two patterns which switch randomly in time if the distance between their centres is in an appropriate range. For the distance at which the two patterns occurred at the same probability, the interval of time during which each pattern prevailed was a random variable whose probability density distribution was approximately the Gamma distribution. The average time interval for two patterns was much larger than both the vortex-shedding period and the time scale of the free-stream turbulence, being strongly dependent on the turbulence intensity.

Introduction

At sufficiently high Reynolds numbers there are two patterns of flow around two circular cylinders of the same diameter d which are placed in tandem in a uniform flow of velocity U_∞ [1]. One is the reattaching flow (hereinafter referred to as pattern 1), and the other is the vortex-shedding flow (referred to as pattern 2). Two patterns switch randomly in time; they are illustrated in Fig. 1. The purpose of the present paper is to study statistical properties of the pattern switching. A similar study was made by Kim and Durbin [2] on the pattern switching of flow around a pair of two circular cylinders arranged side by side in a uniform flow. In the present experiment, turbulence-generating grids were installed at the entrance of the tunnel working section in order to counteract possible effects of uncontrollable residual turbulence in the tunnel and external disturbances.

Experimental Apparatus and Method

Experiments were performed in an open-return wind tunnel with a 0.3 m × 0.3 m square and 1.0 m long working section. The wind tunnel was installed in a large room of approximately 28.0 m in length, 17.5 m in width and 9.0 m in height. The free-stream longitudinal turbulence intensity was 0.24% at a speed of 17.8 m/s. Biplane grids were installed at the entrance of the working section to produce the free-stream turbulence with different intensities and length scales. Two circular cylinders of diameter d = 0.285 m, were machined brass cylinders. The cylinders were set normal to the main flow, spanning the central part of the working section. No end plates were employed. The aspect ratio of the cylinders was 10.5; the blockage ratio was 9.5%. Each cylinder had a pressure tap at the midspan which was connected to a semi-conductor strain-gauged pressure transducer with a small cavity. The pressure transducer of the upstream cylinder was located at the time-mean rear stagnation point, while that of the downstream cylinder was located at the time-mean front stagnation point.

The longitudinal velocity fluctuations were measured by an I-wire probe operated by a constant-temperature hot-wire anemometer. The free-stream turbulence was measured at a central position 0.46 m downstream of the grids in the empty tunnel; this position was to be the time-mean front stagnation point of the upstream cylinder. The experiments were performed in a Reynolds-number range $U_\infty d/\nu$ = (2.0 -3.7) × 10^4, where ν denotes the kinematic viscosity.

Results and Discussion

The distance l between centres of two cylinders was set at a particular value without operating the wind tunnel. Then the tunnel was operated to study the pattern-switching property at the same distance. By this procedure the flow was expected to lose its memory about its previous patterns. Figure 2 shows a sample of time histories of the pressure fluctuation at the rear stagnation point of the upstream cylinder p_1. Pattern 1 and pattern 2 are associated with time intervals in which the pressure fluctuation p_1 is quiet and violent, respectively. The time intervals during which pattern 1 or pattern 2 is recorded is denoted by t_1 or t_2, and the fraction of time for pattern 1 or pattern 2 is denoted by γ_1 and γ_2. In the present study the distance l = $l_{0.5}$ where $\gamma_1 = \gamma_2 = 0.5$ was chosen as a representative distance at which properties of the pattern switching were studied. At this particular distance, the average time interval of pattern 1, T_1, is equal to that for pattern 2, T_2.

The probability density distributions of the time interval t_1 or t_2 were approximately represented by the Gamma distribution $f(x)=[\lambda/\Gamma(r)](\lambda x)^{r-1}e^{-\lambda x}$, where $x = t_i/T_i$ (i = 1 or 2), and r and λ are constants which are related to the mean value x_m (= 1) and the variance x_{var} of x by x_m = r/λ and $x_{var} = r/\lambda^2$.

The average time interval T_1 is plotted in Fig.3 as a function of Reynolds number for the main flow without turbulence grids. The time interval T_1 has a tendency to decrease with increasing Reynolds number. It deserves noting that values of the non-dimensional time interval T_1U_∞/d are of the order of 2 × 10^4 ~ 5 × 10^5, which is much larger than both the vortex-shedding period $U_\infty T_v/d \approx 5$ (where T_v is the average period of the vortex shedding from the upstream or downstream cylinder) and the time scale of turbulence $U_\infty T_{turb}/d \approx 360$, where $T_{turb} = L_x/u'$, L_x and u' being the longitudinal length scale and the r.m.s longitudinal velocity fluctuation.

The pattern-switching property obtained in the working section might possibly be affected by the residual turbulence in the wind tunnel and by uncontrollable external disturbances. This situation led to an idea of introducing the free-stream turbulence of sufficiently high intensity to counteract effects of the external and residual disturbances on the pattern switching. Results of this measurement showed that the time interval T_1 and the distance $l_{0.5}$ were relatively independent of Reynolds number but strongly dependent on the free-stream turbulence intensity. The effects of the length scale of the free-stream turbulence, which was in a range $L_x/d = 0.15 - 0.85$, on T_1 and $l_{0.5}$ were found to be relatively weak. Figure 4 presents the the non-dimensional time interval $T_1 U_\infty/d$ as a function of the turbulence intensity u'/U_∞, Reynolds numbers being in the range of $2.6 \times 10^4 - 3.8 \times 10^4$.

The T_1-u' curve has two different branches. One is for smaller turbulence intensities, and the other is for higher turbulence intensities. The former might be affected by the residual turbulence in the wind tunnel. This was because the autocorrelation function of the longitudinal velocity fluctuation (not shown) had a long tail for the smaller turbulence intensities. The long tail is expected to be associated with low-frequency disturbances in the wind tunnel. On the other hand, the autocorrelations for the higher intensities had no such a long tail; that is, the grid-generated turbulence counteracted the residual turbulence. Thus the second branch of the T_1-u' curve in Fig. 4 can be interpreted as being a 'true' bahaviour of the time interval associated with the grid-generated turbulence. Figure 5 demonstrates that the distance $l_{0.5}$ plotted against the turbulence intensity has the similar bahaviour as the time interval T_1. An interesting point in Figs. 4 and 5 is that there exists a large scatter of the data at a particular turbulence intensity $u'/U_\infty \approx 0.18$. This is because both T_1 and $l_{0.5}$ had a relatively strong dependence on Reynolds number at this turbulence intensity.

Conclusion

The present study has shown that the average time interval of the pattern switching is a rapidly decreasing function of the free-stream turbulence intensity, whereas the turbulence length scale has only an insignificant effect on the time interval. The time interval is much larger than the time scale of the free-stream turbulence L_x/u' and the period of vortex shedding from the circular cylinders.

References

1. Zdravkovich, M.M.: Review of low interference between two circular cylinders in various arrangements. ASME J. Fluids Eng. 99 (1977), 618.

2. Kim H.L.; Durbin, P.A.: Investigation of the flow between a pair of circular cylinders in the flopping regime. J. Fluid Mech. 196 (1988), 431.

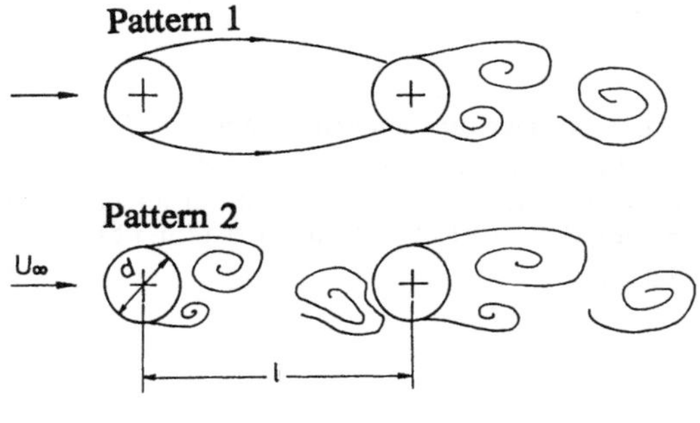

Fig. 1 Two patterns of flow around two circular cylinders of same diameter; pattern 1 : reattaching flow, pattern 2 : vortex-sheddingflow.

Fig. 2 Sample record of pressure fluctuation at time-mean rear stagnation point of upstream cylinder p_1. $Re = 2.44 \times 10^4$, $u'/U_\infty = 0.00158$, $L_x/d = 0.16$, $l_{0.5}/d = 3.47$.

Fig. 3 Average time interval of pattern 1, T_1, and distance $l_{0.5}$ as functions of Reynolds number for main flow without turbulence-generating grid. $u'/U_\infty = 0.0024$, $L_x/d = 0.855$. Closed symbols for $T_1 d/U_\infty$, open symbols for $l_{0.5}/d$ and different symbols for different experimental runs.

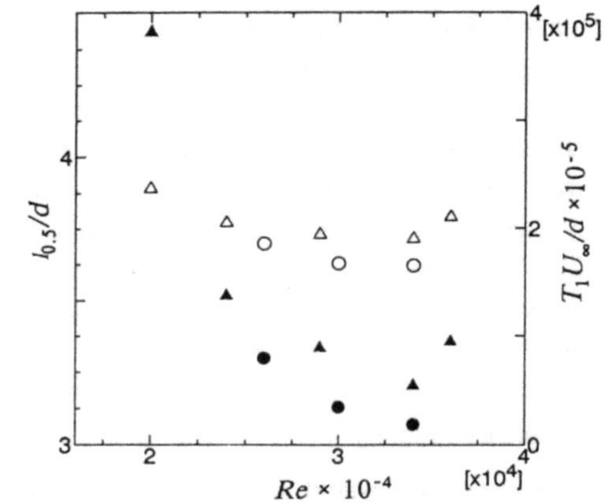

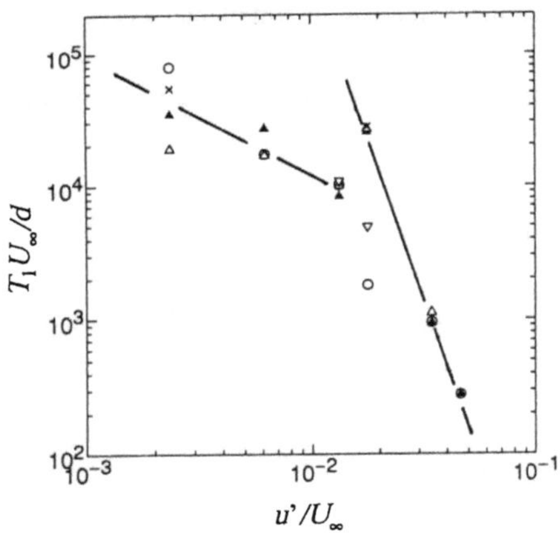

Fig. 4 Average time interval of pattern 1, T_1, as function of free-stream turbulence intensity. Different symbols for different Reynolds numbers in range of $Re = (2.6\text{-}3.8) \times 10^4$.

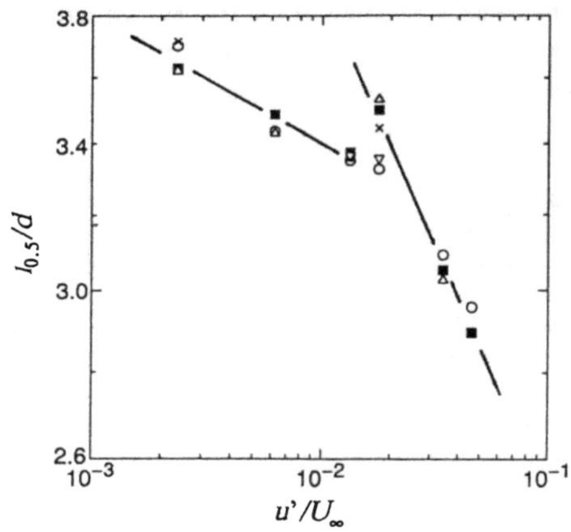

Fig. 5 Distance $l_{0.5}$ as function of free-stream turbulence intensity. Different symbols for different Reynolds numbers in range of $Re = (2.6\text{-}3.8) \times 10^4$.

The Effects of Waves on Vortex Shedding from Cylinders

R.H.Arkell and J.M.R. Graham
Department of Aeronautics,
Imperial College, London SW7 2BY, UK

Abstract

This paper describes the results of an investigation into the effects of an in-line combination of mean flow and free surface waves on the shedding of vortices behind a circular cylinder and the subsequent long term development of the wake. The investigation is carried out both numerically and by experiment. The numerical method is based on a viscous vortex (particle) method. The experimental investigation involves the towing of a cylinder through waves in a laboratory flume, and comparison of the transverse force with a quasi-steady model.

Introduction

Vortex particle methods based on the vorticity-streamfunction form of the unsteady Navier Stokes equations have been used successfully to predict 2-D flow over a circular cylinder and other bluff body sections (Stansby and Smith [1], Graham [2], Benson et al [3]). These methods have an advantage of low numerical diffusion when modelling the convection of complex evolving features of separated flow. In the present paper all the flow simulations are limited to 2-D, although vortex wakes of circular cylinders in uniform incident flow are normally three-dimensional for Reynolds numbers above 140. In the case of a vertical cylinder in waves the effect of vertical motion of the wake is also neglected. For higher Reynolds numbers a particle based k-ε model has been developed. Experimental data for a cylinder experiencing a combination of mean flow and free surface waves is somewhat limited; Matten [4] towed a vertical cylinder through waves, and oscillated a cylinder whilst towing it. Others (Mercier [5], Obasaju et al. [6]) performed experiments in which cylinders were oscillated in currents. The present paper is concerned with the effects of in-line free surface waves on the vortex wake of towed cylinders, for which the mean flow velocity component is greater than the particle velocity amplitude of the wave (u_W)

Wake evolution (Numerical Results)

The evolution of a vortex street interacting with a periodic progressive wave idealised by a point vortex array indicates that break-up is characterised by vortex pairing with ejection from the core (Graham, Arkell, Zhou [7]); here a more representative study of wave/vortex street interaction using a viscous particle based 2-D vortex method is reported. In this method vorticity is carried on particles and interpolated to a mesh for calculating convection and diffusion (Graham[2]). The present calculations use irregular grids so that mesh densities far downstream of the cylinder are chosen for good resolution of the developing wake. To study the effects of waves, a velocity field $u(t)=u+u_w.\sin(\omega t-kx)$ was added to the mean flow to represent the horizontal component of the wave (k is the wave number and ω the angular frequency). In addition to its self-induced velocity, the vortex street is subject to a temporal-spatial variation of velocity due to the progressive wave motion.

Fig.1 λ=20D (D~cylinder diameter); τ_w=25.06 (period); $u_w=0.5u_{cyl}$

Some vortex coupling is present, but full vortex ejection is stifled; coupled pairs of finite core vortices interact so as to diminish their strengths (cancellation). Shearing and collision lead to a loss of structure and tend to diminish vortex strength and subsequent interplay (Fig.1). We quantify a global transverse 'wake spreading' factor for various wave cases:

$$A(t)=\left(\sum_{p}^{ncell}\left(\Gamma_p.y_p\right)^2\right)^{\frac{1}{2}}$$

where Γ_p, y_p are circulation and transverse coordinate of cell p

Fig.2 is a plot of A(t) for various wave cases up to t*=70. Fig.2 shows that for wave cases with small particle velocity amplitudes (u_w=0.2), A(t) is not strongly dependent on u_w; vortex/vortex interactions are weak and wake break-up is performed substantially by a process of mixing and cancellation. There are three distinct regions in the wake: (a) **near wake region**; shedding and vortex structures are very similar to the Von-Karman street, (b) **transition region**; vortex/vortex interaction commences, (c) **far wake region**; the resultant structures from the end of the transition region adjust to the forcing wave. The point at which transition begins is dependent on the wavelength/period of the wave; the shorter the wavelength the closer the transition point lies to the cylinder, so that for very short wavelength waves (~10D) the transition point approaches the back of the cylinder and no 'normal' shedding region is apparent, the shedding becoming nearly symmetric. For higher particle velocity amplitude wave cases (u_w=0.5) vortex/vortex interactions are more dynamic due to the greater range of vortex strengths and greater convective effects of the perturbing wave. Mixing, shearing, collision and coupling are all much in evidence. The transition region starts near to the cylinder and occupies a relatively long region; high energy interactions within the transition region decay with distance downstream, leading to structures spatially consistent with the wave as in the lower particle velocity cases.

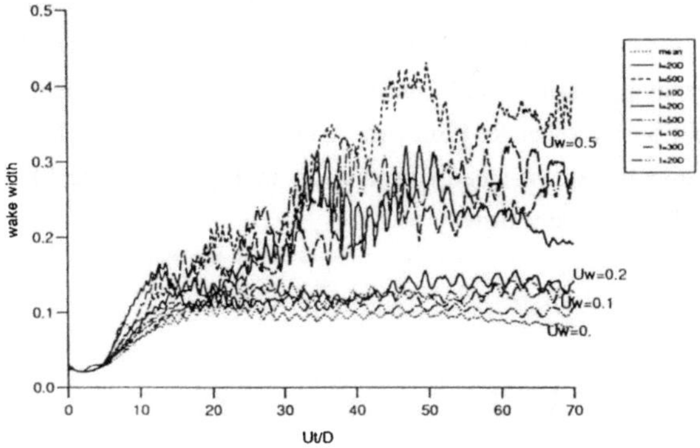

Fig.2

A modified k-ε turbulence model with wall functions was incorporated into the code for higher Reynolds number flows. The method employed extends the vortex particle method so that k and ε are interpolated onto particles for convection. Results for steady incident flow past cylinders indicate that most turbulent kinetic energy (k) is located at vortex centres; levels of k increase in the street up to about 10 D behind the cylinder where they reach a maximum and then decay downstream. The plot of turbulent kinetic energy contours below (Fig.3) shows concentrations of k at vortex centres; however, the highest levels exist in areas of vortex collisions and coupling.

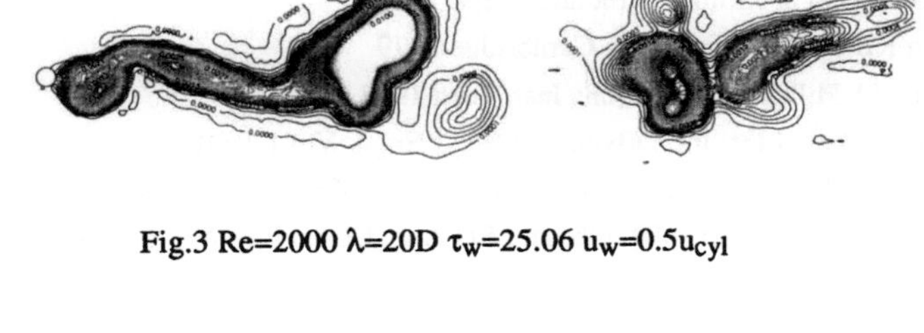

Fig.3 Re=2000 λ=20D τ_w=25.06 u_w=0.5u_{cyl}

Transverse forces (Experimental Results)

A vertical cylinder fitted with pressure tappings at one depth station was towed at steady speeds through in-line waves in a flume. Examples of a transverse force time history obtained by integrating the pressures are shown in Fig.4 for a case of steady towing at 25.9cm/s through waves of period τ_w=2.0s. This is compared with the quasi steady transverse force equation of Bearman, Graham and Obasaju [8], which has also been shown to agree with numerical predictions at lower Re (Graham, Arkell and Zhou [7]). The difference between the measurements and the quasi-steady prediction in Fig.4 is more likely due to the three-dimensionality of the incident flow.

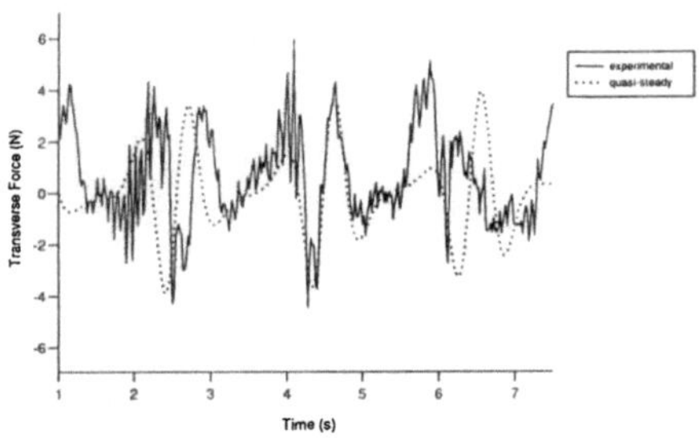

Fig.4

Acknowledgment

This work was supported by the United States Office of Naval Research under grant number N00014-90-J-4122

References

[1] Stansby PK, Smith PA Proc. 8th Int. Conf. Offshore Mech. & Artic Engg. The Hague, Vol 2 1989
[2] Graham JMR Numerical Methods for Fluid Dynamics III 1988
[3] Benson MG et al J.Fluids Structures Sept. 1989
[4] Matten RB PhD Thesis, Univ. Cambridge 1979
[5] Mercier JA PhD Thesis, Stevens Inst. Tech 1973
[6] Obasaju ED et al Int. Jnl. Offshore & Polar Engg. Vol 1 1991
[7] Graham JMR, Arkell RH, Zhou C Y Behaviour of Offshore Structures, 445 1992
[8] Bearman PW, Graham JMR, Obasaju ED App. Ocean Res. Vol 6 p166 1984

Calculation of Unsteady Separation from Stationary and Moving Walls

A. T. DEGANI AND J. D. A. WALKER

Department of Mechanical Engineering and Mechanics
Lehigh University
Bethlehem, Pennsylvania 18015, USA

Summary

The influence of a moving wall on boundary-layer separation at high Reynolds number is considered. For a translating and rotating cylinder, it is shown that separation is inhibited with increasing rotation rate and ultimately completely suppressed at relatively moderate wall speeds.

Introduction

Unsteady separation at high Reynolds numbers occurs in many situations, ranging from the relatively large-scale shedding of vorticity from bluff bodies to small-scale phenomena, such as "bursting" in transitional and turbulent boundary layers. Here the modern terminology for separation is used to imply a process wherein a viscous boundary layer, that is initially thin and attached to the surface, starts to develop strong outflows locally and thereby separates from the surface. In two dimensions, the prelude to separation can involve a wide variety of developing complex flow patterns in the near-wall flow. However, most occurrences at high Reynolds numbers evolve toward a generic terminal state, characterized by an abrupt focussing of the vorticity field into an erupting "spike" that moves rapidly away from the surface. This is the fundamental process in which vorticity leaves the vicinity of the wall at high Reynolds numbers; a narrow plume containing elevated levels of vorticity erupts into the mainstream, subsequently provoking a viscous-inviscid interaction that often culminates in a vortex roll-up. In three-dimensional unsteady boundary layers, a similar (but more complex) generic eruptive process occurs [1]. Through careful experiments [2], it has been possible recently to observe the narrow-band eruptions that are characteristic of boundary-layer separation at large Reynolds numbers. It may be noted, for example, that the principal mode of breakdown (or bursting) of the turbulent wall layer is due to such tightly focussed eruptions that lead to the production of new hairpin vortices [2].

In most circumstances, separation takes place as a particular fluid particle is compressed to zero thickness in the streamwise direction, thereby giving rise to a "spike-like" response in the displacement surface [3]; furthermore the most commonly observed situation is one in which the separation is moving in a direction opposite to the local mainstream at the instant of the eruption. The latter case has been referred to [3] as "upstream-slipping" separation and a number of examples of the phenomenon have been found [3, 4, 5]. The other possible extreme of "downstream-slipping" separation [3] has been controversial and is usually hypothesized to be associated with a downstream moving wall. Controlling and manipulating the sequence of events in boundary-layer separation is of considerable importance in a number of practical applications. In the present study, the use of moving walls in suppressing separation is investigated for the classic model problem of flow past an impulsively started translating and rotating cylinder in the limit as $Re \to \infty$.

Governing Equations

The nondimensional boundary-layer equations for the flow past a circular cylinder in terms of a standard Eulerian formulation are

$$\frac{\partial u}{\partial t} + u\frac{\partial u}{\partial x} + v\frac{\partial u}{\partial y} = U_e \frac{dU_e}{dx} + \frac{\partial^2 u}{\partial y^2}, \qquad \frac{\partial u}{\partial x} + \frac{\partial v}{\partial y} = 0 \tag{1}$$

The Reynolds number $Re = U_o a/\nu$, where U_o is the speed of the upstream approach flow, a is the cylinder radius and ν is the kinematic viscosity; U_e is the mainstream velocity given by $U_e = 2 \sin x$, where x measures angular distance from the upstream radius. The appropriate boundary conditions are

$$u \to U_e \quad \text{as} \quad y \to \infty; \qquad u = -\beta, \; v = 0 \quad \text{at} \quad y = 0, \tag{2}$$

where β is defined in terms of the cylinder rotational speed by $\beta = \Omega a/U_o$. An impulsive start is assumed at time $t = 0$, and the initial condition for u is given by Ece et al. [6]. It is well known [3], [4] that once the boundary layer starts to develop local regions of strong outflow, computations in the Eulerian system are not able to cope with the severe gradients that occur as the flow focuses into a narrow spike just prior to the onset of eruption. The difficulty may be overcome by recasting the boundary-layer equations in terms of Lagrangian coordinates [3], [4], viz.

$$\frac{\partial u}{\partial t} = U_e \frac{dU_e}{dx} + \left[\frac{\partial x}{\partial \xi}\frac{\partial}{\partial \eta} - \frac{\partial x}{\partial \eta}\frac{\partial}{\partial \xi}\right]^2 u; \qquad \frac{\partial x}{\partial t} = u, \tag{3}$$

and the associated continuity equation is

$$\frac{\partial x}{\partial \xi}\frac{\partial y}{\partial \eta} - \frac{\partial x}{\partial \eta}\frac{\partial y}{\partial \xi} = 1. \tag{5}$$

Here (x,y) are now independent variables representing the coordinates of a particle which at the initial time $t = t_o$ was located at (ξ, η). In the Lagrangian system, the trajectories of a large number of fluid particles are computed as a function of (ξ, η, t). The advantage is that the boundary-layer development may be evaluated accurately up to the time when a stationary point $(x_\xi = x_\eta = 0)$ develops in the field $x(\xi, \eta, t)$, signifying a local thickening and breakdown of the boundary-layer assumptions [3], [4].

The boundary-layer equations were first solved in an Eulerian coordinate system, and the solution of the velocity field at $t = t_o$ was used to initiate the Lagrangian calculations; the results are independent of the particular choice of t_o. The numerical algorithm used to solve the above equations is second-order accurate in time and space and employs an upwind-downwind scheme, as described in detail by Peridier et al. [4]. In this study, a mesh with 201 streamwise and 101 normal points was typically used with a time step of $\Delta t = 0.0005$.

Results and Discussion

The flow development was calculated for various values of the parameter β; the flow topologies are quite complicated and will be reported elsewhere. Note that the initial flow development has been described by Ece et al. [6]. For $\beta = 0$ the results obtained by Van Dommelen and Shen [3] were confirmed; in addition, the results obtained by Shen and Wu [7] for $\beta = 1$ and 1.5 were also reproduced. The focus in this study is on larger values of β and specifically determining the value of β beyond which the "spike-like" boundary-layer eruption is suppressed. Calculations were carried out for $\beta = 1.5$ through 2.0 in steps of 0.05.

Shortly after the impulsive start, two detached eddies develop in the first and fourth quadrant [6] (c.f. Figure 1). With increasing β the streamwise extent of the eddy near the lower surface L diminishes; however, the boundary layer first erupts in the fourth quadrant on the upstream side of this detached eddy. A typical streamwise velocity profile at separation in this range is shown in Figure 2, labeled $\beta = \beta_1$. In terms of $\theta = \pi - x$, the mainstream velocity in the direction of θ increasing is positive in the fourth quadrant, and the separation is moving upstream at the eruption. Note the shape of the profile is indicative of an effectively inviscid zone (where the velocity profile is flat) sandwiched between a viscous layer near the surface and a layer farther aloft [4]. With increasing wall speed, the streamwise location of the eruption moves downstream toward $\theta = 0$, and the speed at separation approaches zero, as for $\beta \simeq \beta_2$ in Figure 2. A further increase in β still results in a focussed eruption but now with the separation moving downstream (as with $\beta = \beta_3$). With increasing β the relative strength of the separation appears to diminish and occur at decreasing values θ_s; thus $U_e(\theta_s)$ decreases and eventually a critical value of $\beta = \beta_c$ is reached where the upper shear layer vanishes, as implied in Figure 2. At this value of β, separation is entirely suppressed; this critical value appears to be in the range $1.85 < \beta < 1.9$. For values of β beyond 1.9, a sharply focussed boundary layer eruption does not occur and the boundary layer appears to thicken gradually with time.

Acknowledgements: Work supported by the AFOSR under Grant No. 91-0069.

References

[1] Van Dommelen, L. L. and Cowley, S. J.: On the Lagrangian description of unsteady boundary-layer separation. J. Fluid Mech. 210 (1990) 593-626.

[2] Smith, C.R., Walker, J. D. A., Haidari, A. H. and Sobrun, U.: On the dynamics of near-wall turbulence. Phil. Trans. Roy. Soc. Lond. A 336 (1991) 131-175.

[3] Van Dommelen, L. L. and Shen, S. F.: The spontaneous generation of the singularity in a separating boundary layer. J. Comp. Phys. 38 (1980) 125-140.

[4] Peridier, V. J., Smith, F. T. and Walker, J. D. A.: Vortex-induced boundary-layer separation. Parts I and II. J. Fluid Mech. 232 (1991) 99-165.

[5] Doligalski, T. L. and Walker, J. D. A.: The boundary layer induced by a convected two-dimensional vortex. J. Fluid Mech. 139 (1984) 1-28.

[6] Ece, M. C., Walker, J. D. A. and Doligalski, T. L.: The boundary layer on an impulsively started rotating and translating cylinder. Phys. Fluids 27 (1984) 1077-1089.

[7] Shen, S. F. and Wu, T.: Unsteady separation over maneuvering bodies. AIAA J. 28 (1990) 2059-2068.

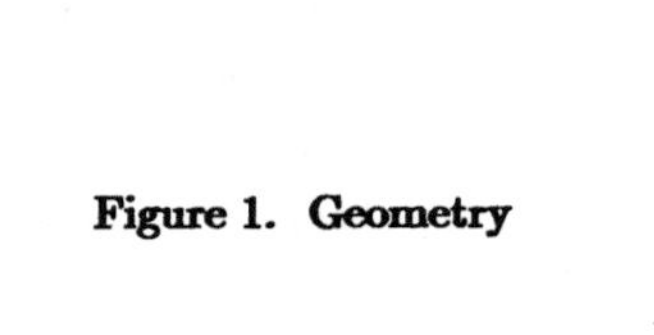

Figure 1. Geometry

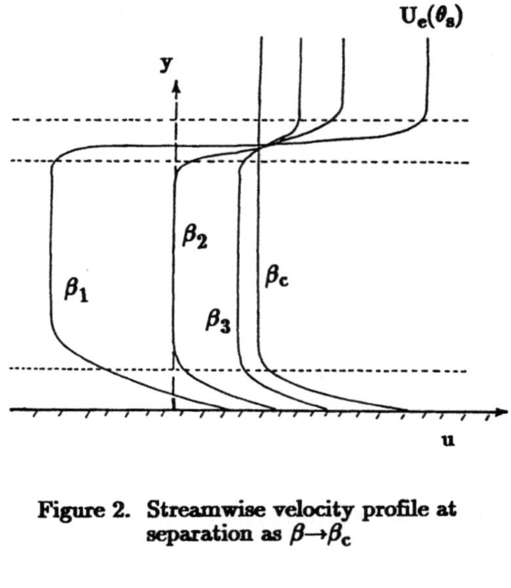

Figure 2. Streamwise velocity profile at separation as $\beta \rightarrow \beta_c$

Unsteady Three-Dimensional Separated Flows Around a Sphere - Analysis of Vortex Chain Formation

U. DALLMANN, H. GEBING, H. VOLLMERS

DLR - Institute of Theoretical Fluid Mechanics,
Bunsenstraße 10, 3400 Göttingen, Germany

Introduction

The near-wake vortex formation process behind blunt bodies is still unclear. While experiments often provide high-resolution of the far-wake vortex flow or the wall-flow (oil-flow visualization) it is still difficult to get insight into the unsteady, three-dimenisonal near-wake flow. Therefore, the aim of our analysis of a low Mach number ($Ma = 0{,}4$) numerical flow simulation around a sphere is to shed some light onto the near-wake flow region. Different types of vortices, vortex chains and vortex sheddings appear in experiments [1] for increasing Reynolds numbers. For the time being, we are limited to a $71 \times 71 \times 71$ mesh with approximately 25 grid points for $Re = 400$ in wall-normal direction within the boundary layer near the point of flow separation. Since the sequence of topological changes of the various quantities defining a flow field in parameter space is not necessarily dependent on the numerical resolution (although the quantitative differences for some fixed parameters may be considerable) we can hope to identify generic features of structural flow changes despite of the numerical resolution limitations given. A review of our earlier results is given in [2].

The following questions have been faced: Does there exist any connection/correlation between periodical topological (qualitative) changes of the vorticity field, pressure field, instantaneous streamlines, streak and timelines and the regions in space where complex eigenvalues of the velocity gradient tensor indicate Galileian invariant vortex flow regions? How does the temporal flow behaviour change with Reynolds number and how is it associated with spatial changes of various flow quantities in the wake formation zone?

The numerical simulation is based on a finite difference method to solve the complete, unsteady, three-dimensional, compressible Navier-Stokes equations. The implicit numerical algorithm used is of the Beam-Warming type and has second order accuracy in space and time. For details see [3].

Results and discussion

For $Re < 200$ the flow field stays steady and shows an axisymmetric closed separation bubble with all the details in good agreement with others' experimental or numerical results [1], [4]. For $Re > 200$ we find three-dimensional unsteady flows and vortex chains begin to form. A spatial left-right symmetry appears at first (its slow rotation reported earlier by Schulte-Werning et al. has been identified to be due to numerical errors) and the closed near-wake bubble with recirculating fluid breaks up. The so-called primary separation line remains almost stationary and circular but connects two critical points, one saddle point and one nodal point, due to the apparent flow symmetry. For $Re = 400$ only the rear stagnation point oscillates within that plane of symmetry with a Strouhal number $St = 0.133$ (Fig. 1). The hot wire technique [1] measures $St = 0.17$ and flow visualization provides $St = 0.114$. This Strouhal number belongs to the shedding of hairpin-shaped vortices. Our numerical flow visualization (Fig. 2) simulates timelines of hydrogen bubbles or smoke released from a circular ring around the sphere. Vortex chains can be seen. No far-field vortices appear in the instantaneous velocity fields in a frame of reference where the sphere is at rest. The appearance of a permanent plane of flow symmetry for $Re < 450$ and, therefore, zero side force is also in good agreement with experimental observation [1].

At $Re \approx 450$ the periodical sequence of topological changes of the skin friction pattern changes. An additional low Strouhal number ($St_2 = 1/4\ St_1$) appears already at $Re = 400$ but at $Re = 450$ it even dominates the spectrum of drag and lift. It is associated with a periodical azimuthal motion of a saddle-node-saddle pattern change (Fig. 1). This wall flow pattern is in agreement with similar numerical simulations [4] for $Re = 500$. The side force time record shows that a quantitative and qualitative change occurs (Fig. 3). For $Re = 400$ the side force coefficient converges slowly to zero, oscillating with St_1. For $Re = 500$, after a transient with almost vanishing side force, an amplification of side force unsteadiness sets in. Lift and side forces reach comparable amplitudes. Their spectral behaviour becomes 'noisy' for $Re \geq 500$ while the drag coefficient still exhibits a single dominating spectral peak even at $Re = 750$. These findings are in qualitative agreement with experiment where one finds an enhancement of a lateral flipping motion of the wake resulting in unsteady normal forces as Re increases further. A comparison with experimental Strouhal numbers fails for the time being since even long-time computations have not reached a stable periodical state yet (Fig. 3). We also identify instantaneous secondary flow separations which occur at first at $Re = 450$ (Fig. 1). We note that secondary vortex flow separations are necessary (of course not sufficient) for axisymmetric vortex shedding to occur. According to [1] almost axisymmetric near-wake vortex shedding has been observed at $Re > 750$ together with a Strouhal number branching. However, for the time being the 71^3-mesh does not seem to be fine enough to resolve the associated high-frequency mode as an instability of the separating high-shear layer.

In Fig. 4 we compare the changes of the instantaneous velocity field (which occur only in the near-wake, Fig. 2) with the corresponding regions of vortical flow (identifyable by regions of complex eigenvalues of the velocity gradient tensor [2]) and indicate locations of instantaneous static pressure minima. The periodical topological changes in Fig. 4 show that fluid and vorticity are driven periodically up and down, toward and away from the plane of symmetry and this is accompanied by convoluted loops of three-dimensional vortical flow regions (Fig. 5) and vorticity structures (Fig. 6). The vorticity field shows streamwise oriented vortex lines with hairpin-shaped, tongue-like loops extending into the far-wake and vortex-line reconnections. Further details of the periodically appearing three-dimensional convolutions of streak surfaces, regions of vortical motion, vorticity and iso-pressure surfaces, etc., will be published elsewhere.

References

[1] Sakamoto, H., Haniu, H., *A Study on vortex shedding from spheres in a uniform flow.* Journal of Fluids Engineering, Vol. 112 (1990) 386-392.

[2] Dallmann, U., Kordulla, W., Vollmers, H., Schulte-Werning, B., *Analysis of the changing topological structures of three-dimensional separated flows.* DLR-IB 221-92 A12 and in: 'Physics of Separated Flows' K. Gersten (Ed.), Vieweg-Series 'Notes on Numerical Fluid Mechanics', 1992 (to appear).

[3] Schulte-Werning, B., *Numerische Simulation und topologische Analyse der abgelösten Strömung an einer Kugel.* DLR-FB 90-43, 1990 (PhD-thesis Munich Technical University, 1990).

[4] Shirayama, S., *Flow past a sphere: topological transition of the vorticity field.* AIAA Journal, Vol. 30, No.2 (1992) 349-358.

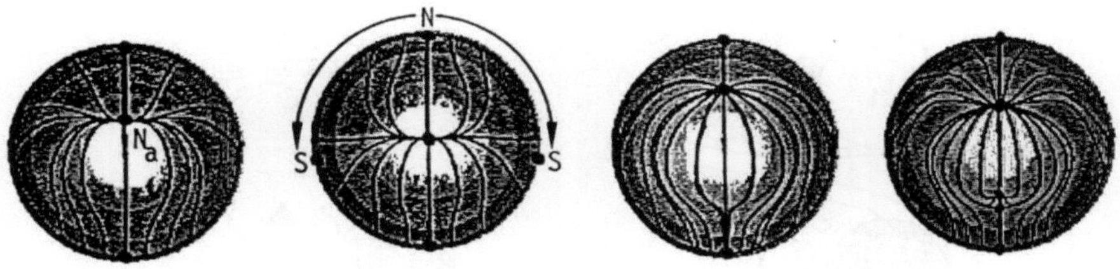

Fig. 1: Time sequence of wall-flow pattern in the separated sphere-flow region for $Re = 450$. The rear attachment node N_a oscillates within the vertical plane of flow symmetry with St_1. The saddle-node-saddle (S←N→S) splitting leads to oscillatory motions of the saddle points S with St_2. For $Re = 400$ only pattern t_1 forms and N_a oscillations occur.

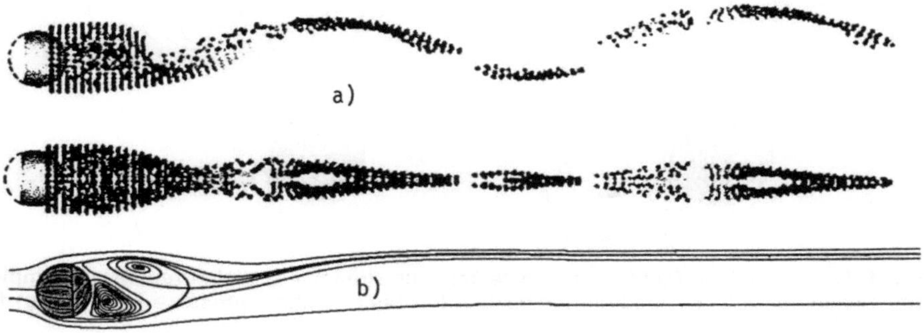

Fig. 2: a) Side view and top view of time lines and streak surface; b) instantaneous streamlines in plane of flow symmetry ($t = t_1 = 393.41$) for $Re = 400$.

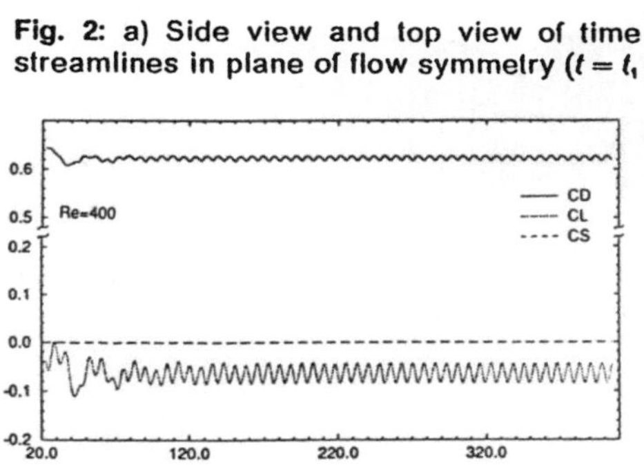

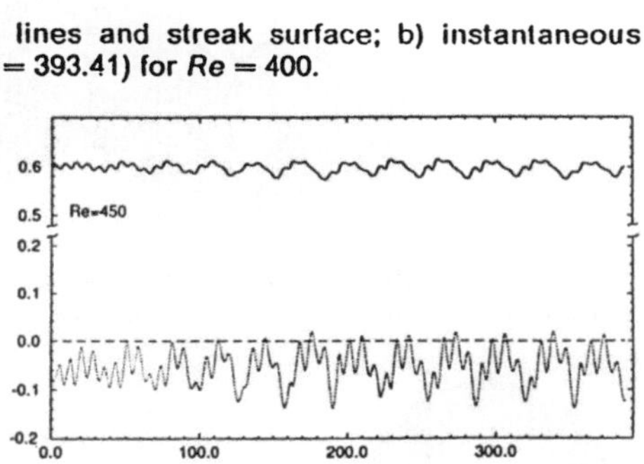

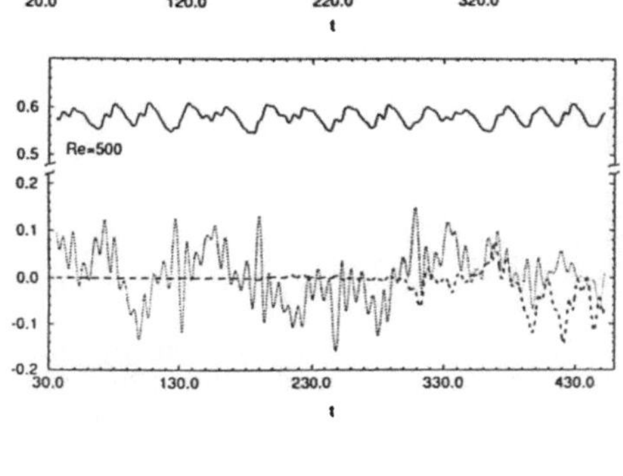

Fig. 3: Time records of drag CD, lift CL and side force coefficient CS for $Re = 400, 450, 500$. The amplification and oscillation of lift and side forces mimic the onset of the experimentally observed lateral flipping motion of the wake.

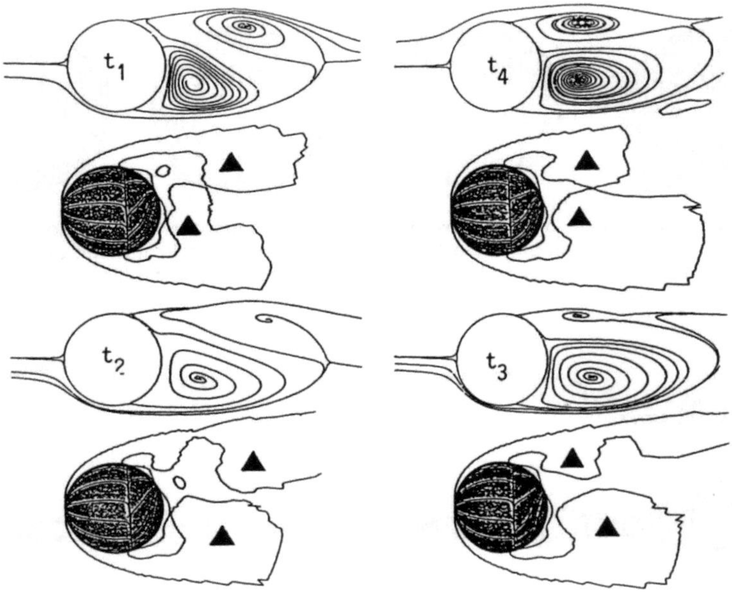

Fig. 4: Time sequence ($t_1 t_2 t_3 t_4 t_1$ ...) of instantaneous streamlines and of regions of complex eigenvalues of velocity gradient tensor and locations of static pressure minima (▲) for $Re = 400$.

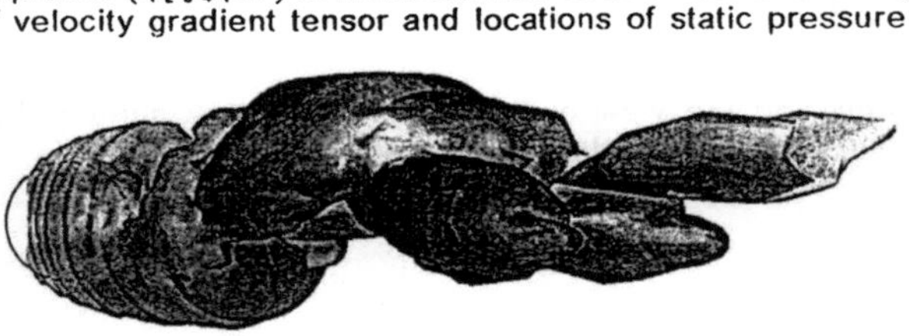

Fig. 5: Perspective view of the three-dimensional surface surrounding the complex eigenvalue region. Vortex chains appear as convoluted loops.

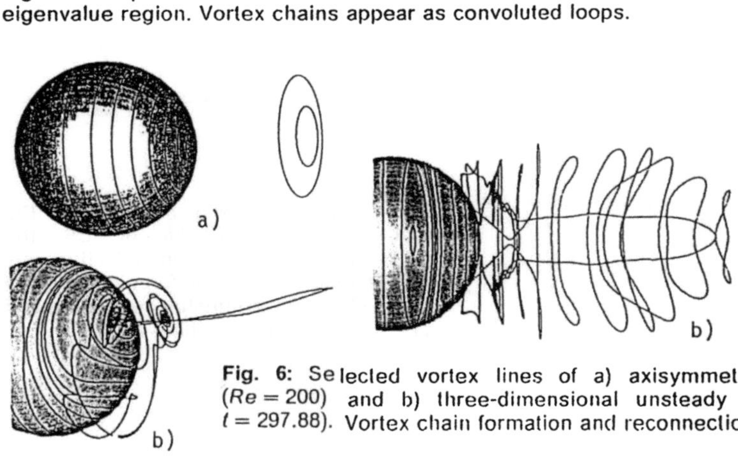

Fig. 6: Selected vortex lines of a) axisymmetric steady flow ($Re = 200$) and b) three-dimensional unsteady flow ($Re = 450$. $t = 297.88$). Vortex chain formation and reconnection can be seen.

Separated Fluid Flows - Steady, Periodical and Transitional Regimes (Numerical Method, Results and Comparisons)

V.A.GUSHCHIN and V.N.KONSHIN

Institute for Computer Aided Desing Russian Academy of Sciences, 2-nd Brestskaya str., 19/18, Moscow, 123865, Russia, phone: /007/095/2509541, fax: /007/095/2508928, telex: 411700 808 ICAD

Introduction

The most part of phenomena taking place in the unsteady and transitional fluid flows around finite bodies is characterised by the existance of the large scale organized structures in them (laminar and turbulent boundary layers, free shear layers, near wake, jets, etc.). Supposing that the dynamics of such structures is described by hydrodynamic equations without any turbulence models it is possible to use Navier-Stokes equations for the description of it's behaviour in the boundary layer and in the near wake for unsteady and transitional flow regimes around finite body using special tranformation of coordinates. To do this it is necessary to constuct numerical method which possesses by some special properties. The finite difference scheme of method must be second order of accuracy in space variables, to possess by minimal scheme viscousity, to be workable in wide range of Reynolds numbers and to be monotonic.

Numerical method and finite-difference scheme

One of the version of the Splitting on physical factors Method for Incompressible Fluid (SMIF) [1], which has been generalized now for three-dimensional problems [2], nonhomogeneous fluids [3] and flows with a free surface [4] is used here for the considered problem.

For the adequate numerical modelling of hydrodynamic flows with large gradients of parameters it is necessary to construct finite-difference scheme with high-order of approximation which has a minimal scheme viscosity and is monotonic and stable over a wide range of Reynolds numbers.

Here we use constructed hybrid finite difference scheme which is based on the combination of modified upwind scheme and modified central difference scheme for the convective terms in the equations of motion with the central differences for the viscous terms.

Some numerical results

At Reynolds numbers $10<Re<40$ steady symmetrical separated flows around circular cylinder is stable. At least two regimes exist at $Re>40$ - unstable symmetrical and stable periodic. The periodic flow which is realized at given fixed Reynolds number doesn't depend on the way of its receiving (influx-outflux of the different intensity, rotation, etc.).

The dependence on time of the total lift coefficient C_y is shown in Fig.1 for $Re=55$, 100 and 225.

At $Re=100$ only the alternative separation from both sides of the cylinder exists. At moderate Reynolds numbers ($200<Re<500$) the unsteady flow is essentially complicated by the existence of the secondary effects: the instability in separated shear layers during some part of the period, the secondary vortexes on the rear part of cylinder surface, the secondary vortexes near reattachment point.

Our calculations, which are based on the adequate reproduction of large scale vortex structures arising in the unsteady separated flows around finite body (boundary layer, free-shear layer, near wake etc.) allow to simulate transitional fluid flows also. The comparison of numerical results (time-average surface pressure distribution) with the experimental data [5] for $Re=4\cdot 10^5$ shown in Fig.2, where also shown experimental data [6] for laminar flows ($Re=10^5$) and experimental data [7] for turbulent flow ($Re=8.6\cdot 10^6$).

The crisis of total drag coefficient and sharp rise of the Strouhal number take place and are simulated numerically (without any turbulence models) for the critical Reynolds numbers ($Re \sim 4\cdot 10^5$), what is in a good agreement with an experimental data [8].

REFERENCES

1. O.M.Belotserkovskii; V.A.Gushchin; V.V.Schennikov. U.S.S.R. J.Comp.Math.Math.Phys. 15 (1975) 190-200.
2. V.A. Gushchin. U.S.S.R. J.Comp.Math.Math.Phys. 16 (1976) 251-256.
3. V.A. Gushchin. U.S.S.R. J.Comp.Math.Math.Phys. 21 (1981) 190-204.
4. O.M.Belotserkovskii; V.A.Gushchin; V.N.Konshin. U.S.S.R. J.Comp.Math.Math.Phys. 27 (1987) 181-191.
5. N.M.Bychkow; V.V.Larichkin. Inst.Teor.and Appl.Mech. SB AS USSR. 1658 (1986) 1-57.
6. A.Fage; V.M.Falkner. ARC RM 1369 (1921).
7. A.Roshko. J.Fluid Mech. 10 (1961) 345-356.
8. V.A.Gushchin; V.N.Konshin. Int. J. Numer. Meth. in Engr. 34 (1992) 469-472.

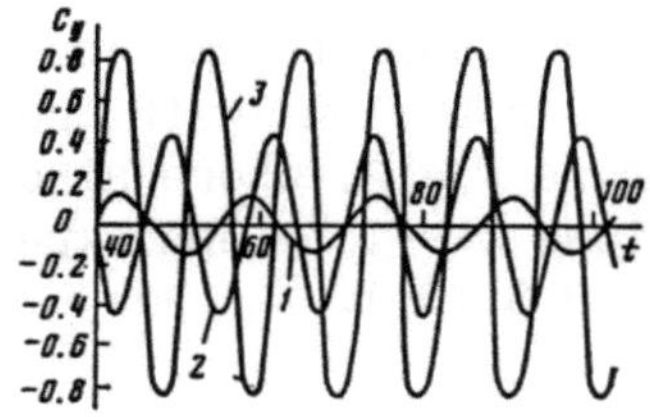

Fig.I. Total lift forth coefficient. I-Re=55, 2-Re=IOO, 3-Re=225

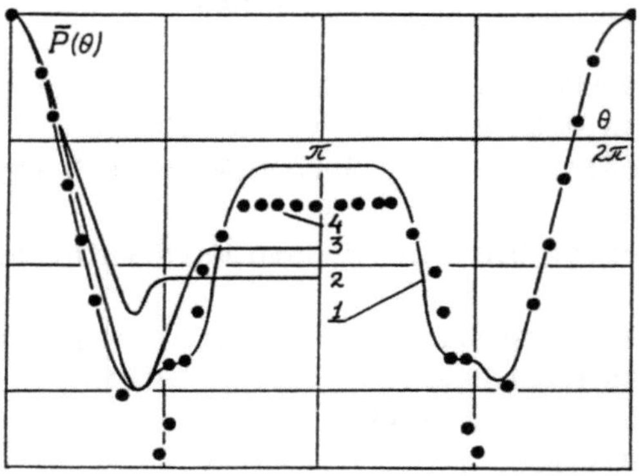

Fig.2. Time-average wall-pressure.
I-present data, Re=400000,
experiments:
2-laminar flow, Re=I00000 [6],
3-turbulent flow, Re=8600000 [7],
4-transitional, Re=400000 [5].

Simulation of Separated and Reattaching Flows Around a Circular Cylinder at Critical Reynolds Numbers

T. Tamura
Ohsaki Research Institute, Shimizu Corporation

T. Otaguro
Institute of Computational Fluid Dynamics

K. Kuwahara
The Institute of Space and Astronautical Science

SUMMARY

We simulate the separated and reattaching flows around a circular cylinder at the critical Reynolds number by the direct finite difference method without turbulence model. The critical flows and the pressure distribution on the cylinder surface are numerically investigated in detail. Their structures and unsteady characteristics are discussed in case the drag crisis of the cylinder occurs.

INTRODUCTION

There have been so far a great deal of numerical studies associated with the circular cylinder flows. However these flows have a undesirable characteristics for computation, such as massive flow separation and wake turbulence in nature. Accordingly the previous studies are almost limited to the simulation of two-dimensional flows at relatively low Reynolds numbers. Recently, the high Reynolds number flows around a cylinder have been calculated by the 3-D direct finite difference method by Tamura et al.[1] and it is concluded that three-dimensional computation is essentially required even for the two-dimensional geometric problem in order to understand the physical mechanism of the flows with separation region. Because the two- and three-dimensional computational results about the wake structures sometimes become very different each other. Here we compute the sub- up to supercritical 3-D flows around a circular cylinder which is assumed to be periodic in the spanwise direction. The third-order upwind scheme[2] is applied to the convection terms of the incompressible Navier-Stokes equations. We bring into focus the dynamic behavior of the separated and reattaching shear layer at the critical Reynolds number where the drag force suddenly drops(the drag crisis). The pressure distribution in the separation bubble is also discussed in order to clarify the mechanism of the drag crisis from a physical point of view.

COMPUTATIONAL MODEL AND BOUNDARY CONDITIONS

Figure 1 displays the grid system for a circular cylinder. The number of grid points is 400×100×40. The O-type of grid system is generated for a cylinder section by solving the

Poisson equation. The circular cylinder is placed a little distance forward of the center of the elliptic computational domain (the upwind and downwind size = 20, 30*D*, *D* being the diameter of the cylinder). The aspect ratio(*H*/*D*, *H* being the span length or the height of the cylinder) is 1.0. The Reynolds numbers(*Re*) computed range over $10^4 \sim 10^6$.

FLUCTUATING FORCES AND PRESSURE IN SEPARATION BUBBLE.

Figure 2 presents time variations of the computed drag and lift coefficients. The Reynolds number is 5×10^5. Owing to the Karman-type vortices, the lift coefficient is periodically changing, but this fluctuation is not clearly sinusoidal. While the drag coefficient is almost constant stably.

Figure 3 shows time variations of the computed pressure coefficients near the separation point at $Re=5\times10^5$. By comparison with the lift fluctuation, we actually can find the fluctuation with the Karman-type vortex shedding frequency. Near the separation point much more high-frequency fluctuation is seen, which is based on small vortices shedding close to the separation point. The level of this fluctuation becomes larger at sometime and smaller at other time, and is changing intermittently by the lower frequency than vortex-shedding. This means the position of the separation bubble is moving by this low frequency.

Figure 4 exhibits the computed instantaneous pressure fields on the cylinder surface at the critical Reynolds number. Coming leeward from the stagnation point on the cylinder surface, the 2-D structure of the initial flow field changes into spanwise wavy structures by the vortex instability(time *A*, *D* and *E*). The checker-board type of pressure distribution appears around the separation region. In the more leeward region, 3-D structures with various scales are clear and flow patterns become more complicated. While, at the other time step, we can find the different kind of patterns, for example, the inclined structure appears in the separation bubble region(time *B*, *C* and *F*). It should be noted that the inclined directions are different between the cases at time *C* and *F*.

RELIABILITY OF DRAG PREDICTION

Figure 5 depicts the drag coefficient as a function of the Reynolds number for a circular cylinder. It includes an experimental typical line based on the data by Cantwell and Coles[3]. The computational drag coefficient is dropping rapidly and the value is very similar to the experimental data. It is discernible that the drag crisis appears in the 3-D computational result.

THREE DIMENSIONAL FLOW STRUCTURES

Figure 6 illustrates the instantaneous 3-D streamlines past a circular cylinder at $Re=5\times10^5$. It is recognized that the separated fluid particles are flowing with rotated by the inclined longitudinal vortices. With great interest, these vortices are found to be also merging. In the figure overlapped with the pressure distribution, it is found that the sharp spiral motion of streamlines appears along these low pressure zone. This wavy or inclined structure is assumed to be a clue of the longitudinal vortices over the critical Reynolds number where the longitudinal vortices are dominant compared to the lateral vortices.

CONCLUSION

Computational results indicate that turbulence phenomena of the flow field including separation region can be captured by the direct finite difference technique. It is concluded that the 3-D computations can give a good solution simulating the flows over a cylinder through the wide range of the Reynolds numbers, especially the critical regime. At the appearance of the drag crisis, we find that longitudinal vortices appear near separation points. In the separation bubble, it is recognized that there are several kinds of three-dimensional patterns of the pressure distribution. We thus conclude that the 3-D structures play an important role in the drag reduction of a circular cylinder.

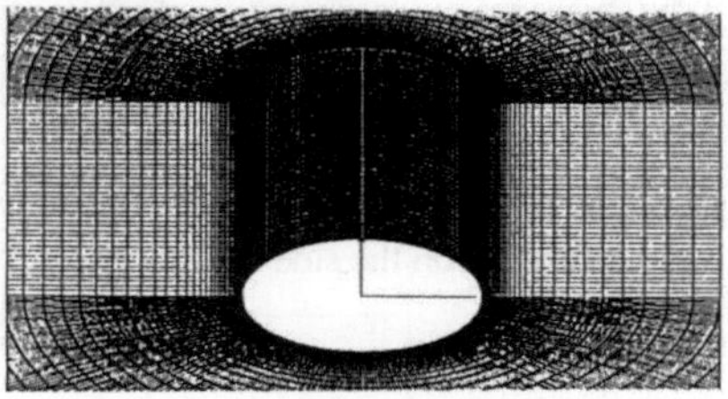

Fig. 1 Grid system for a three-dimensional circular cylinder (H/D=1.0, 400×100×40 points).

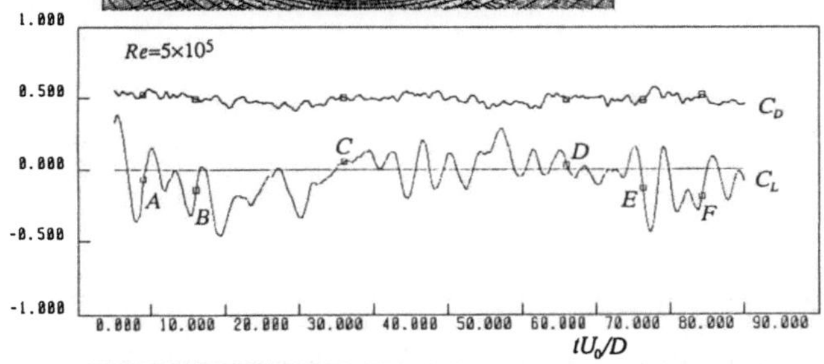

Fig. 2 Time variations of drag and lift coefficients.

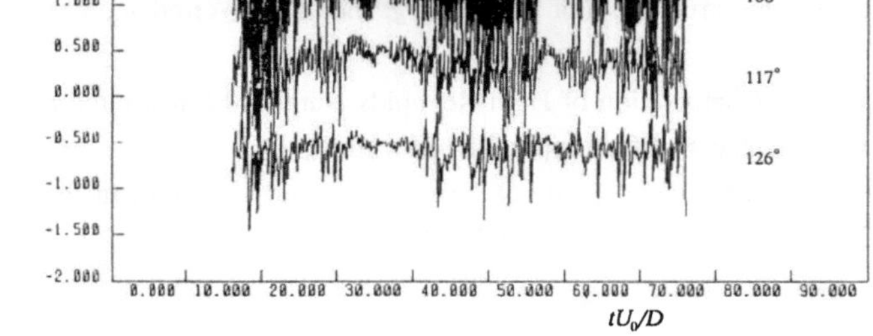

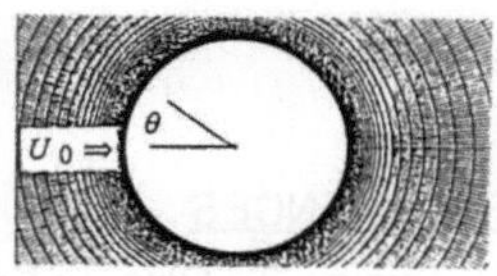

Fig. 3 Time variations of pressure coefficients near the separation point.

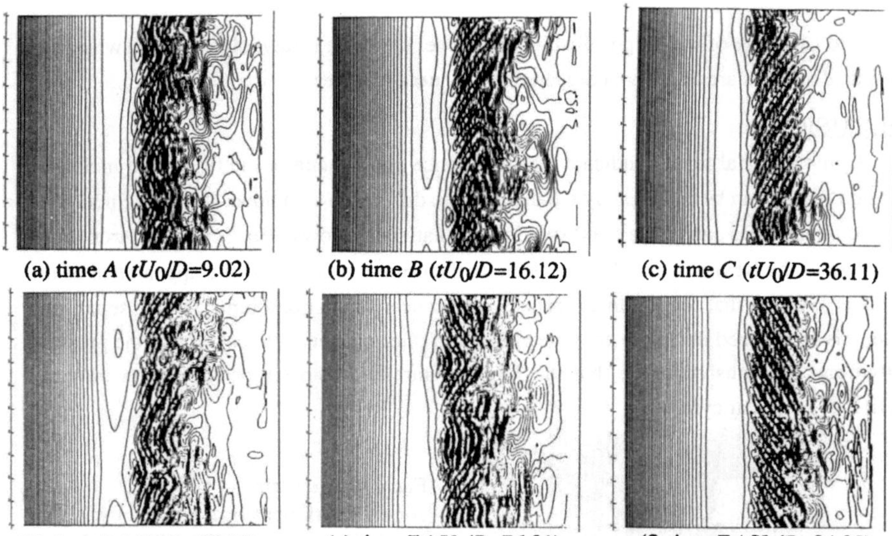

(a) time A (tU_0/D=9.02) (b) time B (tU_0/D=16.12) (c) time C (tU_0/D=36.11)

(d) time D (tU_0/D=66.05) (e) time E (tU_0/D=76.31) (f) time F (tU_0/D=84.30)

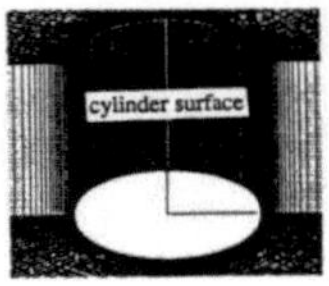

Fig. 4 Computed instantaneous pressure field on the side surface of a cylinder at $Re=5\times10^5$.

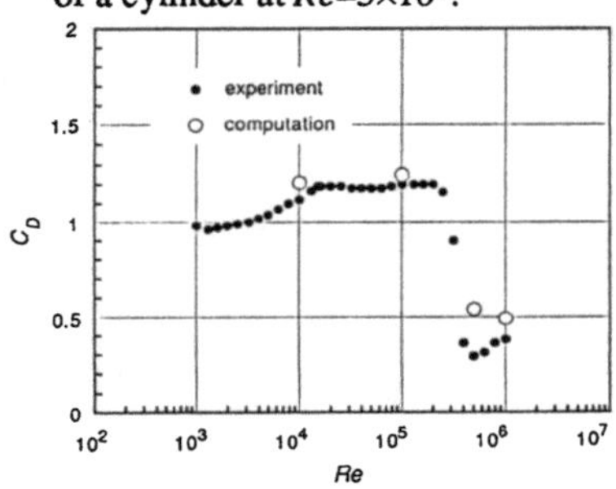

Fig. 5 Drag coefficient as a function of the Reynolds number for a circular cylinder.

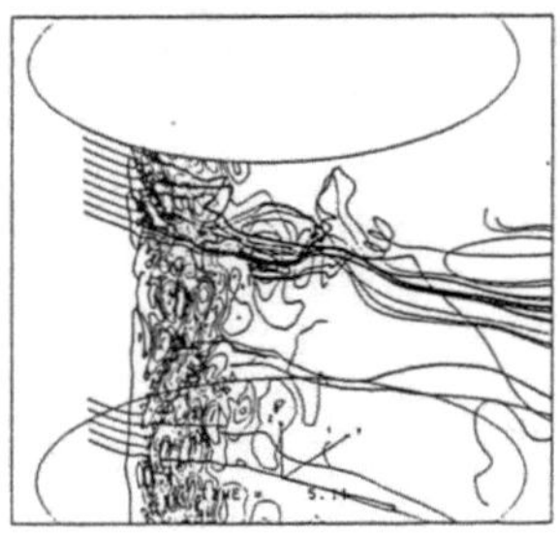

Fig. 6 Instantaneous separated flows with pressure distribution at $Re=5\times10^5$.

REFERENCES

1. Tamura, T.; Ohta, I.; Kuwahara, K.: On the Reliability of Two-dimensional simulation for unsteady flows around a cylinder-type structure. J of Wind Engrg. and Indust. Aerodyn., Vol. 35. 1990.

2. Kawamura, T.; Kuwahara, K.: Computation of High Reynolds Number Flow around a Circular Cylinder with Surface Roughness. AIAA paper, 84-0340, 1984.

3. Cantwell, B.; Coles, D.: An Experimental Study of Entrainment and Transport in the Turbulent Near Wake of a Circular Cylinder. J. Fluid Mech., Vol. 136, 1983.

Unsteady Base Pressure Measurements in the Near Wake of a Cylinder With Imposed Three-Dimensional Disturbances

J. P. BORG and A. A. SZEWCZYK

Department of Aerospace and Mechanical Engineering
University of Notre Dame
Notre Dame, IN 46556, USA

Abstract

The effects of geometrically imposed three-dimensional disturbances that were introduced by a spanwise-periodic trailing-edge splitter plate on the instantaneous base pressure of a circular cylinder were investigated. Measurements indicate a direct coupling to the shedding frequency between the stagnation-line and base-pressure signals. A low frequency component in the base pressure power spectra was found and was directly related to the irregular shedding observed in the wake.

Introduction

It has been previously shown by Roshko [1,2] that the vortex shedding structure can be altered by the introduction of a splitter plate in the near wake of a circular cylinder. A splitter plate has the effect of modifying the region of absolute instability behind a bluff body and in some cases was observed to either enhance or attenuate vortex strength, Roshko[1] and Gerrard [3]. Use of splitter plates delays any direct interaction between the two opposing shear layers and lengthens the regions behind the body before vortices are fully formed, Apelt & Isaacs [4], Apelt, West & Szewczyk [5], Igarashi [6], Unal & Rockwell [7]. A comparison of the measurements in the wake of a circular cylinder with a two-dimensional straight splitter plate to those with a spanwise-periodic trailing-edge splitter plate, Pearson & Szewczyk [8], indicates a variation in wake width, formation length and vortex shedding modes. The present investigation was undertaken to study the effects of a spanwise-periodic trailing-edge splitter plate on the instantaneous fluctuating base pressures.

Experimental Arrangement

All of the experiments were conducted in the University of Notre Dame 61cm. square test section open circuit wind tunnel with a maximum speed of 35 m/s and turbulence intensity of less than 0.1 %. The inlet was 152 cm. and had a 24:1 contraction.

The model was a 5.1 cm. diameter aluminum circular cylinder with a groove along the wake centerline to accommodate a splitter plate. Experiments were performed on two-dimensional splitter plates of varying length as well as spanwise-periodic trailing-edge splitter plates of varying length, amplitude and wavelength. All splitter plates were 1.6 mm thick. The two-dimensional plates had length to diameter ratios of ℓ/D = 0.125, 0.5, 0.625, 1.0, 1.5 and 2.0. Spanwise-periodic trailing-edge splitter plates of varying plate length, ℓ/D = 0.625, 1, 2, and constant amplitude and wavelength, a/D = 0.5 and λ/D = 3 respectively, as well as varying amplitude, a/D = 0.5, 0.875, 1.875, and constant plate length and wavelength, ℓ/D = 2 and λ/D=3 respectively were studied. Peaks and valleys were designated as the splitter plate positions with the most downstream and upstream trailing edge respectively. End plates were used in all experiments and had a 2.5 aspect ratio as recommended by Stansby [9]. The end plates were slotted to accommodate a splitter plate and mounted such that the length-to-diameter ratio was 8.

The model was equipped with flush mounted pressure transducers along the stagnation line and in the base region, at 170 degrees relative to the stagnation line, see Fig. 1. Calibrated pressure transducers had a range of 0-10 inches of H_2O, a sensitivity of 5mv/in H_2O, a frequency response of 50 kHz and a linear response up to 2.5 kPa.

The shedding characteristics from the model were detected with a hot-wire probe using a TSI Flow Analyzer IFA-100. Data acquisition of velocity and pressure signals as well as control of a three-dimensional traverse unit for the probe was performed through a Macintosh IIfx computer with LABView hardware and software. The system had the capability of sampling eight simultaneous signals at a rate of 250 kHz per channel.

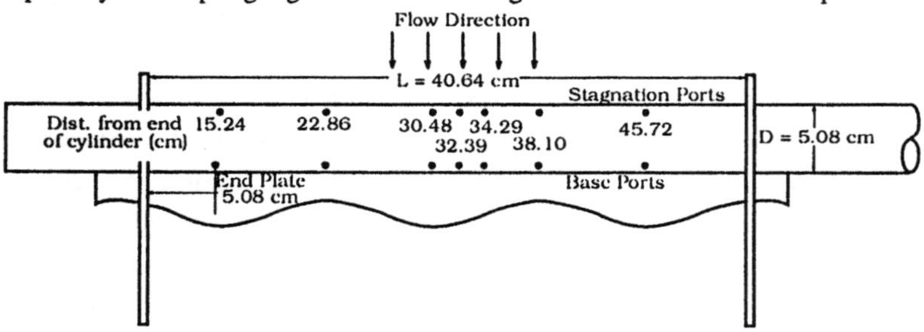

Fig. 1. Cylinder transducer configuration.

Experimental Results

Tests were conducted at Reynolds number of 30,000 in the wind tunnel and with model described above. Unsteady pressure measurements were made at several spanwise locations along the stagnation line and in the base region. Long time averages of the pressure signals were found to be in good agreement with other mean pressure measurements previously reported by Apelt et al.[4, 5] for two dimensional splitter plates and by Pearson &Szewczyk [8] for spanwise-periodic trailing-edge splitter plates.

Selected simultaneously sampled fluctuating base pressure signals acquired from transducers on the cylinder for a spanwise-periodic trailing-edge splitter plate ℓ/D=1.0, a/D=0.5, and λ/D=3.0 and equivalent two-dimensional splitter plates ℓ/D=0.5 and 1.0 are shown in Fig. 2. The transducers are located at the same spanwise positions that correspond to a location of a peak and valley for the periodic splitter plate for all splitter plate configurations. Several observations can be made about the features of the fluctuating pressures. Stagnation-line pressure traces, Figs. 2(a-c), were correlated at the shedding frequency, 38.3 Hz and were always in phase with each other. Spanwise base pressure signals from peak and valley locations Figs. 2(d-i), reveal a strong correlation with the shedding frequency and were always in phase except for phase reversals shown in the Figs. 2(d-i). For the equivalent two-dimensional splitter plates, Figs. 2(f-i) similar phase reversals were also observed. These measurements indicate a direct coupling to the shedding frequency between the stagnation-line and base pressure signals as was observed by Mangalam & Kubendran [9] for a plain circular cylinder. However, no direct feedback of the non-regular phase reversals indicated in Figs. 2(d-i) was found in the stagnation-line pressure signals. For ℓ/D=0.625 the stagnation-line peak and valley pressure signals were found to be 180° out of phase with the base pressure signals at the same spanwise location.

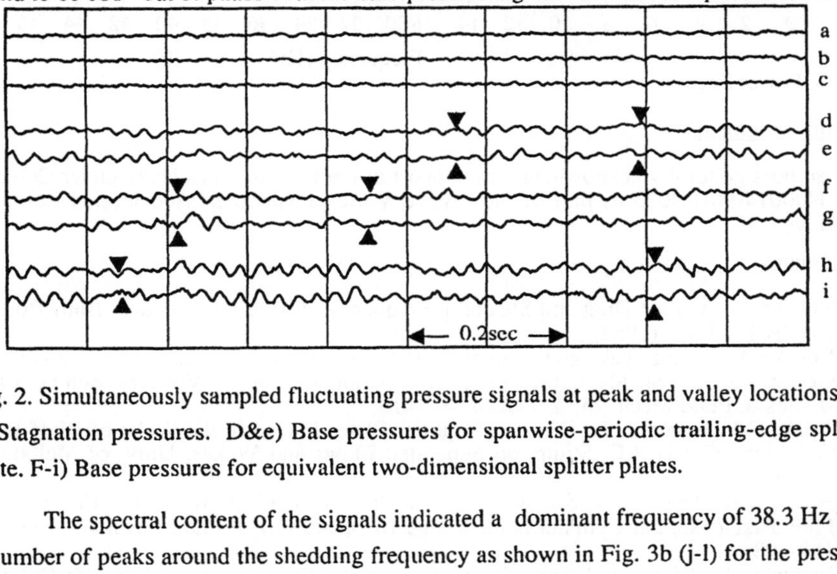

Fig. 2. Simultaneously sampled fluctuating pressure signals at peak and valley locations. A-c) Stagnation pressures. D&e) Base pressures for spanwise-periodic trailing-edge splitter plate. F-i) Base pressures for equivalent two-dimensional splitter plates.

The spectral content of the signals indicated a dominant frequency of 38.3 Hz with a number of peaks around the shedding frequency as shown in Fig. 3b (j-l) for the pressure traces shown in Fig. 2d, 2f, & 2i respectively. The differences in frequency in the peaks around the shedding frequency show up as a low frequency component, 2-5 Hz, in the power spectrum shown in Fig. 3a. Since these are long-time average spectra the peaks in the spectra indicate that the shedding is not regular and different modes of shedding occur in time. As a result the flow exhibits some inherent three-dimensional effects in the shedding process. The results from the pressure measurements low frequency content in the power spectra and irregular shedding, agree with the hot-wire measurements and flow visualization observations of Pearson & Szewczyk[8].

Correlation between peak and valley base pressure signals and hot-wire signals at the peak, valley and mid-point spanwise location both inside and outside the wake was performed. Correlation's inside the wake indicate a phase lag of approximately 90° between the base pressure and hot-wire signals equivalent to the spatial separation of the two signals divided by the convection velocity of the flow for all spanwise locations. Signals from the hot-wire outside the wake were fully correlated to the shedding frequency and indicated no phase lag at any of the spanwise locations.

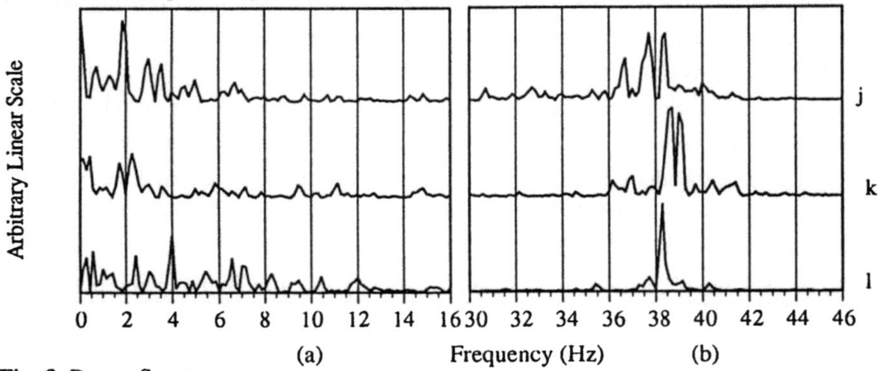

Fig. 3. Power Spectra.

Acknowledgment

The authors gratefully acknowledge the support of the United States Navy under Contract No. N00014-90-J-4083 as part of the "Wake Vortex Dynamics" Accelerated Research Initiative.

References

1. Roshko, A.: On the Drag and Shedding Frequency of Two-Dimensional Bluff Bodies. NACA TN 3169 (1954).
2. Roshko,A.: On the Wake and Drag of Bluff Bodies. J. Aero.Sci. 22 (1955) 124-132.
3. Gerrard, J.H.: The Mechanics of the Formation Region of Vortices Behind Bluff Bodies. J. Fluid Mech. 25, 2 (1966) 401-413.
4. Apelt, C.J.; Isaacs, L.T.: Effects of Splitter Plates Placed in the Wake of Bluff Body Cylinders. C.A.A.R.C. Symp. on Separated Flows and Wakes, Univ. of Melbourne (1970).
5. Apelt, C.J.; West, G.S.; Szewczyk, A.A.: The Effects of Wake Splitter Plates on the Flow Past a Circular Cylinder in the Range $10^4 < R < 5X10^4$.J. Fluid Mech. 61, 1 (1973) 187-198.
6. Igarashi, T.: Effect of Vortex Generators on the Flow around a Circular Cylinder Normal to an Airstream. Bull. JSME 28, 236 (1985) 274-282.
7. Unal, M.F.; Rockwell, D.: On Vortex Formation from a Cylinder. Part 2. Control by Splitter Plate Interference. J. Fluid Mech. 190 (1987) 513-529.
8. Pearson, L. F.; Szewczyk, A. A.: The Near-Wake of a Circular Cylinder with a Spanwise Periodic Trailing Edge Splitter Plate. Proc. Symp. Flow-Induced Vib. ASME-WAM (1992).
9. Stansby, P.: The Effect of End Plates on the Base Pressure Coefficient of a Circular Cylinder. Aeronautical Journal, 78 (1974) 36-37.
10. Mangalam, S. M.; Kubendran, L.R.: Experimental observations on the relationship between stagnation region flow oscillations and eddy shedding of circular cylinder. Instability and Transition, Vol. 1, (1990), 372-386

Development of the Near Wake Structure on a Cantilevered Circular Cylinder with a Free-End

Jean-Louis CHAMPION and Madeleine COUTANCEAU

LABORATOIRE DE MECANIQUE DES FLUIDES (Composante de l'URA CNRS 191)
40, avenue du RECTEUR PINEAU , 86022 POITIERS - FRANCE

ABSTRACT

The paper describes the experimental study of the initial phase of the near wake development behind a finite circular cylinder fixed normal to a flat plate, for an aspect ratio 2≤Al≤5 and a Reynolds number Re=1000. The experiments have been carried out in a vertical water tank using two different visualization methods : tiny discrete solid particles and dye filaments. As a result, the time-evolution of the flow structure is determined and models are proposed. Thus, for the case taken as reference, namely Al=5, three different zones are differenciated along the span of the cylinder and their modifications with the variation of the aspect ratio are established as well as that of the vortex shedding process.

OBJECT OF THE WORK

The object of the present work is to bring information about the near wake structure behind a cantilevered circular cylinder having a flat nose as free-end and to show how the flow sets up with time after an impulsive start and how it varies with the cylinder length for an aspect ratio ranging from 5 to 2. Thus, considering the case where Re=1000 and Al=5 as a reference case, the spatial- and time-evolution of the vortex structure in the recirculating zone are observed up to $t^*=10$ ($t^*=V_0t/D$) by considering both velocity field and vorticity propagation using complementary flow visualizations. Consequently, models of the flow structure are deduced at various stages of the flow establishment and their modifications, with a decrease of Al up to 2, are analysed.

Examination of literature shows the present increasing interest for three-dimensionnal flows around obstacles having a simple geometry, like a circular cylinder. Indeed, they are susceptible both of schematizing realistic situations (tower, chimney,...) and of being calculated. However, it appears that all the investigations devoted to the flow induced by such a short cylinder have been carried out for the established regime, and most of them for high Reynolds numbers ($Re \sim 10^4$-10^5). These investigations concern more especially the flow structure (Kawamura et al. [1] : $Re=3.210^4$, 1≤Al≤8 ; Zdravkovich et al. [2] : $6.010^4 \leq Re \leq 2.610^5$, 2≤Al≤10), the vortex shedding process (Okamoto et al. [3] : $1.310^4 \leq Re \leq 1.510^5$, 1≤Al≤12.5), the unsteady lift and drag forces (Baban et al..[4] : $Re=4,610^4$, Al=2). Although, experiments have been also carried out for lower Reynolds numbers (Re≤300), they correspond to high values of the aspect ratio (Slaouti and Gerrard [5] : 60≤Re≤200, 25≤Al≤230), (Gerich and Eckelmann [6], 150≤Re≤250, 70≤Al≤280). These investigations yielded very interesting data, however, they often remain fragmentary and, in particular, do not cover the domain considered in the present study.

EXPERIMENTAL SET-UP AND TECHNIQUES

The experiments have been realized with the same experimental set-up and analogous techniques as those described in Polidori et al [7]. In particular, complementary types of visualizations have been made with Rilsan particles dispersed in the fluid (water) and dye emitted from the cylinder surface in different cross sections distributed along the span. Here, a 5 cm in diameter cylinder has been used and the visualizations with particle method have been made both in the median longitudinal and several cross sections. Thus, taking account the high complexity and unsteadiness of the flow, 18 cross sections distributed along the 25 cm of the span, for Al=5, have been studied and a short time step (Δ t*=0.5) has been chosen to follow precisely its time-evolution.

RESULTS

The collation of the data deduced from the different photographs, obtained for a same time, permits us to build schematic models of the 3D flow structure. Two examples, corresponding to t*=2.0 and t*=3.5, are proposed in fig.1. Fig. 1a shows that the initial Bénard-von-Karman vortices form a single U-shaped vortex fixed on the plate by its extremities (see also the presence of this structure near the free-end in figs. 2a and b). At this time, (t*=2.0), this vortex has a quasi-2D

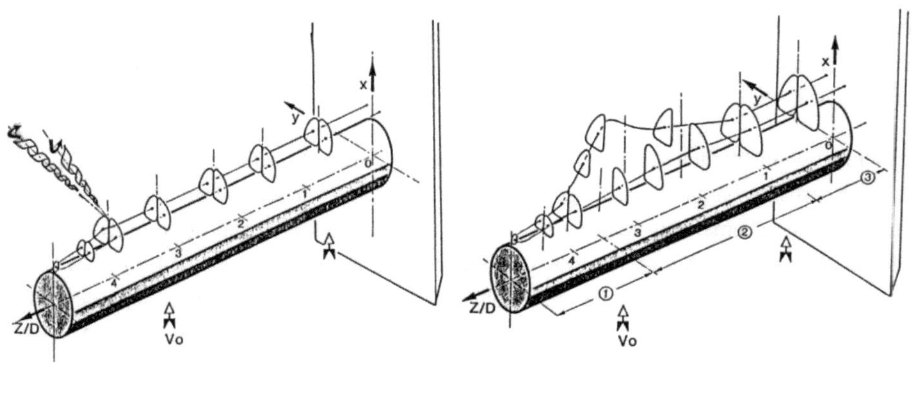

Fig. 1 : Model of the flow structure deduced from visualizations ; 1a : t*≈2.0 ; 1b : t*≈3.5.

development for $0.0 \leq Z/D \leq 3.5$ (fig.3a : the zero-velocity line is parallel to the cylinder axis) and a strongly 3D confined development, due to the spanwise flow, for $Z/D \geq 3.5$, (fig.2a). Moreover, a pair of inclined contra-rotating vortices, induced by the pressure difference between the free-end and the sides of the cylinder, linked to the zero-velocity line, is shed at $t^* \approx 2.5$ in the main flow. With increasing time (fig.1b), the recirculating zone being fed by the spanwise flow in the symmetrical plane (x,z), in a part of the span ($1.0 \leq Z/D \leq 3.5$, for $t^*=3.5$), the vortices are blown away from the cylinder surface, and relatively spaced (fig.3b). On the contrary, near the free-end, the vortices are confined, as described above, and no vortex shedding can occur (fig.3c). Near the plate ($0.0 \leq Z/D \leq 1.0$), the strongly 3D vortices stay on the rear side of the cylinder (fig.4a) and are rapidly disorganized. Thus, three spanwise zones may be identified. Furthermore, the visualizations with dye filaments evidence that the vorticity fluxes, induced both by the free-end and the plate, propagate in opposite directions in the spanwise direction in an helicoidal movement (fig.4), until their collision in the cross section $Z/D=2.0$ (fig.5).

The main effects induced by decreasing the aspect ratio on these phenomena are summarized in the following conclusion and given on the poster. More details will be presented in a further paper, with the influence of the Reynolds number and that of the free-end shape.

CONCLUSION

The results of the present study can be summarized as follows :

1°) for Al=5 :

- the initial pair of Bénard-Von-Karman vortices form a single U-shaped vortex.
- a pair of starting contra-rotating vortices, which develop on the cylinder end, is rapidly shed.
- the spanwise stream, which turns around the free-end, and the perturbation induced by the plate provoke the formation of 3 zones along the span.

2°) for Al<5 :

- the length of the recirculating zone increases with the decrease of the aspect ratio.
- the stretching speed of the recirculating zone is a strongly non linear decreasing function of the aspect ratio.
- the location of the collision between the opposite vorticity fluxes, coming from the free-end and the plate respectively, is a linear function of the aspect ratio.

ACKNOWLEDGEMENTS

This work is supported by the "Ministère de la Défense - Délégation Générale de l'Armement" under Contract No 911507/A000/DRET/DS/SR2.

REFERENCES

1. Kawamura T. ; Hiwada M. ; Hibino T. ; Mabuchi I. ; Kumada M. : J.S.M.E., Vol. 27, pp. 2142-2151, (1984)
2. Zdravkovich M.M. ; Brand V.P. ; Mathew G. ; Weston A. : J.F.M. pp. 557-575, (1989)
3. Okamoto T. ; Yagita M. : J.S.M.E., Vol. 16, pp. 805-814, (1988)
4. Baban F. ; So R.M.C.; Otügen M.V. : Experiments in Fluids 7, pp. 293-302, (1989)
5. Slaouti A. ; Gerrard J.H. : J.F.M., vol. 112, pp. 297-314, (1981)
6. Gerich D. ; Eckelmann H. : J.F.M. ,Vol. 122, pp. 109-121, (1982).
7. Polidori G. ; Pineau G. ; Abed-Meraim K. ; Coutanceau M. , IUTAM Symposium Göttingen (1992)

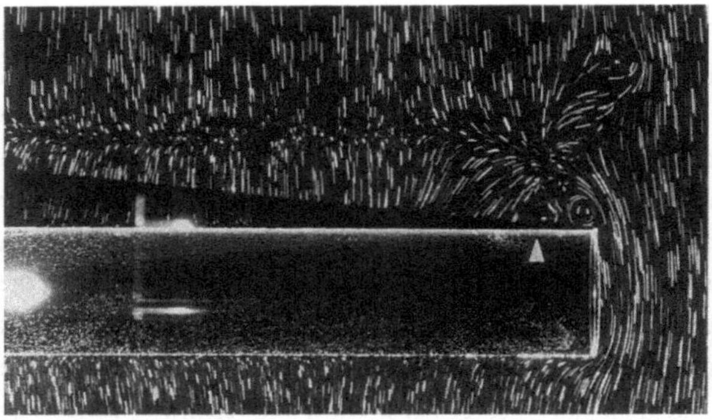

Fig. 2a : t*=2.0 , Symmetrical plane

Fig. 2b : t*=2.0 , Z/D=4.5

Fig. 3a : t*=3.5 , Z/D=1.0

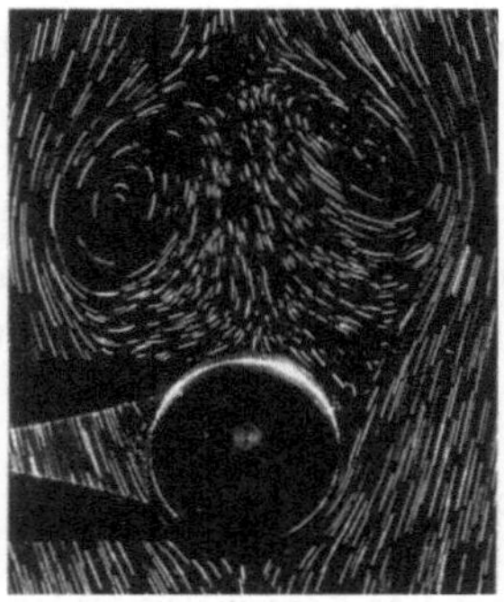

Fig. 3b : t*=3.5 , Z/D=3.0

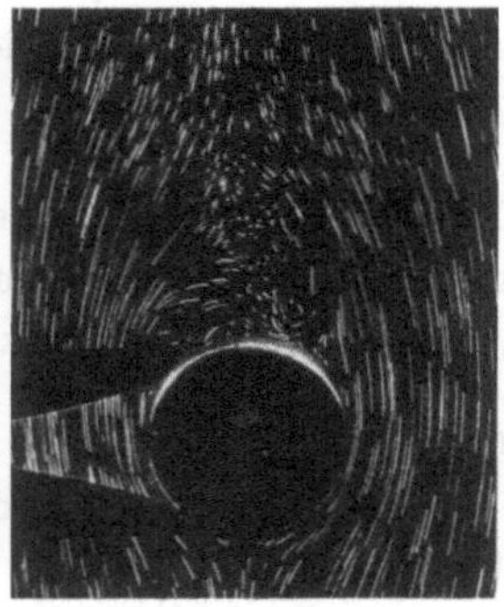

Fig. 3c : t*=3.5 , Z/D=4.0

VISUALIZATIONS BY DISCRETE PARTICLES ; Re=1000, Al=5

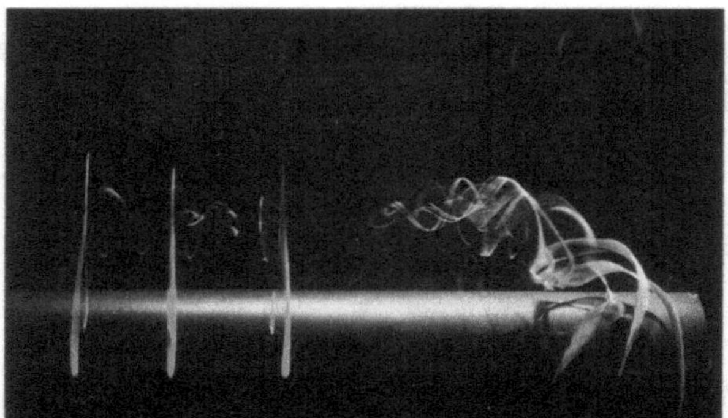

Fig.4 : t*=3.5

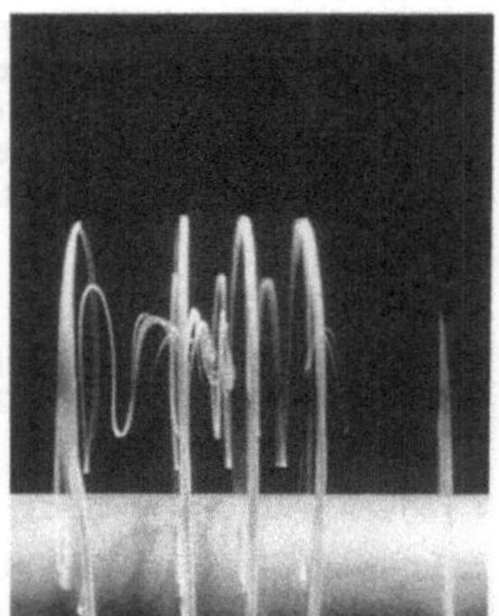

Fig. 5 : t*=2.5

VISUALIZATIONS BY DYE METHOD ; Re=1000, Al=5

Boundary Condition Influence on the Flow Around a Circular Cylinder

P.G. Esposito*, R. Verzicco+ and P. Orlandi+

**I.N.S.E.A.N.*
Via di Vallerano 139 – 00128 Roma, Italy

+ *Università di Roma "La Sapienza"*
Via Eudossiana 18 – 00185 Roma, Italy

Summary

Two-dimensional simulations of the flow around a circular cylinder, performed at several values of the Reynolds number, revealed that when a constant velocity is assigned at the outer boundary, accurate solutions are obtained only for large distances of the boundary from the body. On the other hand, at a reasonable distance from the cylinder, radiative boundary conditions produce results which compare well with the experimental observations, also if the external boundary is quite close to the cylinder.

1 Introduction

The wake past a circular cylinder is still very attractive for the large number of fundamental questions arising from the flow complexity. The study of this flow was performed experimentally and numerically giving rise to some questions for which even now, a satisfactory answer has not been given. For example the experimental relationship $S(Re)$ presents discontinuities whose origin did not find a consensus.

The experimental approach is very useful to emphasise the importance of three dimensional effects. In our opinion for a quantitative description of the 3-D vorticity field, the experiments require a cumbersome work producing informations that are not comparable to those obtained by the direct numerical simulations. It is well known that the direct simulations can not be used at the high Re numbers important for practical applications. On the other hand, for the wake past a cylinder, the changes in the vortical structures occur at Re numbers affordable with the direct simulation. A particular care should be devoted to mesh refinement and boundary conditions, since the outer boundary affects the results, a constant velocity assumption is appropriate only at large distance from the body. Orlansky [7] radiative boundary conditions are very efficient and, in principle, can be located at any distance from the body. The location of the outer boundary at 50 radii from the body is an optimal choice, mainly because our interest is in the evolution of the three-dimensional vortical structure. Two-dimensional tests have been performed with several domain sizes, boundary conditions and mesh refinements as a preliminary study for the development of a code for three-dimensional simulations. Radiative boundary conditions produced results in good agreement with those measured by Coutanceau and Bouard [2] and those of Williamson [8].

2 The Model

2.1 The Governing Equations

The Navier-Stokes equations for an incompressible viscous flow in dimensionless form are

$$\frac{\partial \mathbf{u}}{\partial t} + \nabla \cdot (\mathbf{u}\mathbf{u}) = -\nabla p + \frac{2}{Re}\nabla^2 \mathbf{u} \tag{1}$$

where $\mathbf{u} = \mathbf{u}(\theta, r, z) = (u_\theta, u_r, u_z)$ is the three dimensional solenoidal velocity field. Re is defined as $2UR/\nu$ where U is the freestream velocity, R the radius of the cylinder and ν the kinematic viscosity. In 3-D the Navier-Stokes equations in primitive variables are an optimal choice to reduce the storage and the number of numerical operations. In cylindrical coordinates the continuity equation assumes a simple form as in Cartesian coordinates when the quantities $q_\theta, = u_\theta$, $q_r = r{\cdot}u_r$ and $q_z = u_z$ are used; this consideration led us to express the Navier-Stokes equations in terms of the q_i variables.

The numerical method substantially does not differ from that in Ref. [6] used to solve flows in general curvilinear coordinates. We wish to underline that for the flow around a cylinder the radial size of the computational domain is $L_r = 50R$ and an hyperbolic tangent function has been used to cluster the grid near the cylinder. The finest grid was 128×128 in the $r-\theta$ directions, in view of subsequent 3-D computations with $128\times128\times64$ points, which is affordable by an IBM RISC/6000 540 with 128 MB of memory.

A tentative 3-D simulation at $Re = 180$ was started with 32 grid points in the spanwise direction and, although spacing was similar to that in [4], whatever perturbation decayed to reach a 2-D periodic state. Before withdrawing any conclusion about this unexpected behaviour, it is necessary to perform a large number of simulations with several grid spacings and $Re > 180$. This task requires a lot of time and it is still under development.

2.2 Initial and Boundary Conditions

The initial condition is given by an impulsive start of the cylinder in a fluid at rest; at a chosen time, a perturbation is introduced as a clockwise and a successive counterclockwise rotation of the cylinder.

On the surface of the cylinder a no-slip condition has been applied. As mentioned before the choice of the outer boundary condition is crucial in flows around bodies and it was proved that radiative boundary conditions are very efficient. The q_i quantities at the outer boundary are evaluated by the equation

$$\frac{\partial q_i}{\partial t} + c_i \frac{\partial q_i}{\partial r} = 0 \qquad i = \theta, r, z. \tag{2}$$

where c_i is the convection velocity which is obtained at the previous time step by the Eq. (2) itself. The convection velocity satisfies the limitation $0 < c_i < c_{max}$ where

$c_{max} = \Delta r/\Delta t$. Eq. (2) does not preserve the global mass, therefore at each time step the total mass defect was redistributed among all the cells of the outflow to recover the mass conservation.

3 Discussion of results

The wake past a cylinder at $Re < Re_o \simeq 34$ is steady and the separation bubble grows in time up to a size which is a function of the Reynolds number and we checked that the time-growth at $Re = 32$ agrees with the experimental results reported in [2]. At this low Reynolds numbers the outflow conditions are not relevant and equal results are obtained with radiative or Dirichlet conditions. For $Re < Re_o$ the perturbations are dissipated and a steady configuration is recovered. On the other hand at $Re = 78$ the perturbation is no longer dissipated and vortices are shed from the cylinder. The influence of the boundary conditions has been checked in this different flow structure and the top of Fig. 1 shows that the vorticity field is satisfactory in both cases near the body, while in the far field, spurious vorticity is produced with $c_i = 0$ in the Eq. (2). At a higher Reynolds number

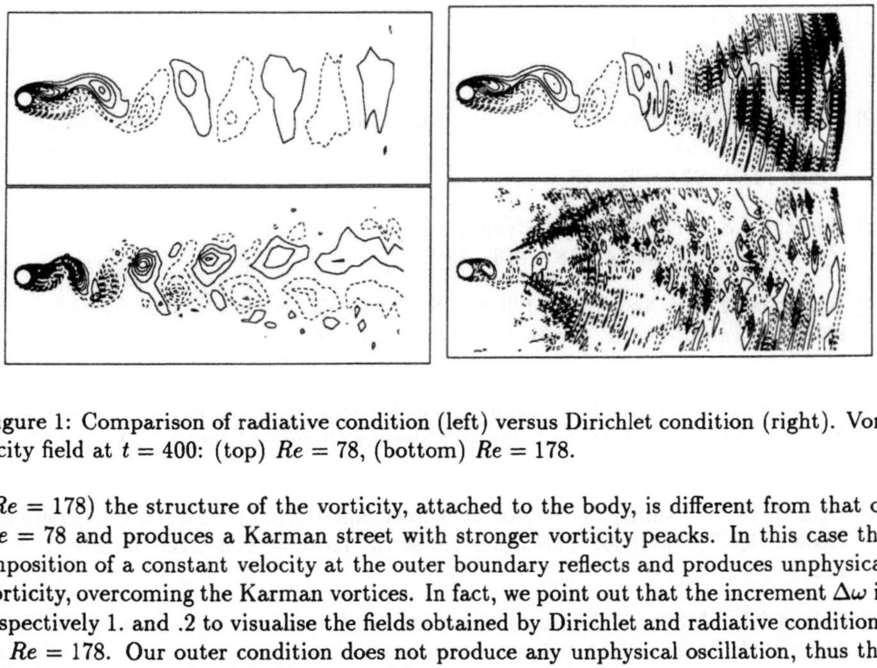

Figure 1: Comparison of radiative condition (left) versus Dirichlet condition (right). Vorticity field at $t = 400$: (top) $Re = 78$, (bottom) $Re = 178$.

($Re = 178$) the structure of the vorticity, attached to the body, is different from that of $Re = 78$ and produces a Karman street with stronger vorticity peacks. In this case the imposition of a constant velocity at the outer boundary reflects and produces unphysical vorticity, overcoming the Karman vortices. In fact, we point out that the increment $\Delta\omega$ is respectively 1. and .2 to visualise the fields obtained by Dirichlet and radiative conditions at $Re = 178$. Our outer condition does not produce any unphysical oscillation, thus the diffusion term introduced by Braza *et al.* [1] is not necessary.

Usually in the studies of a flow around circular cylinders the curve $S(Re)$ is given; in this short paper we do not want to show it once more but we wish to report the values for the Strouhal numbers found at different Re.

Re	64	78	121	178	200	285
S	.134	.148	.174	.192	.197	.189

4 Conclusions

In the present study, two-dimensional simulations of the flow around a circular cylinder show that radiative boundary conditions have to be imposed at high Reynolds numbers. The advantage of these boundary conditions results mainly in 3-D flows where the outflow at 50 radii from the body gives a satisfactory discretization to accurately describe the complexity of three-dimensional vortical structures.

Acknowledgements

P.G. Esposito wishes to acknowledge the the Italian Ministry of Merchant Marine for the support received in the frame of the I.N.S.E.A.N. research plane 1990. The research was also partially supported by the Ministero dell'Università e della Ricerca Scientifica e Tecnologica.

References

[1] Braza M., Chassaing P. and Ha Minh H. "Numerical Study and Physical Analysis of the Pressure and Velocity Fields in the Near Wake of a Circular Cylinder " *J. Fluid. Mech.* **165** (1986), 79.

[2] Coutenceau M., Bouard R. "Experimental Determination of the Main Features of the Viscous Flow in the Wake of a Circular Cylinder in Uniform Translation " *J. Fluid. Mech.* **67** (1977), 85.

[3] Gerrard J.K. "The Wakes of Cylindrical Bluff Bodies at Low Reynolds Number " *Trans. R. Soc. Lond. A* **288** (1978), 351.

[4] Karniadakis G.E. and Triantafyllou G.S. "Three-Dimensional Dynamics and Transition to Turbulence in the Wake of Bluff Objects " *J. Fluid. Mech.* **238** (1992), 1.

[5] König M., Eisenlohr H., Eckelmann H. "The Fine Structures in the Strouhal-Reynolds Numbers Relationship of the Laminar Wake of a Circular Cylinder " *Phys. of Fluids* **A 2** (1990), 1607.

[6] Orlandi P. " A Numerical Method for Direct Simulation of Turbulence in Complex Geometries " *Center for Turbulence Research* **Annual Research Briefs** (1989), 215.

[7] Orlanski I. "A Simple Boundary Condition for Unbounded Hyperbolic Flows " *J. Comput. Phys.* **21** (1976), 251.

[8] Williamson C.H.K. "Oblique and Parallel Modes of Vortex Shedding in the Wake of a Circular Cylinder at Low Reynolds Number " *J. Fluid. Mech.* **206** (1989), 579.

Magnetohydrodynamic Flow Around a Circular Cylinder

G. Gerbeth

Research Center Rossendorf, Inc.
P.O. Box 19, 8051 Dresden, Germany

A. Alemany

Institut de Mecanique de Grenoble
BP 53 X, 38041 Grenoble Cedex, France

1. Introduction

The flow around a circular cylinder is considered for the following particular configuration: The fluid is electrically conducting and the whole system is inside an external magnetic field. Due to Ohms law of electrodynamics the interaction between velocity and magnetic field induces an electrical current in the fluid. This current in turn leads (due to its interaction with the magnetic field) to an electromagnetic body force acting on the fluid. In addition the induced current yields a small perturbation of the external magnetic field.

Main goal of the paper is to point out the interesting features of such a magnetohydrodynamic (MHD) problem in frame of the general research on flow around obstacles.

2. Problem definition

In the following we consider an electrically conducting fluid with constant material parameters (density ρ, kinematic viscosity ν, dynamic viscosity η, electrical conductivity σ, magnetic permeability μ) flowing around an infinitely long circular cylinder of radius R. The non-dimensional system of Navier-Stokes and Maxwell-equations is given by

$$\partial_t \underline{v} + (\underline{v}\ \mathrm{grad})\underline{v} = -\mathrm{grad}p + \Delta\underline{v}/Re + N\,(\underline{j} \times \underline{B}) \tag{1}$$

$$\mathrm{div}\,\underline{v} = 0, \quad \mathrm{div}\,\underline{j} = 0, \quad \mathrm{div}\,\underline{B} = 0 \tag{2}$$

$$\underline{j} = \underline{E} + \underline{v} \times \underline{B} = \mathrm{rot}\,\underline{B}/Rm \tag{3}$$

where R, U_0, ρU_0^2, R/U_0, B_0, $\sigma U_0 B_0$ are used as units of length, velocity $\underline{v}$, pressure p, time, magnetic field $\underline{B}$, and electrical current density $\underline{j}$, respectively. In addition to the usual Reynolds-number $Re=U_0R/\nu$ two independent parameters appear:

magnetic Reynolds-number: $Rm = \mu\sigma U_0 R$

Interaction parameter: $N = \sigma B_0^2 R/\rho U_0$

The interaction parameter is a measure of the ratio of the electromagnetic to the inertia forces whereas Rm determines the magnetic field distortion due to the flow. In many cases

the Hartmann-number: $Ha^2 = N*Re = \sigma B_0^2 R^2/\eta$

is also used to describe the electromagnetic influence on the flow field. It characterizes the ratio between electromagnetic and viscous forces. The induced magnetic field is usually very small in the laboratory, and can be taken into account by a first order expansion in terms of Rm: $\underline{B}=\underline{B}_0+\underline{b}$, $|\underline{b}| \sim Rm*B_0$ ($Rm \ll 1$). In that way Rm disappears from the system and the problem is determined by the two parameters (Re,N) or (Re,Ha).

For the sake of brevity we consider electrically insulating cylinders only. The z-axis of the coordinate system coincides with the cylinder axis. The fluid is taken as unbounded.

3. Flow parallel to the cylinder axis

If the flow velocity is always parallel to the axis of the infinitely long cylinder the non-linear terms in the Navier-Stokes-equation vanish. However, the remaining linear problem is non-trivial in the MHD case. To give a first idea of typical MHD flow structures we present some results here, for further details we refer to /1/. Disappearance of the inertia term means that the resulting flow is a balance between electromagnetic and viscous forces. Therefore, the Hartmann-number is the only relevant parameter.The system (1)-(3) can be solved analytically for this simple geometry. Typical results for the flow field are shown in Fig.1. Fig.1a shows the symmetrical flow field at small Ha: The flow is dominated by viscous forces. Fig.1b shows the case of dominating electromagnetic forces with the characteristic MHD shear layers: The flow is concentrated in a *parabolic wake* parallel to the magnetic field, bounded by the parabolas $|y| \sim |x/Ha|^{1/2}$. Inside the wake there is a *deep core* region (where the velocity is constant) and the *outer core* (where a slow velocity decrease $\sim x^{-1/2}$ occurs). The *Hartmann layers* near r=1, the thicknesses ∂ of which are $\partial \sim Ha^{-1}$, are characterized by high velocity gradients. Within the *tangential layers* in the vicinity of the points r=1, x=0, the velocity is in the opposite direction compared to the wakes and decreases exponentially. Velocity *transition layers*, the thickness of which are of the order of $Ha^{-1/2}$, separate the parabolic wakes from the outer region where the fluid is at rest.

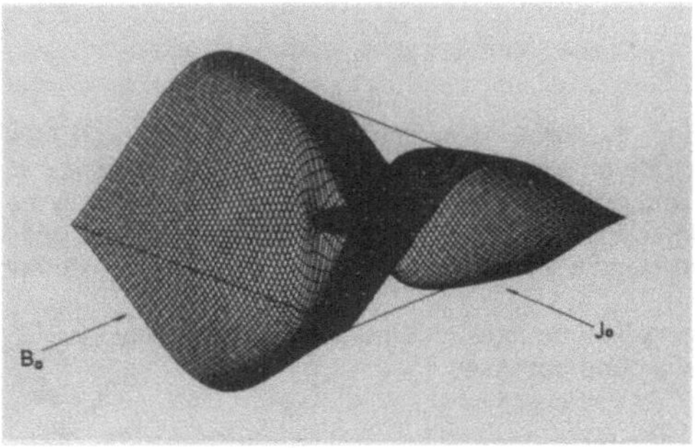

Fig. 1a: MHD flow along the cylinder
weak magnetic field: Ha = 1

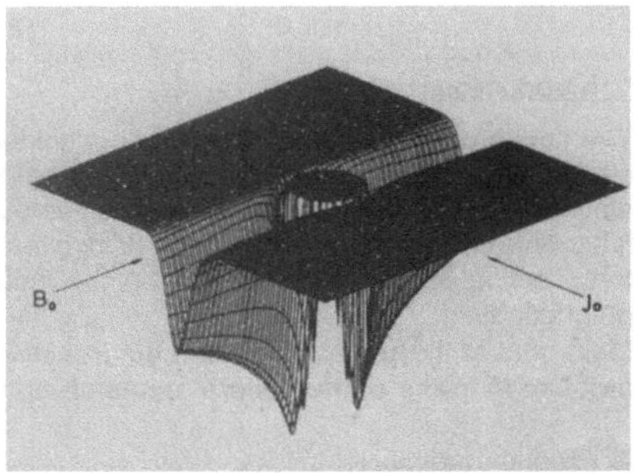

Fig. 1b: MHD flow along the cylinder
strong magnetic field: Ha = 100

4. Flow perpendicular to the cylinder axis

The undisturbed velocity U_0 is parallel to the x-axis whereas the magnetic field is parallel to U_0 (aligned field) or perpendicular to it ($\underline{B}_0=B_0\underline{e}_y$ - transverse field). We distinguish between a non-separated flow without boundary layer separation and a flow with separation. Obviously, the transition between these two regimes must depend on the parameters (Re,N) or (Re,Ha).

4.1 Non-separated flow

If the magnetic field is strong enough the non-linear term of eq.(1) can be neglected, and the flow is determined by Ha only. Again interesting non-trivial flow configurations arise for $Ha \gg 1$ even in this linear case. Such problems were studied extensively in the sixtees and we refer to /2/ for a summary. Main features are:

- Besides the usual transport mechanisms (diffusion due to viscosity, convection due to the main stream) there is a third transport mechanism transmitting the velocity disturbances along the magnetic field lines (Alfvén waves). If the Alfvén-number α=N/Rm exceeds unity not only a downstream but also an upstream wake occurs since the Alfvén wave transport is in both directions from the cylinder. Experimental evidence of that upstream wake was obtained by measurements of the induced magnetic field /4/.
- For $Ha \gg 1$ the flow field is as follows: Inside the wakes there are *dead water* regions in front and behind the cylinder with an extension proportional to Ha. The wakes are separated from the undisturbed free flow regions by *shear layers* emerging at the cylinder poles and growing parabolically towards the up- and downstream directions.

An unusual increase of the cylinder drag coefficient was found recently /3/ resulting in a drag increase C~Ha*ln(Ha) in case of a strong (Ha≫1) transverse field. This is somewhat surprising since the typical drag increase of a cylinder or a sphere is linear with Ha. The increase can be understood by considering the pressure at the cylinder surface. Fig.2 shows a strong increase of the front stagnation pressure proportional to $Ha^{4/3}$. However, the region of that high pressure becomes smaller with increasing Ha and is confined to an angle proportional to $Ha^{-1/3}$. Therefore, the front stagnation pressure gives a drag linear with Ha whereas the integrated pressure over the cylinder surface yields the leading term Ha*ln(Ha). The surprising drag increase is in good qualitative agreement with the available experimental data /3/.

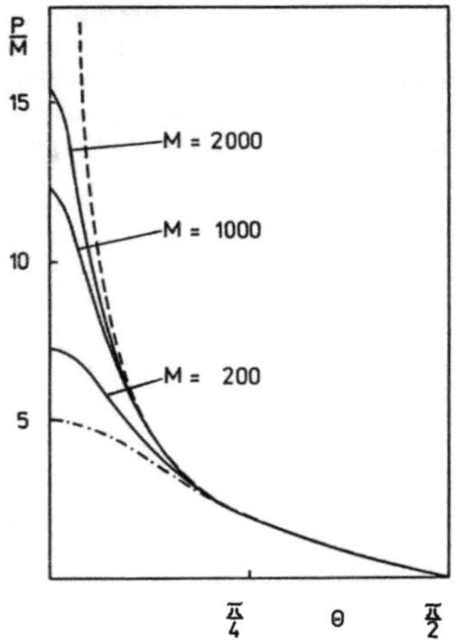

Fig. 2: Surface pressure in a transverse magnetic field
- - exact solution Ha=30, --- asymptotic solution Ha→∞, Θ=0: front stagnation point

4.2 Flow with separation

In which way the magnetic field influences the boundary layer separation, the von Karman vortex street, the downstream wake? There are no numerical simulations available up to now dealing with that problems. In the following we present briefly some experimental results obtained at the Institut de Mecanique de Grenoble. For more details we refer to /4,5/. The following quantities were measured in an aligned magnetic field: mean and fluctuating velocities, the induced magnetic field, the surface pressure. A strong enough magnetic field is able to prevent boundary layer separation even for high Reynolds-numbers. This was proved experimentally for Re=13.000, 17.400, and 34.500 by measuring the induced magnetic field at the position x=5R, y=1.6R downstream of the cylinder. The critical magnetic field strenght above which no separation occurs was shown to depend only on the interaction parameter with the critical value N_c=0.55. Previous theoretical analysis of the bondary layer equations /6/ gave N_c=1, but did not take into account the magnetic field influence on the inviscid flow. Further results are:

* The Strouhal-number decreases slightly with increasing N.
* There is a strong coupling of velocity and magnetic fluctuations with a phase shift of appr. π/2.
* The separation point is shifted downstream with increasing interaction parameter.
* There is no Bernoulli-law for the inviscid outer flow region due to the electromagnetic Joule losses.

The measured surface pressure is shown in Fig.3. The downstream pressure first increases with N and than decreases. This can be attributed to two competing effects: Pressure increase due to the decrease of the inviscid tangential velocity and the downstream shift of the separation point.

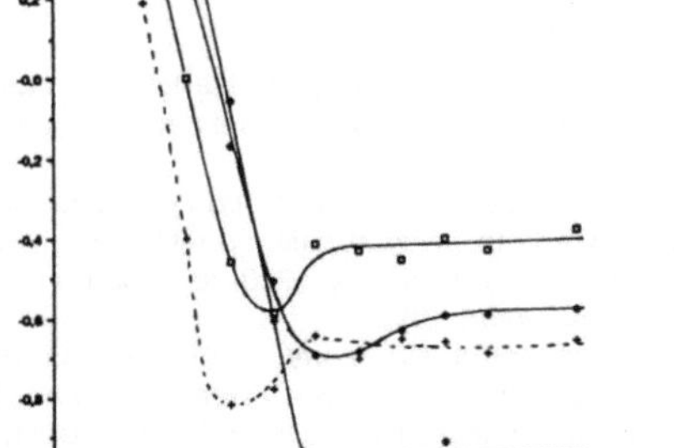

Fig. 3: Surface pressure in an aligned magnetic field
+ N=0, □ N=1, * N=4.1, ♦ N=8.2

Finally an energy spectrum obtained on the axis of the downstream flow is shown in Fig.4. Here the kinetic energy spectrum exhibits a k^{-3} law. One possible explanation of this result is that due to the influence of the magnetic field the flow configuration is nearly two-dimensional. Then the Karman eddies (which give a frequency readily observed in the spectrum) appear like an external forcing on the two-dimensional turbulence resulting in the k^{-3} enstrophy cascade.

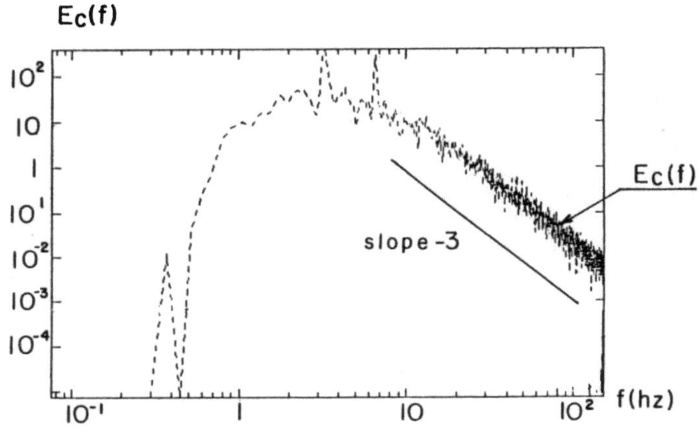

Fig. 4: Energy spectrum at the downstream position x/d=2.5, y/d=0
Re=13.043, Ha=49.5, N=0.19

5. Discussion

We presented a brief review on recent results on MHD flow around a cylinder. Compared to usual hydrodynamics the MHD approach to study the cylinder wake exhibits the following unique advantages:

* A variety of flow regimes in the cylinder wake (including flow without separation at high Reynolds-numbers) can be realized simply by changing the magnetic field strength.
* Three-dimensional disturbing effects (like cylinder end-effects or oblique vortex shedding) are suppressed by the magnetic field.
* The external magnetic field is perturbed since the flow itself induces a magnetic field. This induced field represents an additional interesting quantity.

The magnetic field can be considered as a comfortable control parameter of the flow field.
Significant contributions to modern aspects of wake dynamics research (transition between absolute and convective instabilities, chaotic behaviour in open flow systems, formation of the vortex structures) can be expected by further studying the MHD cylinder wake.

References

/1/ G. Gerbeth, A. Thess, Ph. Marty: Theoretical study of the MHD flow around a cylinder in crossed electric and magnetic fields. European Journal of Mechanics, Vol.9, No.3, 1990, 239-257

/2/ A. Tsinober: MHD flow around obstacles (in russ.). Zinatne, Riga, 1972

/3/ G. Gerbeth: New results on MHD drag coefficients. The Sixth Beer-Sheva Int. Seminar, Jerusalem, February 1990, to appear in: Progress in Astronautics and Aeronautics, Washington 1992

/4/ J. Josserand, Ph. Marty, A. Alemany, G. Gerbeth: MHD flow around a cylinder in an aligned magnetic field. The Sixth Beer-Sheva Int. Seminar, Jerusalem, Febr. 1990, to appear in: Progress in Astronautics and Aeronautics, Washington 1992

/5/ J. Lahjomri, Ph. Marty, A. Alemany: Cylinder wake in a magnetic field aligned with the velocity. Accepted for publication in J. Fluid Mech.

/6/ J. Buckmaster: Separation and magnetohydrodynamics. J.Fluid Mech., Vol. 38, 1969, 481-498

Development of Vortex Structure Behind a Circular Disk

H. Higuchi and H. Balligand

Department of Mechanical and Aerospace Engineering
Syracuse University, Syracuse, NY 13244, U.S.A.

Summary

Vortical structure behind a circular disk was investigated experimentally in water as the flow past the disk was accelerated from rest and successively moved at constant speeds. Initial vortex ring growth and its distortion through subsequent vortex shedding were documented. Flow field was monitored until the steady state wake oscillation was established. The velocity time traces in the wake taken with a LDV were analyzed using a wavelet analysis technique, and time scales of vortical structures were identified and correlated with the flow visualizations.

The disk of 10.16 cm diameter was either towed in a water tank or placed in a water channel started from rest. Maximum model blockage in the test section was 2- 2.6% in all cases. An acceleration parameter, $A_p=D(dU/dt)/U_0^2$ was varied between 0.1 and 4.7. The flow motions were visualized with fluorescent dye together with neutrally buoyant particles as needed. Illumination by a laser sheet or UV lights was employed and the image sequence was digitized in real-time by a microcomputer for analysis.

As the flow was initiated, formation of an axisymmetric vortex ring and secondary Helmholtz shear layer instability waves were clearly identifiable. Cross sectional views taken in the towing tank are shown in Figs. 1ab (all the visualization images in this paper are from the video shown at the conference presentation). During the acceleration stage, the main vortex structure and its trajectory among various acceleration rates scaled well using the instantaneous circulation based on a slug flow model (as defined in, e.g., [1]) as a parameter. During the subsequent constant speed phase, the vortex trajectories with respect to the parameter $T=U_0t/D$ were independent of the Reynolds number after similar accelerations. On the other hand, the scaling using the slug flow model was less satisfactory partly because the model overestimates the variation of circulations.

With continued motion of the disk at constant speed, the vortex ring started to tilt without any preferential direction, and the rear stagnation point on the disk shifted off-center

(Fig.1c). In other cases, when a significant asymmetry in the vortex core diameter developed in the circumferential direction, a sudden appearance of a secondary structure was often observed at the edge of the disk. The disturbance having a reverse flow component subsequently destructed the original vortex ring and triggered the vortex to be shed downstream as documented in Fig. 2.

In order to study the path from the axisymmetric to the three-dimensional structure, subsequent distortion and shedding of the vortex ring were studied in a water channel where the flow was started from rest. The acceleration parameter was approximately 0.1. Figure 3 shows a sequence of events starting with the formation of the initial toroidal structure. When the distorted vortex ring is stretched away from the disk, a large asymmetric bound vortex appears on one side of the disk and the near wake is driven to one side. A similar phenomenon was described for a sphere wake in [2]. When this structure is convected downstream, the axisymmetric near wake structure is gradually established. Shear layer instability waves in the immediate wake and the low frequency asymmetric oscillation in the far wake appear. Observed steady-state vortex oscillations were consistent with measurements by others ([3, 4]).

The velocity time traces in the wake taken with a LDV were analyzed using a wavelet analysis technique [5, 6]. Figure 4 shows representative result corresponding to the measurement in the near wake region. Two bands of nearly periodic signal can be seen: the low frequency oscillation of approximately 3 second (g2 wavelet) duration and 10 second period or $St \cong 0.14$ if deemed periodic; the second is somewhat more intermittent and has approximately 0.8 second wavelet duration. The first corresponds to asymmetric low-frequency wake oscillations and the second corresponds to the axisymmetric shear layer waves. Some phase relation between the two structures appears to be present. Closer to the disk, the low frequency oscillations were not as clearly discernible, indicating the axisymmetric nature of the immediate wake as discussed in [3]. From the conventional power spectra, Strouhal number for the shear layer instability waves was higher than previously reported, and the availability of the momentum thickness data of the shear layer would be valuable for comparison. Nonetheless, by applying the wavelet analysis technique to the velocity signals at various axial stations, time dependent nature of the oscillations and different time scales of vortical structures could be elucidated and correlated with the real-time flow visualizations.

Acknowledgement
This work was supported in part by the Sandia National Laboratories and the Naval Air Warfare Center.

References

1. Shariff, K. and Leonard, A., "Vortex Rings", Ann. Rev. Fluid Mech. (1992) 24, 235-279.

2. Viets, H., "Accelerating Sphere-Wake Interaction," AIAA J., Vol. 9, No. 10, Oct. 1971, 2087-2089.

3. Berger, E., Scholz, D. and Schumm, M., "Coherent Vortex Structures in the Wake of a Sphere and a Circular Disk at Rest and under Forced Vibration," J. of Fluids and Structures, (1990) Vol. 4, 231-257.

4. Fuchs, H. V., Mercker, E. and Michel, U., "Large-scale coherent structures in the wake of axisymmetric bodies," J. Fluid Mech. (1979) Vol. 93, Part 1,185-207.

5. Farge, M., "Wavelet Transforms and Their Application to Turbulence," Ann. Rev. Fluid Mech. (1992) 24, 395-457.

6. in cooperation with Dr. J. Lewalle, Syracuse University.

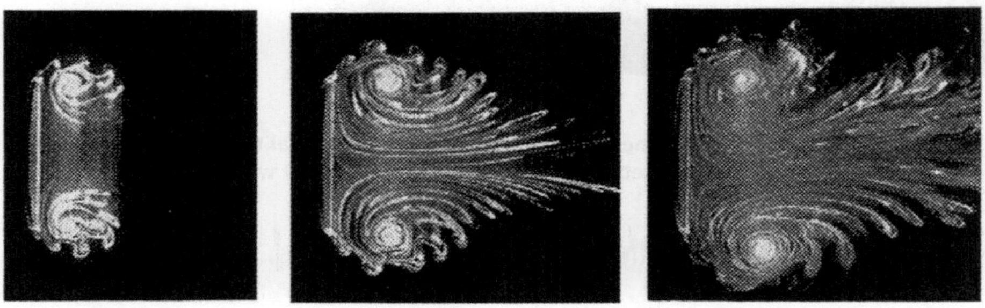

(a) T=1.8 (b) T=4.0 (c) T=6.7

Fig. 1 Formation of vortex ring and its deformation (A_p=1, Re=7300)

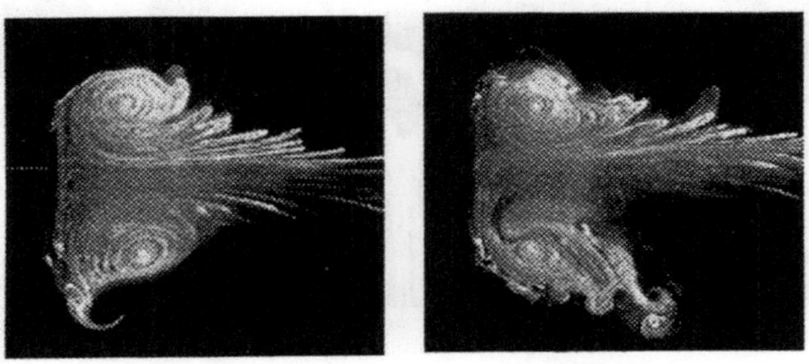

(a) T=6.0 (b) T=7.6

Fig. 2 Destruction of the vortex ring caused by sudden formation of secondary structure at the edge (A_p=3, Re=4200)

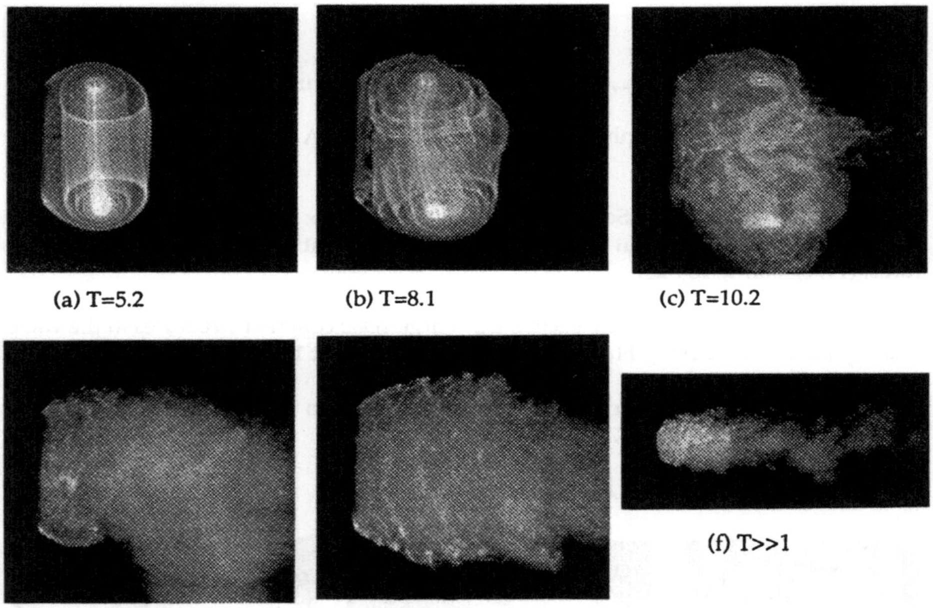

(a) T=5.2 (b) T=8.1 (c) T=10.2

(f) T>>1

(d) T=13.4 (e) T=29.9

Figs. 3 Water channel experiment showing a gradual transformation of axisymmetric vortex into steady-state 3-D wake (Re=6000)

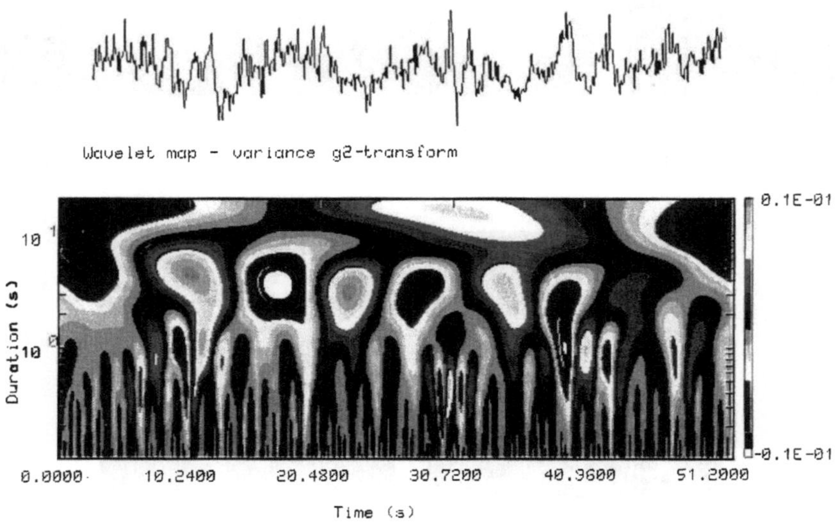

Fig. 4 Wavelet analysis of velocity at x/D=1, r/D=1 Re=6000

End Effect on a Yawed Circular Cylinder in Uniform Flow

Tsutomu HAYASHI*, Fumio YOSHINO**, Ryoji WAKA**,
and Tetuya KAWAMURA***

* Department of Mechanical Engineering, University of Nottingham, University park, Nottingham, NG7 2RD, UK. Present address: **

** Department of Mechanical Engineering, Tottori University, Minami 4-101, Koyama, Tottori 680, Japan.

*** NASA Ames Research Center, Moffett Field, CA 94035, USA.
Present address : Department of Computer and Information Science, Chiba University, Yayoi 1-33, Inage-ku, Chiba 263, Japan.

Summary

An attempt was made to determine clearly the effects of the endplates on the pressure distributions on the surface of the yawed circular cylinder. It can be concluded that the endplates act to decrease the base pressure and operate to enhance the pressure gradient along the cylinder axis with increasing yaw angle. The pressure behind the cylinder decreases greatly towards the upstream endplate of the cylinder, although there is a strong wedge-shaped recovery of pressure in the region near the rear stagnation. The strong secondary flow behind the yawed circular cylinder is responsible for the recovery region.

Introduction

Flow around the yawed circular cylinder can be said to be quasi-two-dimensional if the flows have a similar flow pattern and the same pressure distribution in each section parallel to the free stream and along the cylinder's axis. If quasi-two-dimensional characteristics are assured, we can discuss the general characteristics of the flows around the yawed cylinder in a uniform flow. As finite length cylinders occur frequently in engineering, it is important to investigate, and to understand end effects of the yawed circular cylinder.

An attempt was made to determine clearly the effects of the endplates on the pressure distributions on the cylinder surface, and on the power spectrum for two cases : when the endplates were installed inside the shear layer region of the jet, and when the endplates were removed.

An arrangement of a circular cylinder in the jet and the coordinate system is illustrated in Fig. 1. The wind tunnel has an exit section 204 mm width and 284 mm height. A specimen

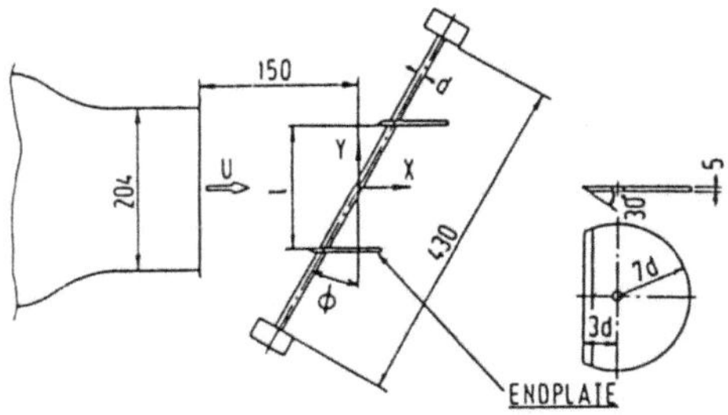

Fig.1. Configuration of apparatus

circular cylinder was constructed of a hollow circular cylinder of 10 mm outer diameter and 430 mm length, whose outer surface was chromium-plated. The model cylinder was arranged horizontally in the jet and extended beyond the shear layer flow of the jet edges. A constant Reynolds number of 2 x 10^4, which was based on the diameter of the cylinder and the mean velocity, was maintained throughout the experiment.

Pressure Coefficient

The experimental results demonstrate that for yaw angles ϕ varying from 0° to 30° the stagnation pressure coefficients obey the independence principle and distribute almost uniformly along the yawed circular cylinder. The base pressure coefficients differ from the value expected by this principle, however, and these spanwise distributions have a gradient along the axis of the yawed cylinder.

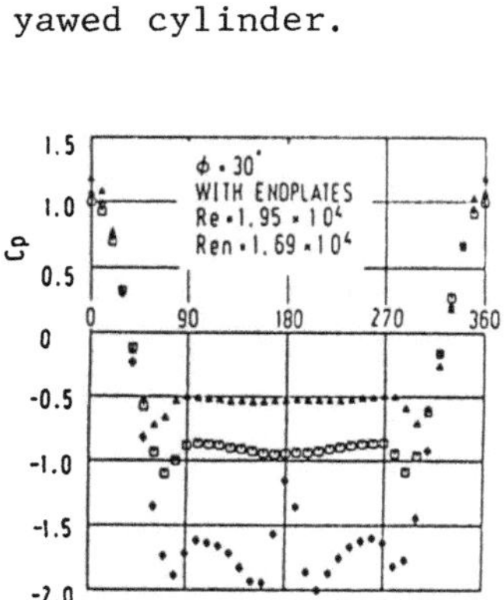

Fig.2. Examples of measured pressure distribution

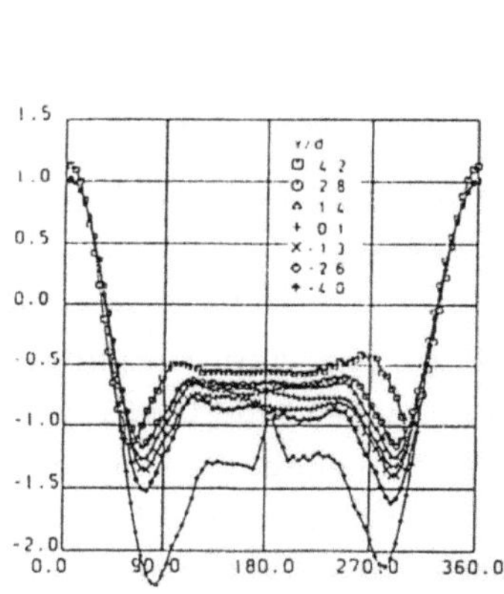

Fig.3. Example of calculated pressure distribution

Some examples of measured pressure distributions on the surface of the yawed circular cylinder are illustrated in Fig. 2 for three sections along the cylinder axis, the sections at the yawed upstream endplate side, the middle section and the yewed downstream side, for the case with endplates. In this case, it can be concluded that the endplates act to decrease the base pressure and operate to enhance the pressure gradient along the cylinder axis with increasing yaw angle. It is especially worthy of note that the pressure behind the cylinder decreases greatly towards the upstream endplate of the cylinder, although there is a strong wedge-shaped recovery of pressure in the region $160° < \theta < 200°$.

These experimental results are also compared with the results calculated by the direct numerical simulation and shows good agreement, as shown in Fig. 3. It was also found from the calculated flow pattern that the strong secondary flow behind the yawed circular cylinder was responsible for the wedge-shaped pressure recovery region near $\theta = 180°$.

Influence of Aspect Ratio

An influence of the aspect ratio (l/d) on the pressure distribution along the cylinder axis is also investigated in this study. Distributions of the stagnation pressure coefficient and the base pressure coefficient along the cylinder axis are shown in Fig.4. C_{pnst} is a constant 1.0 except near the endplates for the aspect ratio between 5 to 15. The stagnation pressure coefficient shows the similar distributions in spite of

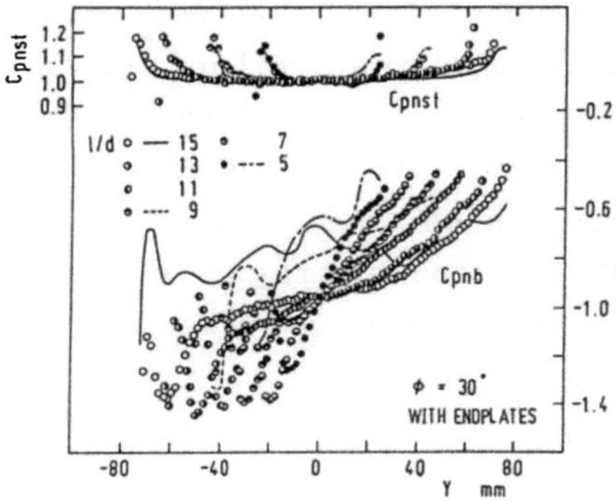

Fig.4. Influence of the aspect ratio on the pressure distribution

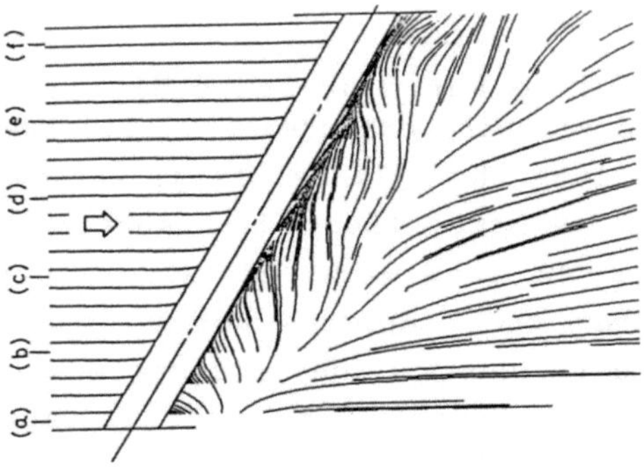

Fig.5. Time averaged streamlines and particle path lines by the numerical calculation.

different aspect ratios, and has good agreement with the results of the numerical calculation which are shown by lines in the figure. On the other hand, every distribution curve of C_{pnb} has a similar shape and the same minimum value -1.4 in the region near the upstream endplate and the same maximum value 0.48 at the end of the downstream side. It is worthy of special mention that C_{pnb} has the same value 0.97 at the middle of span independent of the aspect ratio (l/d).

Secondary flow in wake

A picture in which the streamlines and the particle path lines are averaged over T = 25 - 37.5 using the result of the numerical calculation is shown in the horizontal plane including the center axis of the yawed circular cylinder between the constraining side walls in Fig. 5. In the upstream side of the cylinder, however, time averaged streamlines are reproduced to explain the manner of their curvature, and in the downstream side of the cylinder particle path-lines which are trace of the particle moving with the mean velocities are reproduced to explain in detail the secondary flow in the wake.

This study also proposes a way to infer the turbulent structure in a vortex wake being shed from a finite-length yawed circular cylinder in a uniform flow, using the power spectra and the cross correlation of velocities at two points along a spanwise line parallel to the cylinder axis.

References

1. Smith, R.A., Moon, W.T. & Kao, T.W., J. Basic Eng. Trans. ASME, (1972), 771
2. Ramberg, S.E., Journal of Fluid Mechanics, 1-28 (1983), 81.

Causal Time-Periodic Solutions for Spatially Developing Media

By R. E. Hunt

Department of Applied Mathematics and Theoretical Physics,
University of Cambridge, England

Abstract

Model equations are of immense use in the study of the spatiotemporal evolution of instabilities in wakes, shear flows and so on. The response of such equations to time-periodic oscillatory forcing is of great interest, as such forcing is common in physical situations. An important stage in finding this time-periodic response is concerned with ensuring that the *causal* solution is obtained. For parallel flows, Brazier-Smith & Scott [1], amongst many others, have developed a method which was originally used in the field of plasma physics, which involves manipulating the Fourier inversion contours which are used in solving the problem.

However, for non-parallel flows, where the coefficients of the model equation vary with the spatial location x, this method is useless since the technique of taking Fourier Transforms cannot be applied. It may often be possible to find solutions to the equation using other techniques; but determining which of the possible solutions is the correct causal one requires a new method. We present here a theorem which gives a criterion for determining when a given solution is indeed the causal one. As a prototypical example of how this theorem may be applied in practise, we shall use it for the linearized Ginzburg–Landau equation, which is a good simple model of instability growth in many fluid-dynamical situations. The cases of linear and quadratic variation with x of the control parameter μ of the equation will be covered, and in both cases the *exact* causal time-periodic solution will be presented.

Background and History

Methods involving Fourier Transforms have been used to solve partial differential equations in fluid dynamical situations for many decades now. The problem of *causality* has vexed researchers for almost as long. To illustrate the difficulty, consider a PDE subjected to time-periodic forcing $\exp(-i\Omega t)$. This problem will typically have several possible solutions, corresponding to several different ways of passing the Fourier inversion contour around the singularities in the complex wavenumber plane.

Only one of these solutions, however, can be the causal one, that is to say the one which corresponds to forcing $\exp(-i\Omega t)\,\theta(t-t_0)$ in the limit $t_0 \to -\infty$. (I.e., the causal solution is the reponse to an oscillatory forcing which started at some particular instant a very long time ago.) To obtain this solution, we must be sure that we take our inversion contour along the correct route in the wavenumber plane.

Early research suggested that a suitable method to obtain this route was to make Ω slightly complex: $\Omega \mapsto \Omega + i\varepsilon$. This simulated a small viscosity or friction in the system, and had the effect of moving any singularities which happened to lie precisely on the real wavenumber axis into either the upper or lower half-plane. The inversion contour could then unambiguously be taken along the real wavenumber axis, passing above those singularities which had moved into the lower half-plane and below those which had moved into the upper half-plane.

As the problems being tackled have become more sophisticated, however, it has become clear that this method is too simplistic. A technique which was originally developed to overcome these difficulties in the field of plasma physics by Briggs [2] and Bers [3] has been brought into use in fluid-dynamical problems during the last decade. This technique requires that, in order to correctly obtain the inversion contour, the solution must be made to be analytic as a function of Ω in the entire upper half Ω-plane, $\operatorname{Im}\Omega > 0$. Hence it is necessary to "track" the movements of the poles in the complex wavenumber plane as $\operatorname{Im}\Omega$ is reduced from an extremely large value down to zero. As the poles move, the inversion contour must be deformed to avoid being crossed by them; if this can be achieved all the way down to $\operatorname{Im}\Omega = 0$ then the solution will be continuous for all Ω in the upper half Ω-plane, and it must therefore be the causal one.

Non-parallel Flows

For non-parallel flows, the coefficients of the equation governing the evolution of waves depend on the spatial co-ordinate x. Fourier Transforms therefore cannot be used, and other techniques must be applied to obtain the possible solutions. How can we make sure we find the *causal* one?

The answer is that we must again study the behaviour of the solution for complex forcing frequencies Ω in the upper half Ω-plane. It turns out that not only must the solution be analytic as a function of Ω in the whole of this region, but **it must also decay to zero as $|\Omega| \to \infty$ there**; and this gives us enough information to uniquely choose the causal solution.

The precise details are contained in the statement of the following theorem.

Theorem: let $P(x,t;\Omega) \equiv p(x;\Omega)\exp(-i\Omega t)$ be a solution of the equation

$$\mathcal{L}\left(-i\frac{\partial}{\partial x},\, i\frac{\partial}{\partial t}\right) P = \exp(-i\Omega t)\delta(x), \qquad (*)$$

where $\mathcal{L}$ is a linear differential operator whose coefficients may depend on x. Furthermore, suppose that $p(x;\Omega)$ is analytic as a function of Ω in the region $\operatorname{Im}\Omega \geq -\Delta$ for some $\Delta > 0$; and that for each fixed value of x, $p(x;\Omega) \to 0$ as $|\Omega| \to \infty$ in $\operatorname{Im}\Omega \geq -\Delta$ (with at least algebraic decay in $|\Omega|$).

Then $P(x,t;\Omega)$ is the causal solution of $(*)$ for real Ω.

The Linearized Ginzburg–Landau Equation

The linearized Ginzburg–Landau equation serves as a good illustration of the technique. It arises in many fluid-dynamical situations as a model of the spatiotemporal evolution of small-amplitude wavepackets. We shall use it in the form

$$\frac{\partial A}{\partial t} + U\frac{\partial A}{\partial x} = \mu(\varepsilon x)A + \gamma\frac{\partial^2 A}{\partial x^2} + \exp(-i\Omega t)\delta(x),$$

where U is real and μ and γ are complex, with $\operatorname{Re}\gamma > 0$. The coefficient μ is a function of the slow spatial variable εx, providing a model of a non-parallel flow.

We shall consider the linear and quadratic cases,

$$\mu(\varepsilon x) = \mu_0 + \varepsilon\mu_1 x$$

and

$$\mu(\varepsilon x) = \mu_0 + \varepsilon\mu_1 x + \tfrac{1}{2}\varepsilon^2\mu_2 x^2$$

respectively, with $\operatorname{Re}\mu_2 < 0$. The latter model has been derived rationally as a general model of global-mode growth in arbitrary spatially developing shear flows (subject to certain conditions) by Huerre & Monkewitz [4]. The former model is interesting precisely because it provides a situation where global-mode theory breaks down.

In the case of linear variation of μ with x, the Ginzburg–Landau equation can be reduced to Airy's equation. The transformations

$$\xi = \alpha x + \beta,$$
$$A(x,t) = -\frac{1}{\alpha^2\gamma}\exp\left(\frac{Ux}{2\gamma} - i\Omega t\right)p(\xi;\Omega),$$

for suitable α and β (which depend on Ω, and are given fully below), yield

$$p_{\xi\xi} = \xi p + \delta(x).$$

The solutions of this can be written as a combination of Airy functions:

$$p = \begin{cases} a^+\operatorname{Ai}(\xi) + b^+\operatorname{Ai}(e^{2\pi i/3}\xi), & x > 0 \\ a^-\operatorname{Ai}(\xi) + b^-\operatorname{Ai}(e^{2\pi i/3}\xi), & x < 0 \end{cases}$$

where the $a^\pm$, $b^\pm$ are constants which may depend on Ω, and satisfy the following connection formulae over the source $x = 0$:

$$(a^+ - a^-)\operatorname{Ai}(\beta) + (b^+ - b^-)\operatorname{Ai}(e^{2\pi i/3}\beta) = 0,$$
$$(a^+ - a^-)\operatorname{Ai}'(\beta) + e^{2\pi i/3}(b^+ - b^-)\operatorname{Ai}'(e^{2\pi i/3}\beta) = \alpha.$$

To ensure that p is analytic in $\operatorname{Im}\Omega > 0$, we need only insist that $a^\pm$ and $b^\pm$ are. This does not help: we still have four unknowns, with only two connection formulae.

To proceed further we must guess a suitable form for the $a^\pm$ and $b^\pm$. Fortunately, the connection formulae provide a very good clue: we shall try a solution in which the coefficients $a^\pm$, $b^\pm$ are given by

$$a^+ = c_1 \operatorname{Ai}(\beta) + c_2 \operatorname{Ai}(e^{2\pi i/3}\beta),$$

and so on, where the c_i are constants, independent of Ω. This allows us to examine the behaviour of p as $|\Omega| \to \infty$, using the known asymptotic expansion of $\operatorname{Ai}(z)$ for large $|z|$. To find the causal solution we now need only insist that $p \to 0$ as $|\Omega| \to \infty$ in the entire region $\operatorname{Im}\Omega > 0$. After considering all possible cases, we find that we are indeed able to obtain a unique solution for the coefficients c_i.

We finally, then, obtain the following *exact* solution:

$$A(x,t) = \frac{2\pi \operatorname{sign}(\operatorname{Re}\alpha)}{\alpha\gamma} \operatorname{Ai}\big(e^{2\pi i\theta(x \operatorname{Re}\alpha)/3}\beta\big) \operatorname{Ai}\big(e^{2\pi i\theta(-x \operatorname{Re}\alpha)/3}(\alpha x+\beta)\big) \times \exp\left\{\frac{Ux}{2\gamma} + \frac{i\pi}{6} - i\Omega t\right\},$$

where

$$\alpha = \varepsilon\mu_1(-\varepsilon^2\mu_1^2\gamma)^{-1/3}e^{2\pi i/3},$$

$$\beta = \left(-\frac{U^2}{4\varepsilon\mu_1\gamma} + \frac{\mu_0 + i\Omega}{\varepsilon\mu_1}\right)\alpha,$$

and θ is the Heaviside unit step function.

The method of solution for the quadratic case, $\mu(\varepsilon x) = \mu_0 + \varepsilon\mu_1 x + \frac{1}{2}\varepsilon^2\mu_2 x^2$, proceeds in a very similar fashion. This time, parabolic cylinder functions $D_\nu(z)$ are involved. The exact solution is found to be

$$A(x,t) = \frac{1}{\alpha\gamma}\sqrt{\frac{1}{2\pi}}\Gamma(-\nu)D_\nu(-\beta \operatorname{sign} x)D_\nu((\alpha x + \beta)\operatorname{sign} x)\exp\left(\frac{Ux}{2\gamma} - i\Omega t\right)$$

where

$$\alpha = (\gamma^{-1}(-2\varepsilon^2\mu_2\gamma)^{1/2})^{1/2},$$

$$\beta = \alpha\mu_1/\varepsilon\mu_2$$

and

$$\nu = \frac{1}{\alpha^2\gamma}\left(-\frac{U^2}{4\gamma} + \mu_0 + i\Omega - \frac{\mu_1^2}{2\mu_2}\right) - \frac{1}{2}.$$

References

1. Brazier-Smith, P. R. & Scott, J. F.: Stability of fluid flow in the presence of a compliant surface. *Wave Motion* **6** (1984), 547–560.
2. Briggs, R. J.: Electron-stream interaction with plasmas. *Research Monograph No. 29.* Cambridge, Mass.: M.I.T. Press, 1964.
3. Bers, A.: Linear waves and instabilities. In *Physique des Plasmas* (ed. C. DeWitt & J. Peyraud), pp. 117–213. New York: Gordon and Breach, 1975.
4. Huerre, P. & Monkewitz, P. A.: Local and global instabilities in spatially developing flows. *Ann. Rev. Fluid Mech.* **22** (1990), 473–537.

Transverse Vorticity Measurements in the Wake of an Inverted Oscillated D-Shaped Cylinder

B. WALTER, M. M. KOOCHESFAHANI, J. F. FOSS
Mechanical Engineering Department
Michigan State University

Summary

An oscillating inverted D shaped obstruction has been used to provide a well defined vortex street at Reynolds number (U_0d/ν) equal to 23,000. U_0 and d are defined in Fig 1 which also presents a schematic of the flow based upon phase averaged results.

The purpose of the investigation was to compare time resolved transverse vorticity ($\omega_z(t)$) measurements with the phase (α) averaged velocity [$u(x,y,\alpha)$,$v(x,y,\alpha)$] measurements and flow visualization observations. It was expected that the latter two would exhibit good agreement. Prior experience with the $\omega_z(t)$ measurements in planar shear layers, see Foss (1), has shown that flow fields with a single sign of the mean vorticity ($\approx \partial u/\partial y$) possess vorticity distributions in which the rms values (ω_z) are large with respect to the corresponding mean value. Contrary to this is the plausible expectation that vorticity measurements in the large scale vortex motions would exhibit a clearly recognized distribution during the passage of a large scale coherent motion. The subject flow was selected as a paradigm test to examine the relationships between time resolved vorticity and coherent vortex motions.

Experimental Equipment and Procedure

The d=11.4 cm wide cylinder was used in both wind and water tunnels of common dimensions ($61x61x244cm^3$). It was rotationally oscillated with an amplitude of 1° (see Fig. 1). The forcing frequency equalled the natural shedding frequency ($St=f_{osc}d/U_0=0.275$). Hot wire measurements in the wind tunnel utilized the MSU compact four wire array (Foss and Haw (2)). The micro circulation domains used to evaluate $\omega_z(t)$ were $\leq 1mm^2$. A low data rate (500Hz), multi point ($1 \leq x/d \leq 5, 0 \leq y/d \leq 1.5$) survey was executed to define the phase averaged character of the flow field. Complementary, high rate (10,000Hz), data were acquired in a 16 point traverse ($0 \leq y/d \leq 1.5$) to characterize the vorticity field at x/d=4.2.

Timed releases of hydrogen bubbles at the trailing edge of the D shape revealed the vortical and non-vortical regions of the wake. The latter are inferred to be those locations in which a regular sheet of bubbles remained intact. The vortical region is shown by the cross hatched domain in Fig. 1. Two different dye colors were used to mark the separating shear layers.

The flow visualization records were used to determine the portion of the near wake flow field for which the flow angles (predominantly) exist within the $\pm 36°$ acceptance range of the probe. It was thereby determined that x/d=4.2 should provide acceptable results. Walter (3) provides a complete record of the present experiment and its results.

Results and Discussion

The two color dye visualization observations strongly support the mechanism proposed by Gerrard (4) in which fluid from, for example, $y>0$ migrates across the wake and "releases" the formed vortex motion on the $y<0$ side in order that it may proceed downstream. Phase averaged (u,v) data were used with the flow visualization results to create the composite representation shown in Fig. 1. (Such data plots and schematic representations are available for eight phase points in the driving cycle (3).) The large scale coherent motions are evident in both the dye visualization and the phase averaged (u,v) data. The dotted region represents the domains of noticably higher fluctuations (with respect to the balance of the shaded region) connecting the vortex core regions. These domains are revealed by phase averaged $(\partial Q/\partial t)_{rms}$ data. (Q=velocity magnitude as indicated by a single hot-wire probe.)

The coherent motions, that are clearly shown in this representation, are swept

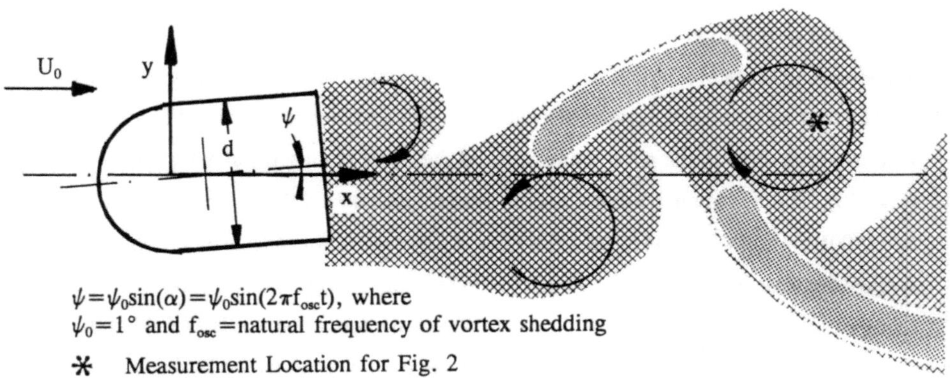

$\psi=\psi_0 \sin(\alpha)=\psi_0 \sin(2\pi f_{osc} t)$, where
$\psi_0=1°$ and f_{osc}=natural frequency of vortex shedding
* Measurement Location for Fig. 2

Fig. 1: Schematic Representation of the Investigated Flow Field at Phase Angle $\alpha=135°$

downstream at the convection velocity: $U_c/U_0=1\text{-}\exp(-0.7x/d)$, for $2 \leq x/d \leq 5$). The particular phase location $\alpha=135°$, that shows the center of a negative sign vortex just upstream of the observation plane ($x/d=4.2$), was selected as the initial location for detailed surveys within a negative sign vortex region; see * of Fig. 1. The continuing samples through this region, which end at $\alpha=180°$ or $\lambda/8$ (=0.4d) upstream of *, are shown in Fig. 2. These figures compare the phase averaged vorticity with the time resolved values during the same

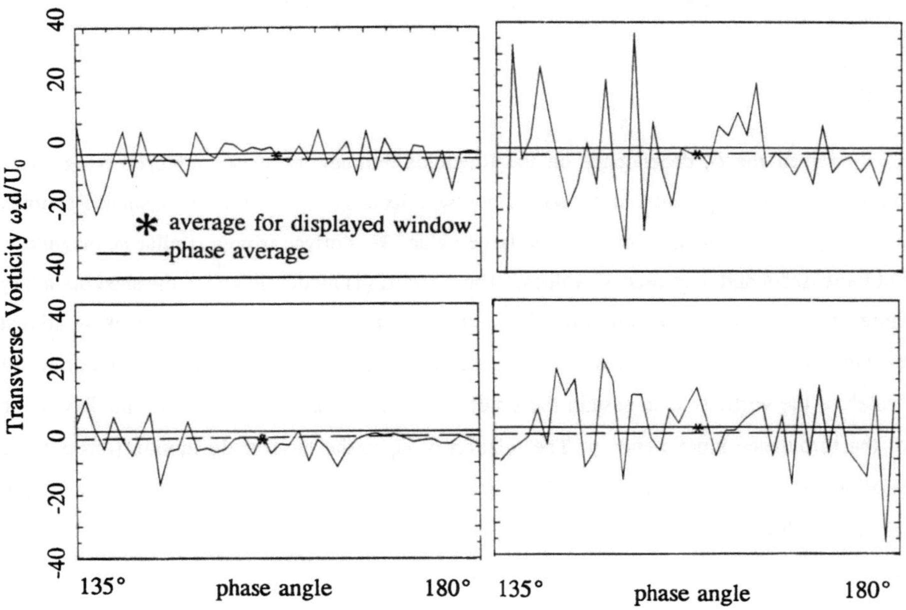

Fig. 2: Vorticity Time Signatures for Four Consecutive Cycles, $x/d=4.2$, $y/d=0.3$

part of the cycle. These plots include the phase averaged values from the complete data set.

It is apparent that there is no characteristic signature which defines the passage of the vortex, albeit the local average ($135° \leq \alpha \leq 180°$) for each realization is surprisingly close to the ensemble mean.

A crude comparison of the present measurements with those made earlier in planar shear layers can be based upon the inferred Reynolds number for the present half wake. Note that the flow for $0 \leq y/D \leq 0.7$ and $135° \leq \alpha \leq 180°$ is considered to be similar to that in a two stream shear layer. Also, long time averaged values can be used to define V_c and θ for the half-width profile; these values are included below.

	Shear Layer		Inverted D-shape	
	single stream ($V_c=U_0$)	two stream ($V_c=\Delta U$)	y/d=0.3, $\alpha=135°$ (V_c=U(0.7)-U(0))	y/d=.3, time avg. (V_c=U(1.2)-U(0))
$\frac{V_c\theta}{\nu}$	78,000	4,800	350	130
$\frac{\tilde{\omega}_z\theta}{V_c}$	7.1	0.78	6.1	39

Given the ω_z dependence on the Reynolds number for these self preserving shear layers, it is quite apparent that the wake flow (with its large ω_z values at these small Reynolds numbers) is far from equilibrium. (The large value: 39, derives from a similar ω_z magnitude but increased θ and decreased V_c values.) The Gerrard (4) model, in which the shed boundary layers from the two sides are strongly mixed near the separation lip, provides a rational explanation for these observations. The suggestion, from the present results, is that a principal "role" of the vorticity is to sustain the small scale motions that are related to the dissipation of the turbulence kinetic energy. The relation of ω_z to the large scale motion (through the relation $\Gamma = \int \omega_z dA$) is problematic. The nonequilibrium condition also predicts a strong decay of $\tilde{\omega}_z$ with downstream distance.

References

(1) Foss, J.F., (1991), "Vorticity Considerations and Planar Shear Layers", Experimental Heat Transfer, Fluid Mechanics and Thermodynamics (1991), Ed. Keffer, J.F., Shah, R.K. and Ganic, E.N., Elsevier Co, pp. 240-251

(2) Foss, J.F. and Haw, R.C., (1990), "Transverse Vorticity Measurements Using a Compact Array of Four Sensors", The Heuristics of Thermal Anemometry, ASME-FED, Vol. 97, p 71

(3) Walter, B., (1992), "Transverse Vorticity Measurements in the Wake of an Inverted D-shaped Prism", Dipl.-Ing. Thesis, Rheinisch-Westfälische Technische Hochschule Aachen

(4) Gerrard, J.H., (1966), "The Mechanics of the Formation Region of Vortices Behind Bluff Bodies", Fluid Mech., Vol. 26, part2, p 401

Two- and Three-Dimensional Flow Past Circular Cylinder(s)

LUO, S.C., Department of Mechanical & Production Engineering, National University of Singapore, 10 Kent Ridge Crescent, Singapore 0511

INTRODUCTION

The main objective of the present paper is to investigate the flow past two tandem finite length circular cylinders with the spacing between the cylinders (l) and the aspect ratio (length to diameter ratio (h/d)) as the main parameters. However, some measurements on two-dimensional flow past two tandem cylinders and flow past a single finite length cylinder were also carried out for the purposes of calibrating the test rig and later the comparison of results.

EXPERIMENTAL SET-UP

All the measurements were carried out in an open loop wind tunnel with a 1 m (height) x 2 m (width) x 16 m (length) working section. A constant 10 m/s free stream velocity (U) was employed in all the measurements. For the part on two-dimensional flow past two cylinders, each vertically installed cylinder is 880 mm in length, 100 mm in diameter and the corresponding Reynolds number (Re) is 6.29×10^4.

For measurements involving finite length single and dual cylinders, the cylinders used have a smaller diameter (d) of 53 mm. Three different lengths (h, = 4d, 6d and 8d) of cylinders were used. The cylinder(s) was(were) also installed vertically, on a mounting board which was 100 mm above the floor of the working section and was therefore outside the wind tunnel boundary layer. Each cylinder used has 8 pressure tappings along a generator. Their distances from the mounting board were equal to 5%, 15%, 30%, 50%, 65%, 75%, 85% and 95% of the length of the cylinder respectively. Pressure distribution at various spanwise locations were measured by rotating each cylinder about its own axis. Because of the smaller diameter, the Reynolds number is only 3.33×10^4. When comparison between two- and three-dimensional flow data is made, the assumption that there is no significant variation in the flow between Reynolds number of 3.33×10^4 to 6.29×10^4 is therefore implied. For both two- and three-dimensional flow, the pressure tappings were connected by 1 m long PVC tubings to Setra 239 $\pm$ 0.1 p.s.i. pressure transducers located just outside the working section. The output of the pressure transducers was digitised by an IBM PC AT based data acquisition system. Mean pressure and drag force (pressure drag only) could then be calculated. FFT was also preformed on the pressure data in order to identify the dominating (shedding) frequency for the estimation of the Strouhal Number (S). The spacing between the cylinders where measurements were conducted corresponded to $l/d = 1$ (cylinders touching), 1.64, 3 and 5.

RESULTS AND DISCUSSIONS

Two-dimensional flow past two tandem cylinders

The results obtained were very similar to those reported in the literature (Zdravkovich [1]). A critical spacing $(l/d)_{cri}$ was found to exist at $l/d = 3.5$ to 3.8. Reattached and co-shedding flow structures were identified at l/d smaller and larger than l/d_{cri} respectively. The above thus confirms that the present experimental set-up which includes the computerised data acquisition/analysis is functioning correctly.

Flow past a single finite length cylinder

The pressure distribution (C_P versus Θ) is similar to its two-dimensional flow counterpart (c_p versus θ, note that lower case and capital letters symbols refer to 2- and 3-D flow respectively) except that the wake pressure (ie. pressure downstream of the flow separation positions which includes the base pressure C_{PB}) is generally higher (less negative). Exceptions to the above occur near the free end of the $h/d = 6$ and 8 cylinders where C_{PB} is about the same or even slightly less negative than c_{pb}, and near the mounting board where the stagnation pressure is less than one because of the presence of the mounting board boundary layer. While the pressure upstream of the flow separation position is relatively insensitive to the spanwise position, the wake pressure is spanwise position dependent in that it is more negative near the free end. The decrease of the wake pressure near the free end is thought to be caused by the presence of a third shear layer which separates from the front top edge of the finite length cylinder. Shear layer is known to entrain fluid and sustain the low wake pressure of a cylinder in a flow. C_P versus Θ for the $h/d = 8$ cylinder at $y/h = 0.15$, 0.5, 0.75 and 0.95 are shown in Fig. 1. Because of brevity, C_P versus Θ at other y/h are omitted and only curves rather than individual data points (collected at an $\Delta\Theta$ of 5° from $\Theta = 0°$ to 105°, and at larger $\Delta\Theta$ there after) are shown in Fig. 1. The pressure distribution at the four spanwise position that are closest to the mounting board were found to be fairly similar. The variation of C_D in the spanwise direction (ie. w.r.t. y/h) is mainly determined by the variation of C_{PB}. This leads to two consequences. 1. C_D is generally smaller than c_d (~ 1.2 in the sub-critical Reynolds number regime). The mean (over the span) C_D is therefore also smaller than 1.2, with the longer cylinder having the larger C_D. 2. C_D is larger near the free end of the cylinder.

The regularity of vortex shedding was found to be suppressed by the turbulence associated with the three-dimensionality of the flow. Vortex shedding frequency can be detected only at the lower half of the $h/d = 8$ cylinder. The corresponding Strouhal number is 0.19. Some of the present data were compared with similar data in the literature (eg. Farivar [2]) and the agreement is generally good.

Flow past two tandem finite length cylinders

The pressure distribution on the upstream cylinder is generally fairly similar to that of a single finite length cylinder. At $l/d = 1$, the pressure distribution of the downstream cylinder suggests that two different flow structures exist at different part of the span. At $y/h = 0.95$, the shear layers which separate from the upstream cylinder reattach onto the downstream cylinder. Away from the free end, reattachment does not appear to occur, the variation of C_{P2} with Θ_2 (note that subscript 1 and 2 refer to upstream and downstream cylinder respectively) is small and the downstream cylinder appears to be simply in the wake of the upstream cylinder. The latter flow structure will be referred to as the "wake flow structure" in the present paper. Similar situation applies to $l/d = 1.64$ except that the reattached flow structure applies to a larger part of the span near the free end of the downstream cylinder. At $l/d = 3$, the pressure distribution of the downstream cylinder suggests that three flow structures exist. Co-shedding flow near the free end, reattached flow near mid-span and wake flow near the mounting board end. At a even larger spacing of $l/d = 5$, the pressure distribution of the downstream appears to be a further extension of the $l/d = 3$ case in that the extent of the span that is subjected to co-shedding and reattached flow both extend downwards (ie. towards the mounting board). For the $h/d = 4$ cylinder, almost the entire downstream cylinder is subjected to co-shedding flow. For the $h/d = 6$ and 8 cylinders, the free end side of the downstream cylinder is subjected to co-shedding flow and the lower part of it is subjected to reattached flow. A lot of data are available here. Due to space constraint, only C_{P1} and C_{P2} versus Θ at $y/h = 0.15$, 0.5, 0.75 and 0.95, $l/d = 5$ and $h/d = 8$ are shown as curves in Fig. 2 and 3 respectively. Some comparison were made with similar data from Taniguchi et al [3] and agreement was found to be quite good.

In terms of drag coefficient, C_{D1} varies with y/h in a way that is similar to C_D of single finite length cylinder versus y/h. For C_{D2}, at small l/d, C_{D2} is negative almost throughout the entire span except near the free end, where C_{D2} increases rapidly and becomes positive. As l/d increases, the whole C_{D2} versus y/h curve moves upwards in the increasing C_{D2} direction so that minimum C_{D2} which occurs at the lower part of the downstream cylinder is near zero. The extent of the span where C_{D2} is positive increases as l/d increases. Again a lot of data had been collected but due to space constraint, only C_{D1} and C_{D2} versus y/h for the case of $l/d = 5$ and $h/d = 8$ are shown in Fig. 4. Mean (over the span) value of C_{D1} and C_{D2} is always smaller than the two-dimensional flow sub-critical Reynolds number value of 1.2.

Regular vortex shedding does not exist throughout the span. For most of the cases investigated it can only be detected at the lower part (ie. nearer to the mounting board) of the downstream cylinder. When regular vortex shedding exists, the corresponding Strouhal number varies from about 0.12 to 0.17 which is in fairly good agreement with published data.

REFERENCES

1. Zdravkovich, M.M.: Review - Review of flow interference between two circular cylinders in various arrangements. Trans ASME J. Fl. Mech. 99(1977) 618-633.

2. Farivar, Dj.: Turbulent uniform flow around cylinders of finite length. AIAA J. 19(1981) 275-281.

3. Taniguchi, S.; Sakamoto, H.; Arie, M.: Interference between two circular cylinders of finite height vertically immersed in a turbulent boundary layer. Trans ASME J. Fl. Mech. 104(1982) 529-536.

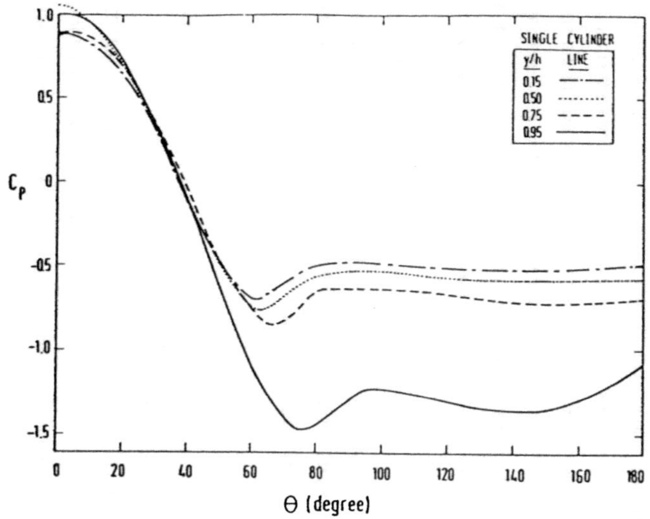

Fig.1 C_P versus θ for the h/d = 8 cylinder.

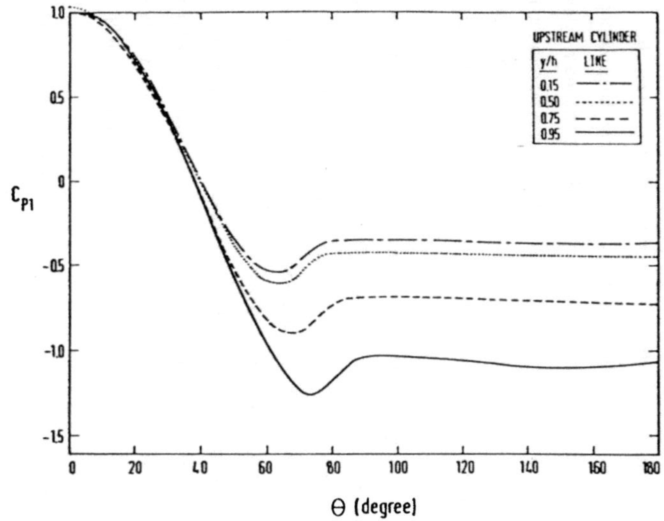

Fig.2 C_{P1} versus θ, ℓ/d = 5, h/d = 8.

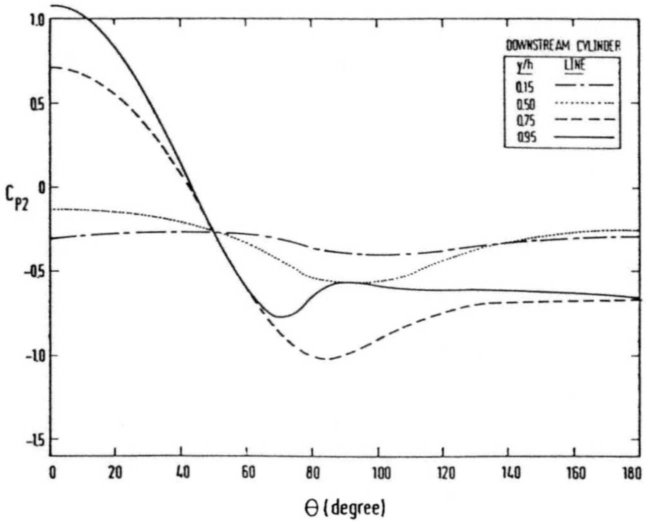

Fig.3 C_{P2} versus θ, ℓ/d = 5, h/d = 8

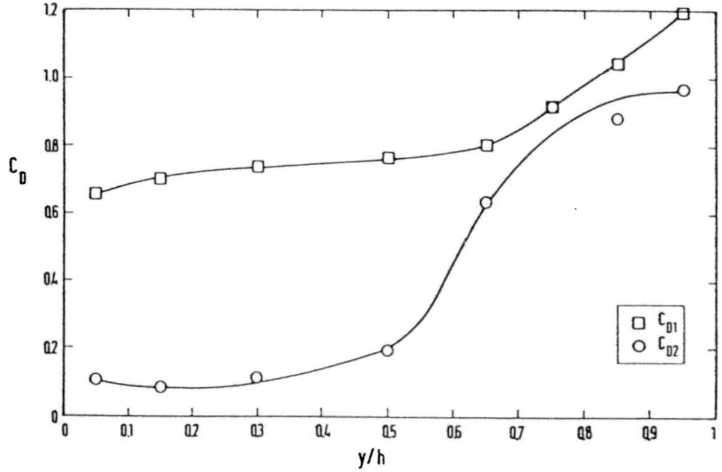

Fig.4 $C_{D1,2}$ versus y/h, ℓ/d = 5, h/d = 8.

Passive Control of Wake from a Circular Cylinder with a Splitter Plate

A. MISIRLIOGLU, M. F. UNAL
Faculty of Aeronautics and Astronautics
Aeronautical Engineering Department
Istanbul Technical University
Istanbul, Turkey

and

P. W. BEARMAN
Aeronautics Department
Imperial College of Science, Technology and Medicine
London, United Kingdom

Summary

Passive control of wake formed from a circular cylinder by means of a splitter plate attached to its end has been numerically investigated at Re=500. The Navier-Stokes equations in two dimensions are solved using the operator-splitting technique: Convection and diffusion of vorticity are treated separately. Variations of flow structure, drag and lift coefficients and Strouhal number with respect to the splitter plate length are investigated. Shedding from the tip of the plate is observed as an interesting flow feature.

Effects of Splitter Plate in Wake Formation

Experimental studies reveal that the characteristics of a bluff body wake can be effectively controlled by a proper placement of a thin streamwise plate behind the body [1-4] . When the splitter plate is attached to the end of the body [1,3], variations with respect to the plate length of interaction of the vortices formed from the body with the plate reflect into changes in the wake characteristics such as the base pressure coefficient and the Strouhal number. For a circular cylinder with short splitter plates, the vortices do not start forming until after the end of the plate and a marked reduction in the pressure drag with respect to the plain-cylinder value is achieved. For plates longer than approximately the cylinder diameter, vortices form before they reach to the end of the plate, and a further increase in the plate length causes an increase in the pressure drag [2]. On the other hand, a sufficiently long splitter plate inhibits the interaction between the shear layers separated from upper and lower sides of the body and thus eliminates the formation of Karman vortices but does not preclude the onset of shear layer instability downstream of separation [3,4].

Related numerical studies are of limited number and they are based on basically inviscid approaches [5,6]. In these studies Lagrangian discrete vortex method is used in which the bluff body is represented either by means of the image method [5] or an algorithm based on a boundary integral method [6].

Herein, the 2-D Navier-Stokes equations in vorticity streamfunction form are solved for flow around a circular cylinder with a splitter plate attached to its end using the operator-splitting technique:

Convection and diffusion of vorticity are solved separately. When modelling convection, a mixed Eulerian-Lagrangian scheme known as the Vortex-in-Cell (VIC) method is used while the diffusion of vorticity is calculated on a mesh using the finite difference technique [7]. The key to the VIC method is to by-pass the Biot-Savart law and use sequentially the Eulerian frame to calculate the velocities at discrete vortex locations through the use of the Poisson equation for streamfunction and

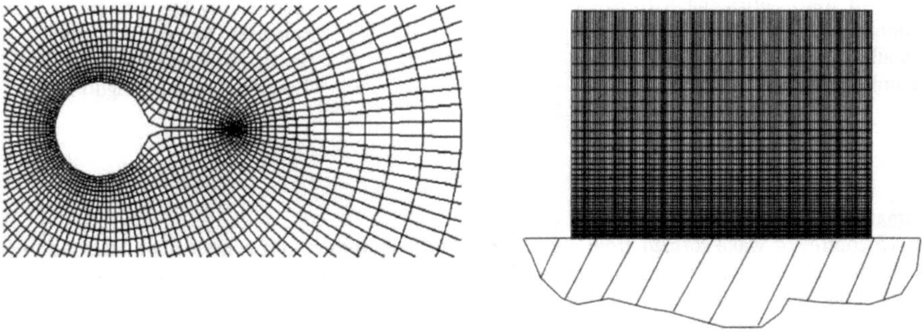

Fig. 1. Grid structures in physical (left) and transformation (right) planes.

Lagrangian frame to track the vortices. The computations are carried out in a rectangular plane obtained by conformal transformation of the domain outside of the cylinder with the splitter plate (Fig. 1). To facilitate the solutions of the diffusion equation for vorticity and the Poisson equation for streamfunction a grid with more crowded points near the cylinder surface is generated in the computational plane. For solution of the Poisson equation periodicity is imposed at the right and left boundaries of the computation plane and the Fast Fourier Transformation is used. The vorticity is created at the wall applying the no-slip condition there. After calculating the diffused vorticity across the mesh and therefore determining the new vorticity field, the Poisson equation is solved and the stream function values at grid points are used to calculate the convection velocities of the vortices. Concequently, the vortices are convected to their new locations at the end of each time step.

The VIC method allows simulation of various flow features that are observed experimentally [2]. Fig. 2 shows the effect of the splitter plate on the vorticity contours. For comparison, these patterns are sampled at the instant when the fluctuating lift acting on the cylinder is minimum. With increasing plate length, the vorticity contours indicate a passage from one case of flow in which vortex formation is delayed until after the end of the plate to another whereas the vortex formation takes place before the end of the plate. For plate lengths $L/D \geq 1.5$, traveling of the vortex across the wake center-line is eliminated. Also evident in Fig.2 is that , up to $L/D \geq 1.5$, the length of the

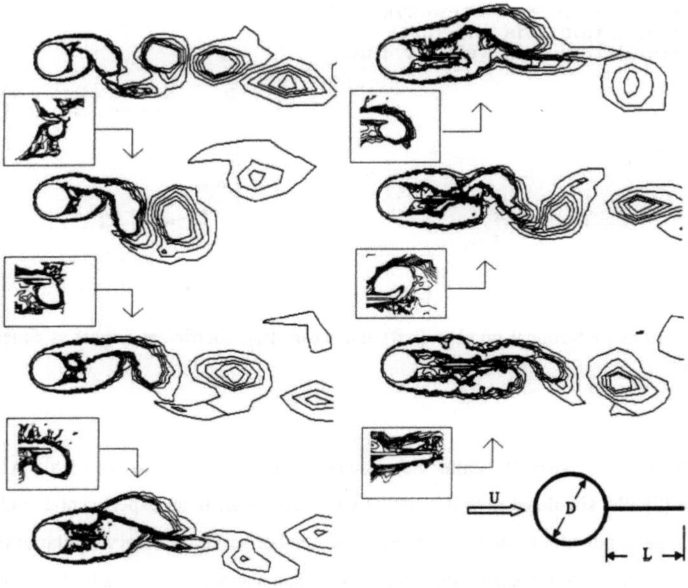

Fig.2. Vorticity contours as function of the splitter plate length. From top to bottom and left to right, L/D=0, 0.25, 0.5, 1, 1.5, 2 and 2.5.

vortex formation increases with increasing plate length. For a further increase from L/D=1.5 to 2.0, the formation length somewhat reduces. However, passing from L/D=2.0 to 2.5 the formation length increases again. An interesting flow feature observed in this study is the cross-flow-induced formation of a small scale vortex from the tip of the splitter plate. This vortex is demonstrated at the insets of Fig.2, which provide enlarged view of the tip region. Evolution of the small scale vortex with respect to that of the large scale vortex formed from the cylinder depends on the plate length. Up to L/D=1.5, a fully developed large scale vortex at the lower side co-exists with a small scale vortex of opposite sign, whereas for L/D=2, vortices of like sign meet at the tip of the plate. For L/D=2.5, the small scale vortex is not observed.

Variation of Strouhal number calculated from the unsteady lift force acting on the cylinder as function of the plate length is shown in Fig. 3. Also included in this figure is the variation determined from velocity fluctuations at a downstream location (X/D=11 ; Y/D=1.5). Similar to the experiment [2], Strouhal number first decreases and passes through a minimum at around L/D=1 and than increases to its plain cylinder value with increasing plate length. However, the change due to splitter plate is much larger than that determined experimentally. At this point, it should be noted that

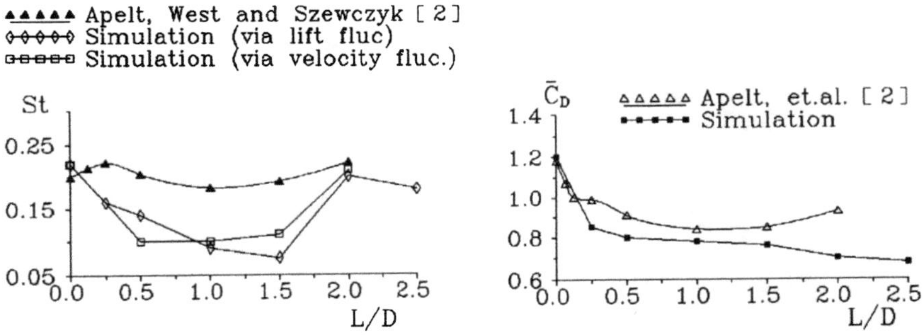

Fig.3. Variations of Strouhal number (left) and mean drag coefficient (right) as function of splitter plate length.

the experiment is at $Re=10^4$ while the present simulation refers to $Re=500$. Also, unlike the experiment [2], the simulated flow is strictly two dimensional. In an experimental study [3] for a flat plate placed perpendicular to free stream, Strouhal number variation is very similar to the case herein, in that it decreases with increasing plate length and than sharply rising to a higher mode of shedding resulting from instability of shear layers immediately after separation.

There is a better agreement in the drag coefficient variation (Fig 3). For a short plate, the mean value of the drag coefficient sharply reduces close to a third of the plain cylinder value; This sharp reduction is followed by a gradual one with increasing plate length, and in contrast to the experiment [2], no recovery of the drag coefficient is predicted for the plate length L/D=2.0.

Acknowledgement
This collaborative study has been made possible by NATO Research Grant CRG No 890427.

References

1. Bearman, P.W.: Investigation of the flow behind a two-dimensional model with a blunt trailing edge and fitted with splitter plates. J. Fluid Mech. 21 (1965) 241-255.
2. Apelt, C.J.; West, G.S.; Szewczyk, A.A.: The effects of wake splitter plates on the flow past a circular cylinder in the range $10^4 < R < 5x10^4$. J. Fluid Mech. 61 (1973) 187-198.
3. Nakamura, Y.; Nakashima, M.: Vortex excitation of prisms with elongated rectangular, H and ⊢ cross-sections. J. Fluid Mech. 163 (1986) 149-169.
4. Unal, M.F.; Rockwell, D.: On vortex formation from a circular cylinder. Part 2: Control by splitter plate interference. J. Fluid Mech. 190 (1988) 513-529.
5. Kawai, H.: Discrete vortex simulation for flow around a circular cylinder with a splitter plate. J. Wind Engrg. and Ind. Aero. 33 (1990) 153-160.
6. Meneghini, J.R.: Bluff body flow simulations using the discrete vortex method with applications to flow control. Imperial College Aero Tech Note 90-106 (1990).
7. Graham, J.M.R.; Cozens, P.D.: Vortex shedding from edges including viscous effects. Fluid Dynamics Research 3 (1988) 111-114.

The Vortical Structure of the Near Wake of a Circular Cylinder

Phuc N. Nguyen[1], Barsam Marasli and James M. Wallace
Department of Mechanical Engineering
Univerisity of Maryland, College Park, MD 20742, U.S.A.

Introduction

For wake flows, much knowledge about turbulence structures has been obtained, but the lack of detailed vorticity data hinders understanding of the dynamics of these vortical structures. Previous attempts to measure vorticity components with X-wires in a cylinder wake have been reported by Hussein and Hayakawa [1], Antonia *et al.* [2] and Bisset et al. [3]. However, to our knowledge the present results are the first simultaneous measurements of all three components of the velocity and vorticity vectors in a wake.

Experimental Facility and Method

The experiment was performed in the open-return wind tunnel of the University of Maryland Turbulence Research Laboratory. The free-stream speed was 5.0 m/s with a 6.3 mm cylinder, giving $Re_d \approx 2050$. A 12-sensor probe, built by P. Vukoslavčević as an extension of the nine-sensor probe described by Vukoslavčević *et al.* [4], was used. The probe consists of three four-sensor arrays which have centroids at the corners of an equilateral triangle. Each array has two pairs of mutually perpendicular V-probes, with prong spacings of about 0.45 mm. The hot-wire probe was calibrated at two speeds within a $\pm 20°$ angle of attack cone. The probe was operated with 12 AA Lab Systems constant-temperature anemometer circuits with a 2000 Hz sampling rate for 90 seconds.

The data were reduced with an iterative error-minimization technique briefly described in Marasli and Nguyen [5], with initial values taken from a V-probe calculation. The solutions, consisting of the three velocity components at the probe centroid and six transverse velocity gradients, are unique. Döbbeling *et al.* [6] showed that using four sensors to measure three velocity components produces only one obviously physical solution to the non-linear sensor cooling equations.

---

[1] Permanent address: Propulsor Technology Branch, David Taylor Model Basin; Bethesda, Md 20084-5000

Discussion of Results

Figure 1 depicts the measured statistics at $x/d = 30$ vs. $\eta \equiv y/L_0$. The wake half-width L_0 is 8.5 mm, and the wake deficit u_0 is 0.82 m/s. The mean vorticity values of $\overline{\Omega}_x$ and $\overline{\Omega}_y$ shown in figure 1(a), are approximately zero as required for this two-dimensional wake flow. The directly measured $\overline{\Omega}_z$ is very close to that obtained by differentiating the measured mean streamwise velocity profile. The maximum rms values of the vorticity components have quite high signal-to-noise ratios, and they peak at the wake centerline as seen in figure 1(b). All three components are distributed about the same across the wake. The skewness and flatness factors are displayed in figures 1(c&d) within $\pm 2.5L_0$, because the intermittent region of the wake produces extremely high values. The skewness factors for ω_x and ω_y are nearly zero across the wake, consistent with the two-dimensional nature of the mean flow. The ω_z skewness factor is positive for the lower wake, crosses zero at the centerline, and is negative for the upper wake, consistent with the sign of $\overline{\Omega}_z$. The flatness factors are approximately the same for all three components, and are about 5.0 in the centerline region.

The vorticity spectra in figures 2(a–c) are for the location $0.12L_0$ from the wake centerline. The ω_z spectrum shows a strong spike at the vortex shedding frequency, $f_s = 168$ Hz which corresponds to a Strouhal number of 0.21. The peak at f_s is noticeably absent from the ω_x and ω_y spectra, demonstrating both the two-dimensionality of the shedding and the ability of the probe to resolve the different components of the vorticity.

The contours of the joint probability distribution functions $P(\omega_x, \omega_y)$ and the covariance integrand $\omega_x \omega_y P(\omega_x, \omega_y)$ are shown in figures 3(a&b) for positions roughly $L_0/2$ from the wake centerline. The contours of the JPDFs are rotated about the horizontal axis by approximately $-45°$ for the lower wake, and about $+45°$ for the upper wake. The contours of the covariance integrand offer the additional information that there is a greater contribution to the covariance from quadrants 2 and 4 than from 1 and 3 for the lower wake, and vice versa for the upper wake. This further illustrates the presence of vortical structures inclined at approximately $\pm 45°$ from the wake centerline, consistent with results of Lasheras and Meiburg [7].

A conditional analysis was carried out using the maximum value of the spanwise vorticity, bandpass filtered at the shedding frequency, as the detection signal. An ensemble average $<>$ includes 14,000 realizations centered around the detection time. This analysis shows that the vortex center, which occurs at maximum values of $<\omega_z>$, is related to low Reynolds shear stress $<uv>$ as seen in figure 4 and consistent with the results of Hussein and Hayakawa [1]. The pattern is reversed for the saddle region, where $<\omega_z>$ is nearly minimum. For other locations, the pattern is similar though less correlated, and exhibits some phase shifts.

References

1. Hussain, A.K.M.F. and Hayakawa, M., "Eduction of Large-Scale Organized Structures in a Turbulent Plane Wake," *J. Fluid Mech.* (1987) **180**, pp. 193-229.
2. Antonia, R.A., Browne, L.W. B. and Shah, D.A., "Characteristics of Vorticity Fluctuation in a Turbulent Wake," *J. Fluid Mech.* (1988) **189**, pp. 349-365.
3. Bisset, D.K., Antonia, R.A. and Britz, D., "Structure of Large-Scale Vorticity in a Turbulent Far Wake," *J. Fluid Mech.* (1990), **168**, pp. 463-482.
4. Vukoslavčević, P., J.M. Wallace, and J.L. Balint, "The Vorticity Field Vector of a Turbulent Boundary-Layer: Part I. Measurement by Hot-Wire Anemometry," *J. Fluid Mech.* (1991) **228**, pp. 25-51.
5. Marasli, B. and P.N. Nguyen, "A Calibration Technique for Multiple-Sensor Hot-Wire Probes," *Bull. Am. Phys. Soc.* (1991), **36(10)**, p. 2623.
6. Döbbeling, K., Lenze, B. and Leuckel, W., "Basic Considerations Concerning the Construction and Usage of Multiple Hot-Wire Probes for Highly Turbulent Three-Dimensional Flows," *Meas. Sci. Technol.* (1990), **1**, pp. 924-933.
7. Meiburg , E. and J.C. Lasheras, "Experimental and Numerical Investigation of the Three-Dimensional Transition in Plane Wakes," *J. Fluid Mech.* (1988), **190**, pp. 1-37.

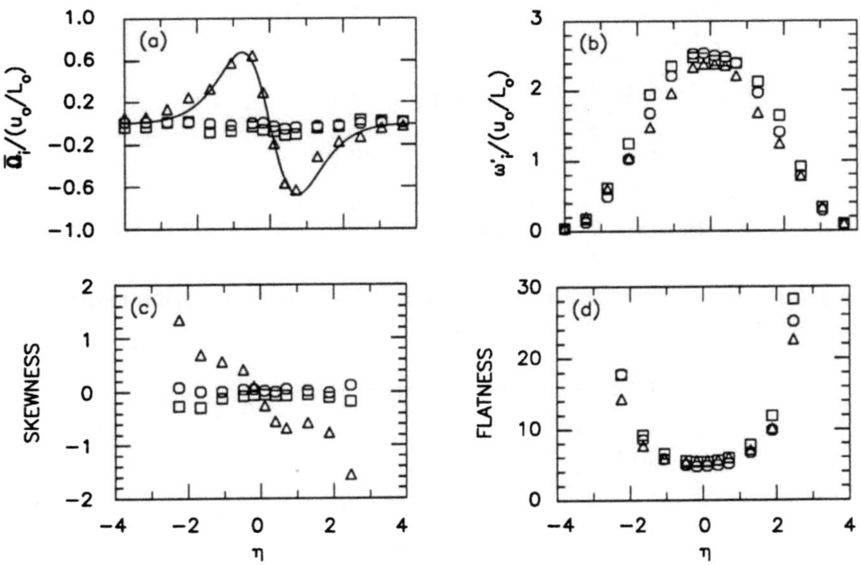

Figure 1. Distribution of (a) the mean vorticity; (b) the rms vorticity components ω_i'; (c) the skewness factor; and (d) the flatness factor across the wake. □, streamwise; ○, cross-stream; △, spanwise components; ——, $-d\overline{U}/dy$.

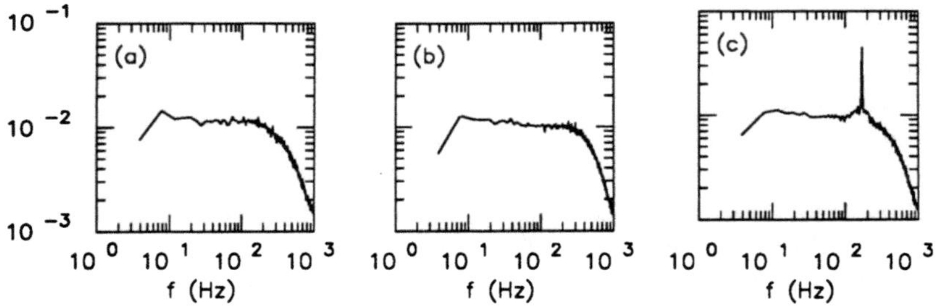

Figure 2. Spectra of the three vorticity components at $\eta = 0.12$: (a) ω_x; (b) ω_y; (c) ω_z.

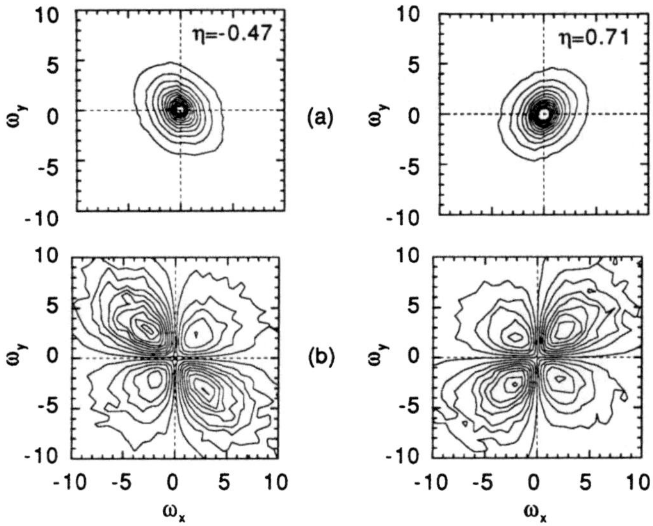

Figure 3. (a) Joint probability density functions and (b) covariance integrand contours on each side of the wake at $\eta = -0.47$ and 0.71.

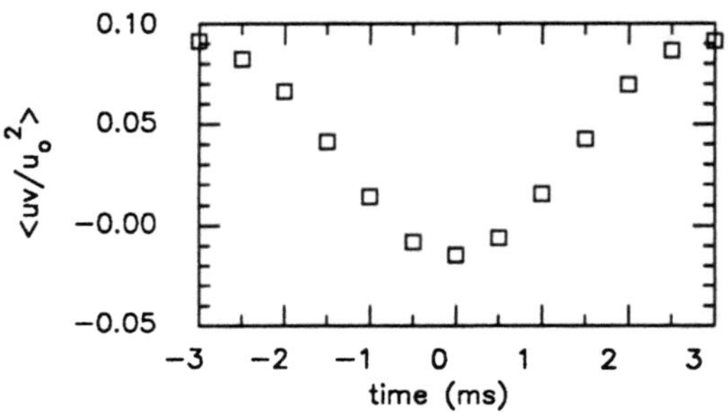

Figure 4. Ensemble averaged time series of the Reynolds stress, $<uv>$, conditioned on the maximum value of ω_z band-pass filtered at the Strouhal (shedding) frequency at $\eta = -0.47$.

Control of Wake Formation and Base Pressure by Transverse Perturbations

J. Sheridan[1], R. Mills[1], K. Hourigan[2], P. Dilin[3], M. Rudman[2], M.C. Thompson[2] and M.C. Welsh[2]

[1] Department of Mechanical Engineering
Monash University, Victoria, AUSTRALIA

[2] CSIRO Division of Building, Construction and Engineering
Victoria 3190, AUSTRALIA

[3] Department of Mechanical Engineering
Huainan Mining Institute, Anhui Province, PEOPLE'S REPUBLIC OF CHINA

Summary

The control of vortex shedding from the leading and trailing edges of long rectangular plates by the application of transverse perturbations is investigated. In particular, the effect on the base pressure coefficient resulting from the external forcing is measured. Evidence is presented for significant modification of the wake formation due to interference between the leading edge vortices and the trailing edge vortex shedding.

Introduction

The correlation between wake formation dynamics and form drag is well recognised (e.g. Roshko [1]; Bearman [2, 3]). For many flows around bluff bodies, an absolute instability exists near the trailing edge, leading to vortex shedding at a Strouhal number almost constant over a wide range of Reynolds numbers. A decrease in the base pressure coefficient, leading to an increased form drag, is observed when the vortex formation region is displaced upstream (Bearman [3]). The studies of Welsh and Gibson [4], Stokes and Welsh [5] and Hourigan *et al.* [6] have shown that vortex shedding from both the leading and trailing edges of long blunt plates can be modified by the application of sound.

In this paper, results are reported of physical and numerical experiments investigating the interference of the vortex shedding from the trailing edge of a rectangular plate by the vortex shedding from the leading edge. The effect of this interference on the base pressure coefficient is also examined.

Experimental Procedure

The flows around long plates with rectangular leading and trailing edges were investigated in an open jet wind tunnel and a water tunnel.

The basic rectangular test plate was 13 mm thick (H), 135 mm span and 130 mm chord (L) and was mounted with endplates in an open jet wind tunnel. The chord of the plate was adjusted by affixing extra plates of identical thickness to the leading edge.

Details of the wind tunnel have been reported by Hourigan *et al.* [6]. The velocity profile across the contraction at the outlet of the wind tunnel was uniform to within 1%. The centreline longitudinal turbulence intensity was 0.3%. The free-stream velocity approaching the test plate for each test was maintained constant within $\pm$ 0.1 m s^{-1}, and the plate was set at zero incidence angle to the flow.

For the wind tunnel experiments, the plates had pressure tappings along the leading and trailing edges in addition to tappings along one of the long sides. The plates were mounted on the centreline of the jet. Loud speakers were placed above and below the plates outside the jet to generate a transverse velocity perturbation at an adjustable amplitude and frequency. The local pressure coefficients along the plate surface were measured for different values of these parameters and for different plate lengths.

For the water tunnel experiments, flow visualization was conducted using hydrogen bubbles in a water tunnel with oscillating walls to provide a transverse velocity perturbation near the plate similar to that produced by the loudspeakers in the wind tunnel.

Computational Procedure

The basic method used is that described in Greengard and Rokhlin [7]. The problems considered in this paper involve flow in an infinite (two-dimensional) domain past rectangular cylinders. To incorporate an obstacle in the flow, its surface is split into M vortex sheet segments and at each iteration of the algorithm the velocity at the centre of each segment due to every vortex in the flow is calculated. Each sheet segment is then assigned the exact circulation per unit length that cancels the tangential component of velocity just *inside* the surface contour. Then, at the beginning of the next timestep, each vortex sheet segment is replaced by a point vortex of the same circulation and placed in the flow at a position *near* the surface such that its influence at the surface segment midpoints remains the same. Vortices thus released are advected by the induced velocity field of all vortices together with the background uniform potential flow. The surface vorticity approach automatically copes with separation and secondary shedding, without the need to resort to using a Kutta condition. Because the vortex shedding releases M vortices into the flow at every time step, the number of vortices can become too large even for the multipole technique to handle efficiently. Thus a vortex merging procedure has been implemented that favours merging close (like-signed) vortices that are distant from the surface of the obstacle.

Results

Physical experiments have been undertaken to date for rectangular plates with chord-to-thickness ratios 10, 13 and 15.

From the water tunnel experiments, the wake consists of large-scale structures originating from the leading edge shedding in addition to shedding from the trailing edge which is more difficult to visualize (Fig. 1). The trailing-edge vortex shedding locks to the sound frequency only over a relatively narrow range of frequencies near the natural Strouhal shedding frequency. The application of sound locks the leading-edge vortex shedding over a wide range of forcing frequencies; these structures shed from the leading edge retain their coherence along the chord of the plate. For a given flow velocity, and applied sound frequency, the phase of the forcing cycle when leading-edge vortices arrive at the trailing edge is a function of the plate length.

a)

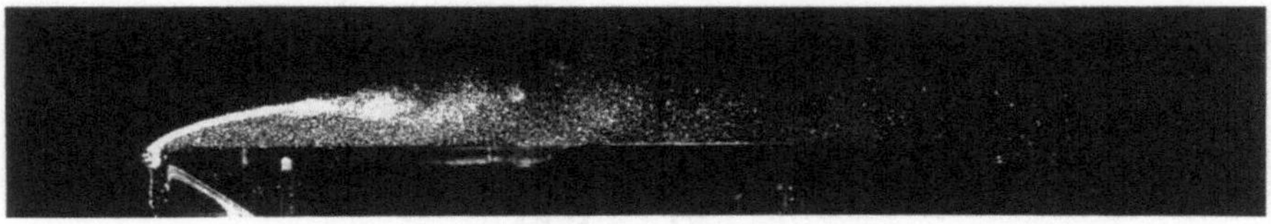

b)

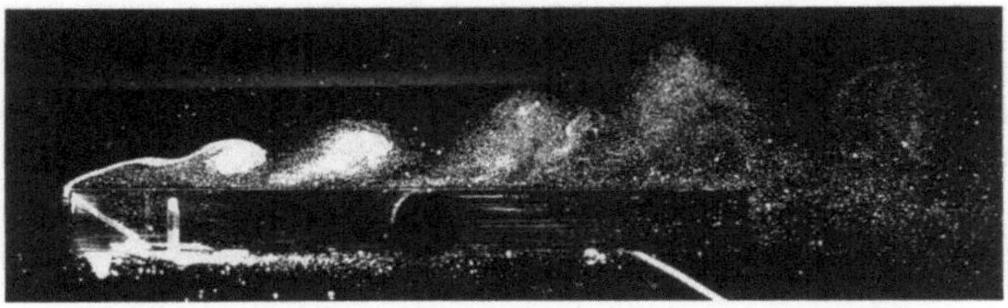

Figure 1: Flow visualization in water tunnel for rectangular plate with chord-to-thickness ratio 10 for: a) unforced flow and b) forced flow. Mean flow is from left to right.

The following results suggest that, for some phases, there can be substantial interference of the trailing edge shedding due to the arrival of vortices shed from the leading edge. In corresponding experiments conducted for plates with aerofoil and semicircular leading edges but not detailed here, no such phase dependency was observed.

From the wind tunnel experiments, the plots of the base pressure coefficient $C_{p,B}$ show a minimum when the sound frequency is close to the Strouhal shedding frequency (Fig. 2) for the chord-to-thickness ratios 10 and 15 but not for ratio 13. This suggests that for the intermediate aspect ratio, the leading edge vortices arrive at the trailing edge at a phase of the forcing cycle that results in interference with the wake development. This interference also results in an increase in the trailing-edge vortex formation length and a subsequent increase in the base pressure coefficient.

Spectra of the fluctuating pressure measured by a probe microphone placed 65 mm downstream of the trailing edge and 20 mm above the top surface of the plate were taken. The acoustic Strouhal number was 0.17 (frequency 140 Hz) in each case. It was found that the amplitude of the fluctuating pressure was lower by a factor of two for the plate with chord-to-thickness ratio 13 compared with the shorter and longer plate. This is consistent with the suggestion of wake interference leading to a higher base pressure coefficient for this intermediate plate length.

At this stage, some preliminary numerical experiments have been undertaken of the shedding from a rectangular plate. Fig. 3 shows the instantaneous vorticity distribution for a plate of chord-to-thickness ratio 10 with transverse forcing at a forcing Strouhal number of 0.2. The dynamics are seen more clearly with video animation, which will be shown at the Symposium. For this configuration, locked vortex shedding both from the leading edge and trailing edge is observed in the predictions, consistent with observations of the shedding in the water tunnel.

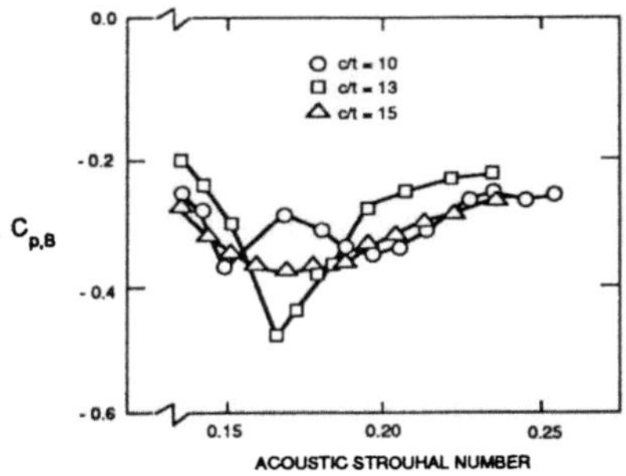

Figure 2: Base pressure coefficients $C_{p,B}$ for forced flow past rectangular plate with different chord-to-thickness ratios: $\times$, $L/H = 10$; $\bigcirc$, $L/H = 13$; $\triangle$, $L/H = 15$.

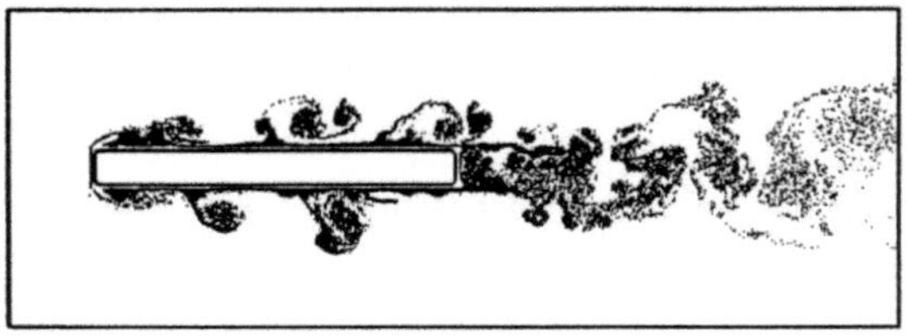

Figure 3: Predicted vorticity distribution for flow, left to right, around a rectangular plate with transverse forcing.

Future numerical investigations will focus on the prediction of the base pressures for plates of different lengths and geometries to determine modification of wake development and the influence on base pressures.

Acknowledgements:
Dr. Sheridan, Dr. Hourigan and Mr. Mills acknowledge the support of a grant from the Australian Research Council and a Monash–CSIRO Grant.

References

1. Roshko, A. (1955) On the Wake and Drag of Bluff Bodies. *J. Aero. Sc.* 22, pp. 124–132.
2. Bearman, P.W. (1965) Investigation of the flow behind a two-dimensional model with a blunt trailing edge and fitted with splitter plates. *J. Fluid Mech.* 21, pp. 241–255.
3. Bearman, P.W. (1967) The Effect of Base Bleed on the Flow behind a Two-Dimensional Model with a Blunt Trailing Edge. *The Aeronautical Quart.* 18, pp. 207–224.
4. Welsh, M.C. and Gibson, D.C. (1979) Interaction of induced sound with flow past a square leading edged plate in a duct. *Journal of Sound and Vibration* 67, pp. 501–511.
5. Stokes, A.N. and Welsh, M.C. (1986) Flow-resonant sound interaction in a duct containing a plate, II: square leading edge. *Journal of Sound and Vibration* 104, pp. 55–73.
6. Hourigan, K., Welch, L.W., Thompson, M.C., Cooper, P.I. and Welsh, M.C. (1990) Augmented Forced Convection Heat Transfer in Separated Flow Around a Blunt Flat Plate, Exp.Thermal and Fluid Science, 4, pp. 182-191.
7. Greengard, L. and Rokhlin, V. (1987) A Fast Algorithm for Particle Simulations, *J. Comput. Phys.* 73, pp. 325–348.

Dynamics of Near Wake of Circular Cylinder at High Reynolds Numbers

Dr. Tapan K. Sengupta** & Raja Sengupta*
Dept. Of Aerospace Engg., IIT Kanpur - 208016, INDIA

Introduction:

The unsteady viscous incompressible flow around an impulsively started circular cylinder is studied numerically and compared with available flow visualization pictures. The problem is solved in stream function - vorticity formulation for the infinitely long circular cylinder, placed in an uniform stream at right angles to its axis. The higher order upwinding scheme used [1] has higher spectral accuracy than other methods which use second order central differencing and/or other forms of upwinding. Results are shown for Reynolds number of 4500 and 10000, for both small and large times.

The stream function & vorticity transport equations are solved in two dimensions subject to the initial and boundary conditions,

at T=0 : The cylinder is started impulsively.

at all times :

i) no-slip condition (which also fixes the boundary vorticity)

ii) as $r \to \infty$: $u \to U_\infty$ & $\delta\omega / \delta\eta = 0$

Results & Discussion:

When the cylinder is started impulsively, all the temporal modes of the fluid dynamical system are excited. It is to be noted that impulsive start only excites the symmetric modes in the wake. The wake bubble remains symmetric for all subsequent times without vortex shedding if an accurate numerical method is used. Computed flows using less accurate schemes show wake bubble asymmetry at large times because of forcing by the asymmetric numerical error. Generally the free and forced vibration characteristics match only in a narrow range of 'synchronization' [2]. With accurate numerical schemes like the present one this necessitates the use of external forcing for inducing asymmetry and subsequent shedding.

Figs. (1a - d) show the comparison of numerical results with the experimental pictures for Re = 4500 which show the higher order eddies clearly. These are reproduced in the computations.

**Assistant Professor *Graduate student

The streamlines obtained satisfy the topological laws of Hunt et al [4].

Figs. (2a-d) show the comparison between the computed results and experiments for Re = 10^4. The experimental results of (2c & d) depict the onset of asymmetry and which has been reproduced by introducing a localized roughness on the leeward side of the cylinder.

It is to be noted that the comparison between experimentally observed path lines and computed streamlines for general unsteady flow is not straight forward. The experimental results depend on the type and quantity of the tracer particles used and the exposure time of the visualization techniques. In spite of the pronounced unsteadiness of the flow field, the closed streamlines of finite dimensions matches well between the experiment and computation. This is not the case for the smaller eddies - observed in the computed results- which are blurred in the visualization pictures because of the above mentioned problems.

Also there are some doubts about the exact nature of starting the flow in an experiment [5,6]. It is our experience that to match the experimental flow visualization pictures we have to work with a particular grid spacing depending on the Reynolds number. A finer mesh fails to show good comparison, implying that the flow is initially more dissipation dominated than what is expected after an impulsive start Ta Phuoc Loc & Bouard [6] also found better match when their initial velocity profile corresponded to Stokes flow rather than an inviscid flow.

Figures (3a -d) show large time results for Re= 10^4.The result for T=20, exhibits the fact that the near wake becomes almost symmetric once again implying the presence of larger time scale superposed on the vortex shedding cycle.

References:

[1] Kawamura T., Takami H. & Kuwahara K.: In Fluid Dynamic Research, vol. 1, pp 145-162 (1985)
[2] Zdravkovitch M.M.: In Z. Flugwiss Weltraumforch., vol. 14, pp 47-58 (1990).
[3] Bouard R. & Coutanceau M. - In J. Fluid Mech.,vol 101 pt 3, pp 583-607 (1980).
[4] Hunt J.C.R., Abell C.J., Peterka J.A. & Woo H.: In J. Fluid Mech., vol. 86, pt. 1, pp 179 -200 (1978)
[5] Sarpakaya T.: In AIAA Jour., vol 29, no.8, pp 1283-1289 (1991)
[6] Ta Phuoc Loc & Bouard R.: In J. Fluid Mech., vol 160, pp 93 -117 (1985)

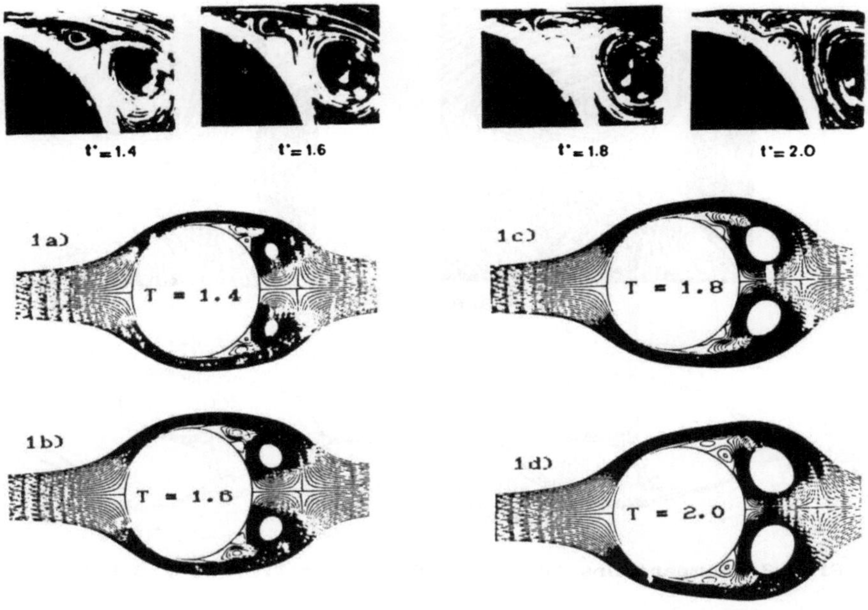

Fig. 1. Comparisons for Re = 4500 at T = 1.4, 1.6, 1.8 & 2.0

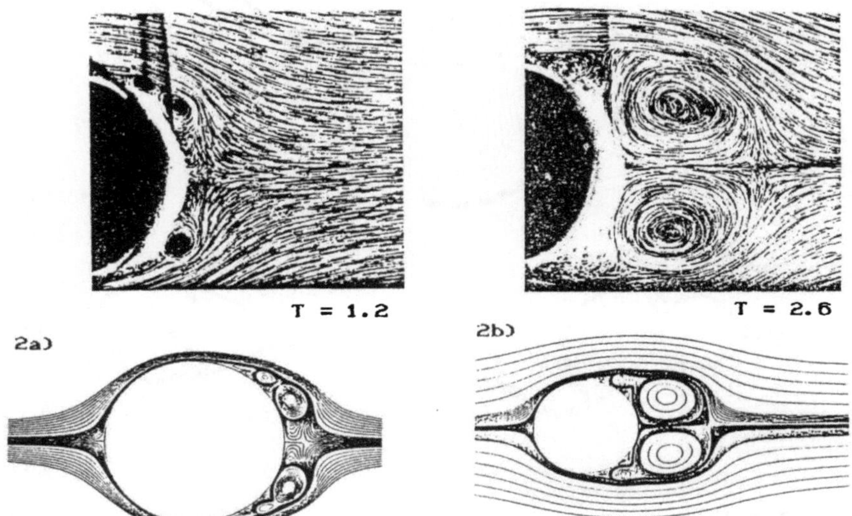

Fig. 2. See next page for legend

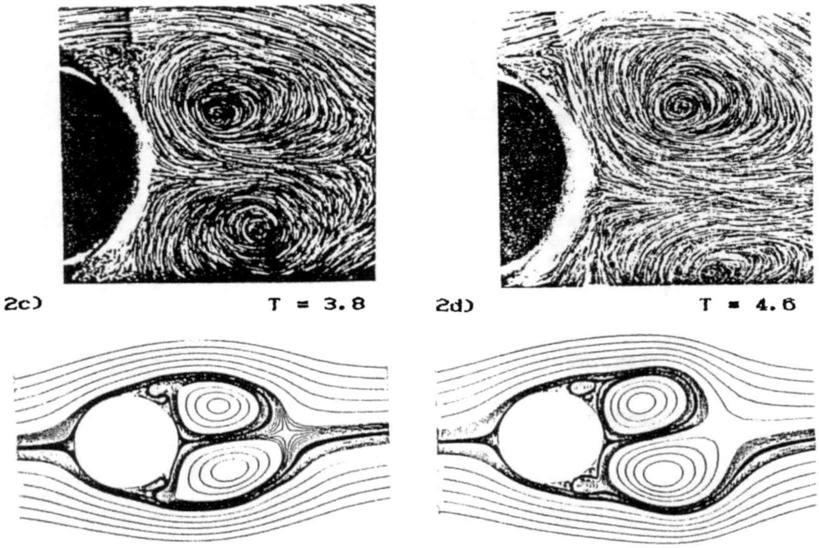

Fig. 2. Comparisons for Re = 10^4 at T = 1.2, 2.6, 3.8 & 4.6

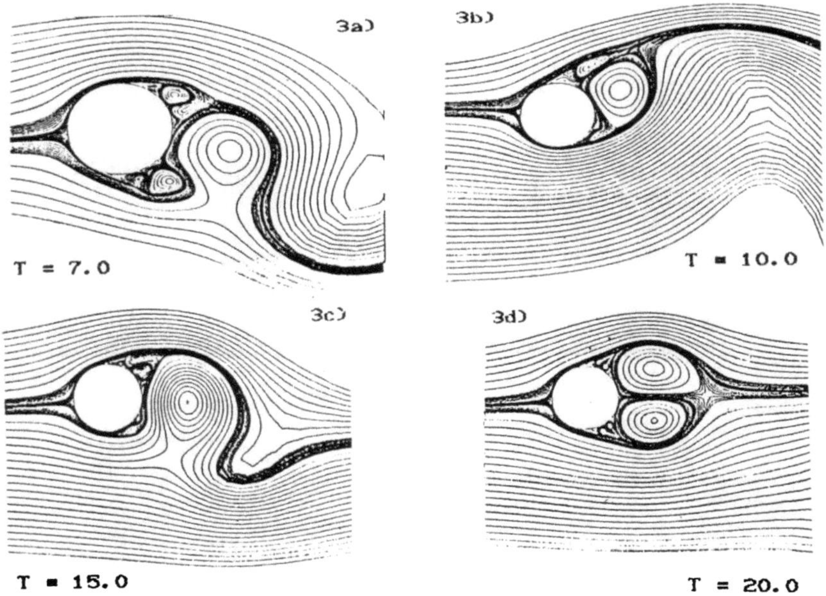

Fig. 3. Large time results for Re = 10^4 at T = 7, 10, 15 & 20

Passive Control of the Wake of a Sphere by Ventilation

G.K. Suryanarayana
Experimental Aerodynamics Division
National Aeronautical Laboratory, Bangalore 560 017, India

H. Pauer, G.E.A. Meier
DLR Institute for Experimental Fluid Mechanics
Bunsenstrasse 10, 3400 Goettingen, Germany

Introduction

The flow field around a bluff body causes a stagnation region in the front which is at a higher pressure and a lower pressure region surrounding the rear. When the two regions are mutually connected or ventilated through an internal duct, an internal flow driven by an effective pressure difference is automatically set up by the external flow. Consequently, additional mass, momentum and energy are introduced in the near wake behind the body and the base pressure mechanism can be passively altered.

Experimental Details

Tests were conducted in water and wind tunnels using wooden and styrofoam spheres. The range of Re was 1700 to 20000 in the water tunnel and 0.6×10^5 to 7.5×10^5 in the wind tunnel. The wooden models were solid and the styrofoam models hollow. Venting was created by making a central hole from the stagnation to the base. In the case of styrofoam spheres, a paper sheet rolled into the necessary shape formed the internal duct.

Results and Discussions

Fig.1 shows variation of the net drag coefficient with Re. In the region Re $< 3.6 \times 10^5$, there is no change of drag due to ventilation. Transition of the external boundary layer begins at nearly the same Re as the basic sphere. In the region 5×10^5 to 7.5×10^5, the vented sphere shows a C_d of 0.05 to 0.06 whereas the basic sphere has a value of the order of 0.12 to 0.14 and increases further with increasing Re. The drag reduction is stable to internal disturbances and unstable to disturbances created on the external flow. A single roughness element on the windward side offsets the drag reduction drastically. When the external flow is tripped using a wire, all the drag reduction vanishes.

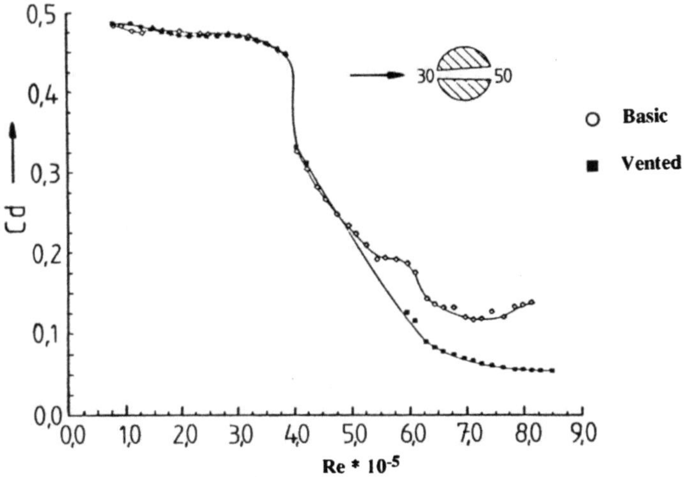

Fig. 1: Variation of Net Drag Coefficient with Reynolds Number

Visualisation at Supercritical Re

Oil-flow visualisations indicate no changes in the locations of separation and reattachment of boundary layer at all Re due to ventilation. Hence, the observed drag reduction is unlikely to be a result of downstream shift of the turbulent separation. The drastic increase of drag due to a roughness element in the external boundary layer is explained by the oil-flow pattern (Fig.2) on the vented sphere at a Re of 5.9×10^5. The disturbance on windward side is a piece of adhesive tape of dimensions 4 x 15 x 0.2mm. It is noted that a pair of steady counterrotating streamwise vortices from specific locations have generated the pattern. Two disturbances result in 4 such vortical signatures and three disturbances (sufficiently isolated) generate 6 signatures.

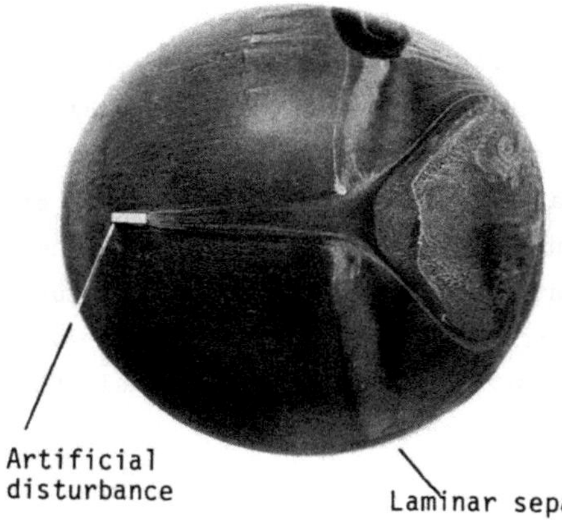

Fig. 2

Generation of streamwise vortices due to partial destruction of the base vortex ring on vented sphere at supercritical R_e

Visualisation at subcritical Re

As seen from experiments in the water tunnel at a Re of 1800, the vent flow travels 1.5 to 2.5 diameters downstream and is then deflected backwards by the reverse flow associated with the near wake. The upstream deflection of the vent flow is in response to the roll-up of the external flow. The deflected vent flow is carried by the reverse flow up to the separation line and emerges along the external separated shear layer. Thus, a large part of the vent flow comes out along with the external shear layer and very little or no vent flow comes out from the center of the near wake even through all the vent mass is introduced at the centre.

At higher Re in the subcritical range, length of the jet is shorter and is turbulent at a shorter distance behind the base. The jet is seen to pulsate and rotate and is invariably deflected upstream, to be carried by the external shear layer. The roll-up frequency remains unaffected by ventilation, as visibly seen and measured with a hot film probe.

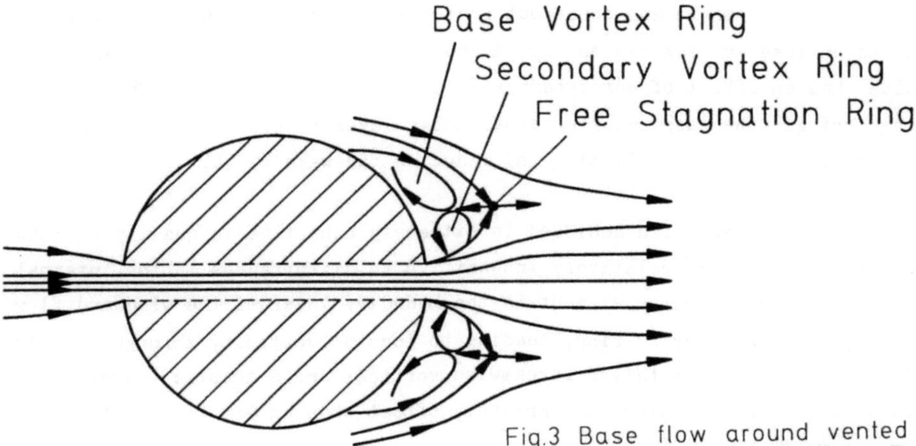

Fig.3 Base flow around vented sphere at supercritical Re

Measurements of Internal Flow

Velocity profiles of the internal flow were measured using a Prandtl tube at 0.75 diameter from the inlet inside the sphere. The mean velocities across the horizontal and the vertical diameters show slightly different profiles at subcritical Re, with the centreline velocity reaching $1.15U_{inf}$. Beyond the critical Re the centerline velocity reduces to $0.9U_{inf}$.

When the external flow is tripped using a wire, the centerline velocity again reaches $1.15U_{inf}$. A notable drop of static pressure occurs inside the duct, and the vent flow stagnates in the near wake just as at lower Re.

Conclusions

Passive ventilation is an effective method of controlling the wake of a bluff body. Wind tunnel tests carried out on smooth spheres show 50% to 60% drag reduction at supercritical Re when 2% of the frontal area of the sphere is vented to the base. Simultaneously, the flow field is symmetrised, stabilised and wake oscillations drastically reduced. Even 1% of venting is sufficient to cause these effects.

Beyond the critical Re, the vent flow breaks through the dead water region of the sphere and comes out from the center of the wake, resulting in an enclosed vortex system beyond the location of turbulent separation, which has an effect of aerodynamically streamlining the base. The vortex ring brings the separated external shear layer towards the center and symmetrises the wake, splitting and reducing the wake width.

The drag reduction achieved is extremely sensitive to quality of the external surface and relatively insensitive to disturbances in the internal flow. A single roughness element in the boundary layer of the windward flow destroys the base vortex ring, leading to loss of vorticity through a pair of localised counterrotating streamwise vortices and a drastic increase of drag. The drag reduction is completely offset when an artificial tripping wire is used.

The Effect of Free Stream Turbulence on a Circular Cylinder for Various Aspect Ratios

S. Szepessy

Department of Thermo-and Fluid Dynamics
Chalmers University of Technology
S-412 96 Gothenburg, Sweden

(during 1992):
Department of Hydromechanics
University of Karlsruhe
D-7500 Karlsruhe, Germany

Abstract

Experiments have been carried to investigate the effect of free stream turbulence on fluctuating lift forces and spanwise vortex shedding correlation of a circular cylinder at Reynolds numbers between $6x10^3<Re<4.6x10^4$, and aspect ratio $0.25<L/d<11$. Turbulence generating grids were used that produced a maximum turbulence intensity of 6.7% and a longitudinal length scale ranging $0.6<\Lambda_x/d<1.1$. Attention is focused on the difference between small and large cylinder aspect ratio. The effect of oncoming turbulence was found to be larger at small L/d.

Introduction

Aspect ratio effects reported by Szepessy&Bearman [9], at $8x10^3<Re<1.45x10^5$, were found to strongly affect the near wake flow behind a cylinder. Reduced aspect ratio made the instantaneous flow more closely two-dimensional, as seen from increased spanwise correlation. This was believed to be due to increased spanwise stability of the vortex shedding, which also increased the amplitude of the vortex shedding associated fluctuating lift. Free stream turbulence influences the cylinder boundary layer, the separation mechanisms, possible reattachment flows, *i.e.* it affects both indirectly and directly the wake vortex motion. As the interaction of turbulence in the boundary layers and shear layers is complex, the effect of turbulence is rarely accurately predictable in bluff body flows, like for a cylinder at subcritical Reynolds numbers which is particularly dependent on the state of the transitional wake shear layers. It is reported that aspect ratio, Reynolds number and free stream turbulence affects the spanwise stability of the vortex shedding. Previously, the effect of free stream turbulence has only been studied for cases at relatively large aspect ratio, whereas this study will attempt to explore the effect of aspect ratio for $0.25<L/d<11$.

Results

Measurements were carried out in a low speed wind tunnel at the Department of Aeronautics, Imperial College, London UK, using a 6cm diameter circular cylinder fitted with moveable end plates (7dx8d). The lift forces were obtained using two spatial pressure averagers connected to eight suitably positioned pressure tappings on the upper and lower half of the cylinder respectively. This technique is well suited for measuring sectional lift forces on a cylinder, as shown in detail in (Szepessy&Bearman [10]). Two different turbulence generating grids were used, with mesh sizes (M) of A=38mm and B=230mm. The cylinder was positioned at four different distances from the grid (x/M) and five different cases were investigated to give a range of turbulence intensities (Tu) and longitudinal length scales (Λ_x/d). At a cylinder Reynolds number of $4\text{x}10^4$ the turbulence parameters were: a) no grid, Tu=0.05%, b) grid A, x/M=100, Tu=2.1%, $\Lambda_x/d=1$, c) grid A, x/M=37, Tu=3.3%, $\Lambda_x/d=0.6$, d) grid B, x/M=17, Tu=5.1%, $\Lambda_x/d=1.1$, e) grid B, x/M=12, Tu=6.7%, $\Lambda_x/d=0.9$.

Figure 1 shows the effect of free stream turbulence on the sectional fluctuating lift (c_L') at a small aspect ratio of unity. For Re<$3.8\text{x}10^4$, the oncoming turbulence has the effect of reinforcing the vortex shedding causing c_L' to increase, and the effect is stronger in the lower Re range. With increasing intensity at a constant length scale (Λ_x/d), c_L' increases as shown by case (e) (6.7%,$\Lambda_x/D\approx1$) exceeding case (b) (2.1%,$\Lambda_x/D\approx1$). Reducing the length scale seems to have a large effect as indicated by case (c) (3.3%,$\Lambda_x/D\approx0.6$), which produced the largest increase of c_L' and was only exceeded by the case with laminar oncoming flow (0.05%) at Reynolds numbers above $4.2\text{x}10^4$. At the upper boundary of the Re range, the grid generated turbulence decreases the vortex shedding associated pressure amplitude and causes c_L' to drop. It is thus possible to divide the present Reynolds number range in two, where the effect of turbulence different on either side of the narrow transitional range ($3.8\text{x}10^4<\text{Re}<4.2\text{x}10^4$) is different. Figure 2 shows that with large aspect ratio the effect of turbulence is smaller than for small aspect ratio. Also the Reynolds number effect is weaker at large aspect ratio, both with and without grid turbulence. At large aspect ratio the imposed turbulence gives an almost constant increase of c_L' throughout the Reynolds number range, with the exception of curve (b) (2.1%, $\Lambda_x/D\approx1.0$) and curve (d) (5.1%,$\Lambda_x/D\approx1.1$) for Reynolds numbers less than $1.6\text{x}10^4$. Those cases exhibit the largest turbulence scale, hence figure 2 supports that the use of a smaller turbulence scale more efficiently affects the flow once the intensity has reached some threshold

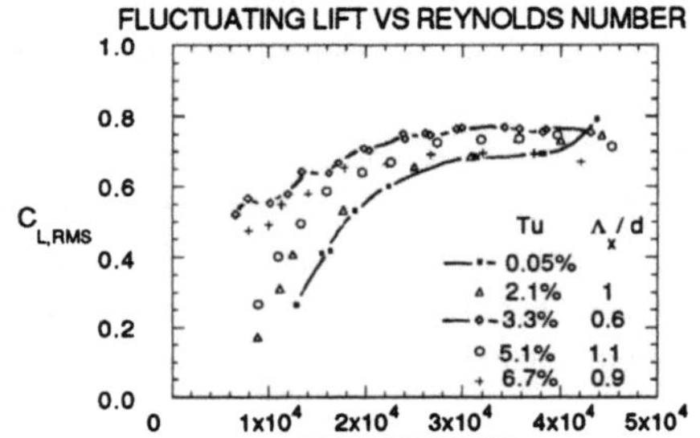

Figure 1) c_L' *vs Reynolds number at L/d=1*

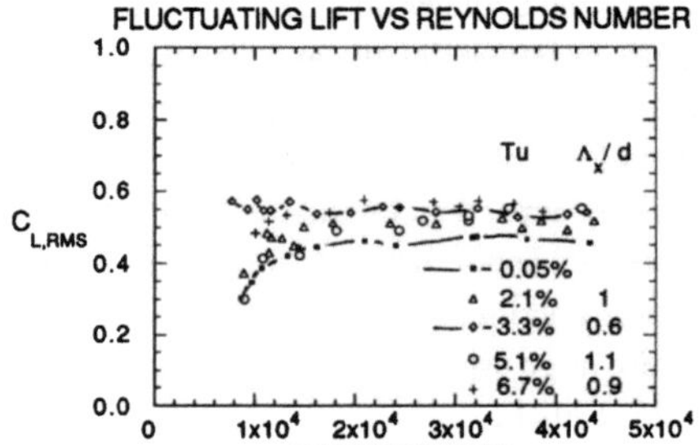

Figure 2) c_L' *vs Reynolds number at L/D=6.7.*

value of say 2%. It is possible that the effect of the intensity saturates and the effect of the scale then prevails. Case (c) (3.3%, $\Lambda_x/D\approx0.6$) shows that also for large aspect ratio the overall largest increase of c_L' occurs with the smallest turbulence length scale, closely followed by the maximum intensity case (e) (6.7%, $\Lambda_x/D\approx0.9$). The increase of c_L' with added turbulence in figure 2 is in agreement with results from the lower Reynolds numbers in Cheung&Melbourne

[3] ($6x10^4<Re<6x10^5$, L/d=4.4&8.3). Figure 3 shows the slightly attenuating effect that turbulence has at $Re=4.2x10^4$, for a range of aspect ratios. The magnitude of the peak at L/d=1 diminishes but the rise in c_L', seen at L/d=6.7 in figure 2, starts at L/d>5. Figure 4 shows that effect of turbulence on c_L' at $Re=1.55x10^4$ is the opposite to that for $Re=4.2x10^4$ and small L/d.

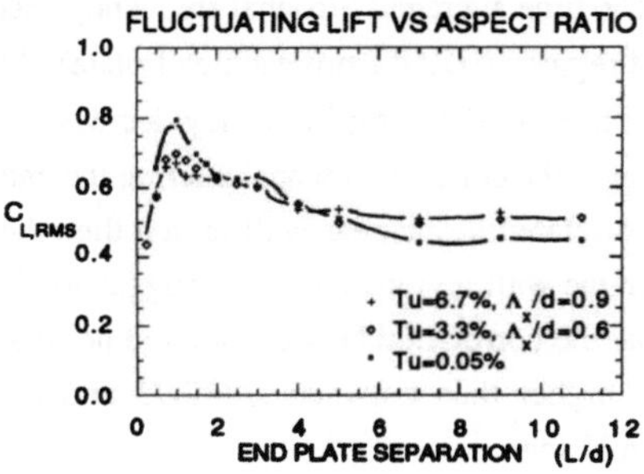

Figure 3) c_L' *vs aspect ratio.* $Re=4.2x10^4$.

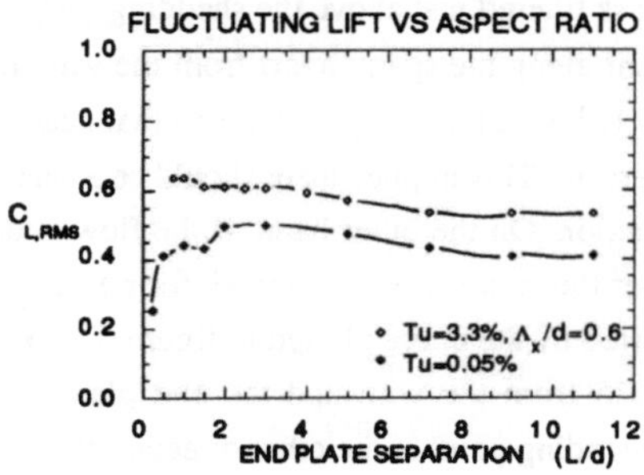

Figure 4) c_L' *vs aspect ratio.* $Re=1.55x10^4$

For laminar oncoming flow the aspect ratio effect is weak, and with added turbulence the flow seems to become even more independent of aspect ratio. From figure 5 it is seen that the spanwise cross correlation at $Re=4.2x10^4$ drops more rapidly than for $Re=1.1x10^4$. This is consistent with other works, which report on increased spanwise correlation lengths when

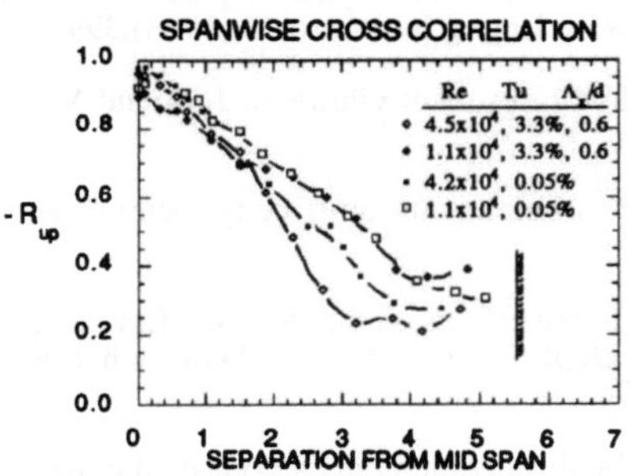

Figure 5) Spanwise cross correlation, L/d=11 Effect of turbulence (Tu=0.05% or Tu=3.3%, $\Lambda_x/d=3.3$*) and Reynolds number;* $Re=4.2x10^4$, $Re=1.1x10^4$

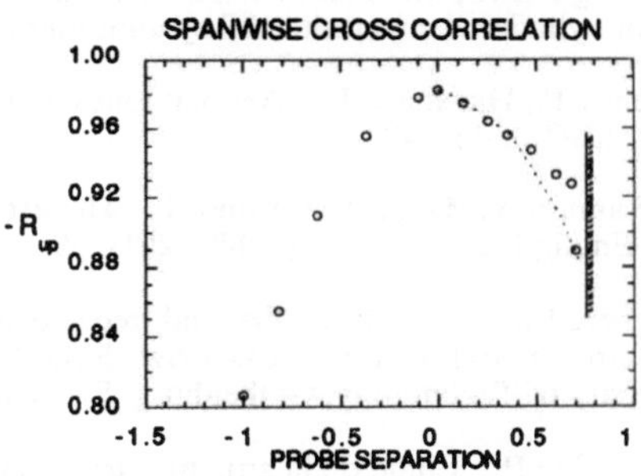

Figure 6) Spanwise cross correlation near the end plate, $Re=4.2x10^4$*, Tu=0.05%; L/d=7.*
---- mirror image of correlation $-0.75<z/d<0$

reducing the Reynolds number from $4x10^4$ to $1x10^4$. (Leehey&Hanson [5], Sonneville [8] and Norberg [7]). For $Re=4.2x10^4$, the correlation does not drop continuously towards zero, but levels out to -0.25 at about z/d=3 (2.5d from the wall) for Tu=3.3% and at 2d from the wall for Tu=0.05%. The sudden flat plateau in correlation, shows that there is a difference in the correlation between the mid span region and the wall region. It suggests the flow is well correlated near the wall, which prevents the correlation curve from rapidly falling off to zero.

The smoothly decaying correlation at Re=1.1x10^4 indicate that the wall proximity influence is weaker at this Reynolds number. Most of the loss of spanwise correlation originates from stochastic vortex shedding phase drift along the span, and not from the superimposed turbulent fluctuations, as seen from agreement in correlations of nonfiltered signals and signals that were low-pass filtered just above the shedding frequency. The time averaged Strouhal frequency was constant along the span , apart from the wall affected regions (≈1.5d from each end plate). At Re=4x10^4, spectral analysis showed that near an end plate the shedding frequency decreases a few percent. This implies there should be a sharper drop of the correlation when entering the end wall region. On the other hand, if the flow is better correlated in the near wall region then the slope of the curve should diminish for positions within the wall region. The latter suggestion is supported by the curve plotted in figure 6, showing that the correlation between a fixed point at z/d=0.75 from the wall and the end plate region is higher than between z/d=0.75 and the corresponding point in the other direction towards the mid span

References

1. Bearman P.W., Morel T.: Effect of free stream turbulence on the flow around bluff bodies. Prog Aero Sci., 20, (1983), 97-123

2. Bruun H.H, Davies P.O.A.L.: An experimental investigation of the unsteady pressure forces on a circular cylinder in a turbulent cross flow. J. Sound Vibr., 40 (4), (1975) , 535-559

3. Cheung J.C.K., Melbourne W.H.: Turbulence effects on some aerodynamic parameters of a circular cylinder at supercritical Reynolds numbers,J Wind Engng Ind. Aero,14,(1983),399-410

4. Leehey P., Hanson C.E.: Aeolian tones associated with resonant vibration, J. Sound Vibr., 13 (4), (1971), 465-483

5. Nakamura Y., Ohya Y., Ozono S.: The effects of turbulence on bluff body mean flow. J Wind Engng Ind. Aero, 28, (1988), 251-259

6. Norberg C.: Reynolds number and free stream turbulence effects on the flow and fluid forces for a circular cylinder in cross flow, PhD Thesis, Dept. Thermo-&Fluid Dyn., Chalmers University of Technology, Gothenburg, Sweden, 1987

7.Sonneville P.: Étude de la structure tridemensionelle des écoulements autour d'un cylindre circulaire. Bulletin de la direction des etudes et recherches, serie A (3), (1976), (in french). "Study of the three-dimensional structure of flow round a circular cylinder"

8. Surry D.: Some effects of intense turbulence on the aerodynamics of a circular cylinder at subcritical Reynolds number. J. Fluid Mech., 52 (3), (1972), 543-563

9. Szepessy S., Bearman P.W., Aspect ratio and end plate effects on vortex shedding from a circular cylinder, J. Fluid Mech., 234, (1992), 191-217

10. Szepessy S., Bearman P.W.: Analysis of a pressure averaging device for measuring aerodynamic forces on a bluff body, to be submitted to Exp Fluids, 1992

On the Three-Dimensionality of Fluctuating Forces Exerted on a Circular Cylinder in a Turbulent Cross-Air-Flow

V. URUBA
Institute of Thermomechanics, Dolejškova 5, Praha 8, ČSFR

Summary

Three-dimensional low frequency broad band force effects on a circular cylinder in turbulent cross-air-flow have been experimentally studied. Experiments were carried out to measure fluctuating global forces and force moments, varying the Reynolds number from $2 \cdot 10^4$ to $8 \cdot 10^4$ and the structure of incoming stream.

An original method for evaluation of local fluctuating force coefficients and correlation lengths from global forces was derived.

Introduction

A bluff body in fluid flow is subjected to fluctuating forces. Chen [1] distinguishes two parts of these: a narrow band component associated with organized vortex shedding and a wide band contribution predominantly at frequencies below the vortex shedding frequency.

In the presented paper the wide band low frequency contributions of fluctuating forces were studied separately. They are often supposed to be caused by turbulence of the incoming stream ([3]).

Model loading in a certain direction is on the fig.1. The length forces in the direction $F_1(z,t)$ produce global force $F(t)$ and perpendicular global moment $M(t)$. The spanwise correlation function r of length forces F_1 is often written:

$$r(z_1, z_2) = exp[-|z_1 - z_2|/(\lambda L)] \qquad (1)$$

where z - spanwise coordinate; λ - dimensionless correlation length; L - length of the model.

We can express the statistical characteristics of the global quantities. Mean values (2) and RMS values (3):

$$\overline{F} = L\,\overline{F_1}\,, \qquad \overline{M} = 0; \qquad (2)$$

$$F' = F'_1 \sqrt{\iint_{LL} r \,dz_1 dz_2}\,, \qquad M' = F'_1 \sqrt{\iint_{LL} r\, z_1 z_2 \,dz_1 dz_2}\,. \qquad (3)$$

Using definitions of fluctuating force coefficients:

$$C'_F = F'/F_r \,, \quad C'_M = 4\,M'/(F_r L) \,, \quad C'_{F1} = F'_1/F_r \,, \qquad (4)$$

where $F_r = 1/2\ \rho\ \overline{U}^2\ L\ D$ - relative force; ρ - air density; U - drag direction flow velocity; D - diameter of the model.

The integrals (3) yield:

$$C'_F = C'_{F1}\ \sqrt{2\ \lambda\{1 + \lambda[exp(-1/\lambda) - 1]\}} \,, \qquad (5)$$

$$C'_M = C'_{F1}\ \sqrt{8/3\ \lambda[12\lambda^3 - 3\lambda + 1 - 3\lambda exp(-1/\lambda)(2\lambda + 1)^2]} \,.$$

The equations (5) represent relation between the global quantities C'_F, C'_M and the local ones C'_{F1}, λ.

The value of $\kappa = C'_M/C'_F$ can vary from 0 for fully-correlated case ($\lambda \to \infty$) to $2/\sqrt{3}$ for absolutely non-correlated case ($\lambda \to 0$). For the values of $\kappa > 1.1$ the equations (5) are not convenient because of great relative errors of evaluated λ and C'_{F1}.

Experimental apparatus

A close-circuit wind tunnel with the working section 0.5 x 0.9 x 1.6 m^3 with several grid generators was used. The parameters of the isotropic turbulent air flow are in the table with respective marks. The definitions of intensity of turbulence I_u and dimensionless dissipative scale λ_e:

$$I_u = 100\%\ \sqrt{\overline{u^2}}/\ \overline{U} \,, \qquad \lambda_e = -\ \overline{u^2} I_u / [L(d\overline{u^2}/dx)] \,, \qquad (6)$$

where u - fluctuation of U, x - drag direction.

The model was of 30mm in diameter and 476mm in length with the circular end plates.

A five-component high frequency aerodynamic balance was utilized ([2]). The signals were filtered by means of 40Hz LP filter to avoid influence of vortex shedding.

Results and conclusion

The measurements in both drag and lift directions of global forces and moments were carried out.

Fig. 2 shows local fluctuating drag (hollow marks) and lift (solid marks) coefficients versus Reynolds number. The turbulence causes increasing of the lift coefficient in the whole investigated range of *Re*. The drag coefficient is increased by the turbulence only up to $Re^* \approx 4 \cdot 10^4$ and decreased for greater Re values. The magnitude of fluctuating drag is higher than that of lift (compare with the case of vortex shedding excitation).

Fig. 3 shows drag and lift correlation lengths versus Reynolds number. The lift correlation length is slightly greater than the drag one. Any significant systematic dependence of the correlation lengths on the length scale of incoming turbulent flow wasn't observed.

Presented results indicate significant importance of the wake and its structure. The value of $Re^* \approx 4 \cdot 10^4$ would be connected with the end of laminar/turbulent transition of fluid motion in shedding vortices ([4]).

References

[1] Chen S.S.: Flow-induced vibration of circular cylindrical structures, Hemisphere, 1987
[2] Uruba V.: Multi-component dynamic balance, to be published in Journal of Czech and Slovak Mechanical Engineering
[3] Vickery B.J.: Wind loads on buildings, University of Sydney, 1968
[4] Zdravkovich M.M.: Conceptual overview laminar and turbulent flows past smooth and rough circular cylinders, Journal of Wind Engineering and Industrial Aerodynamics, vol 33, 1990, pp 53-62

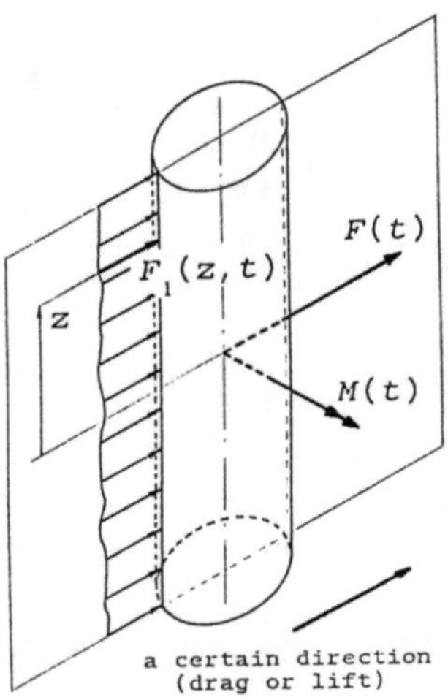

Fig. 1. Model loading in a certain direction

Tab. 1. Flow parameters

	I_u [%]	λ_e [10^{-2}]
☆	0.38	?
○	1.96	2.08
□	3.11	3.33
△	4.21	4.59
◇	7.43	11.1

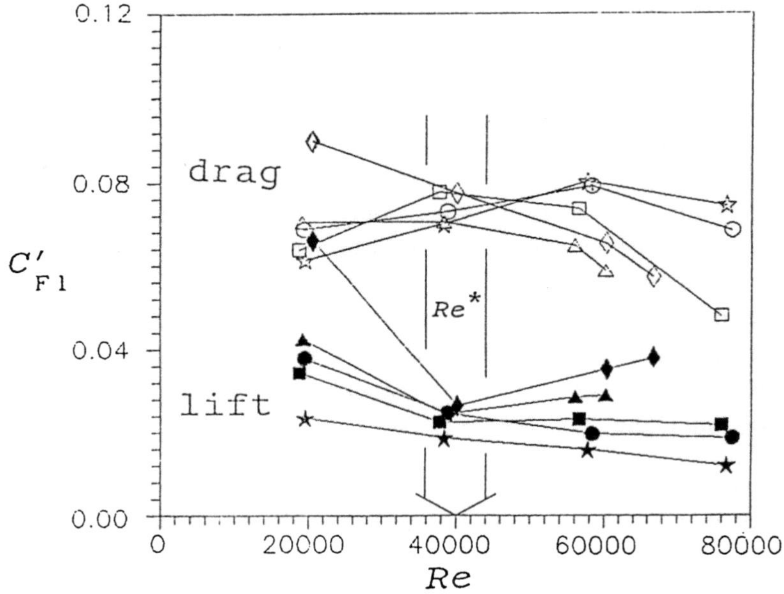

Fig. 2. Local fluctuating drag and lift coefficients as a function of *Re* for various flow parameters

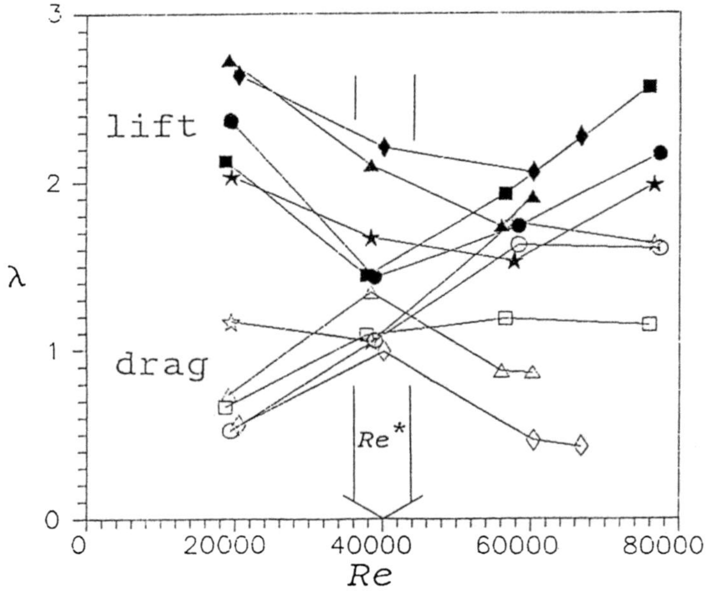

Fig. 3. Dimensionless drag and lift correlation lengths as a function of *Re* for various flow parameters

Numerical Investigation of Instability Waves in the Wake of an Axisymmetric Bluff Body

V. Schwarz, H. Bestek
Institut für Aerodynamik und Gasdynamik, Universität Stuttgart, Germany

H. Fasel
Aerospace and Mechanical Engineering Department, University of Arizona, Tucson, USA

SUMMARY

The stability of a wake behind a cylindrical, axisymmetric bluff body and the subsequent disturbance development are investigated using direct numerical simulations. The simulations are based on the complete Navier-Stokes equations employing the so-called spatial model. The steady axisymmetric base flow was also obtained from Navier-Stokes calculations. Linear stability theory analysis was applied to determine the local stability characteristics of the base flow. For the calculations discussed in this paper, the Reynolds number (based on body diameter and freestream velocity) was Re_D=1000. For this case, the base flow contains a region of absolute instability for the first helical mode (Mode 1). With an integration domain containing only the convectively unstable region of the wake, the spatial evolution of artificially excited Mode-1 disturbances was investigated and compared with linear stability results. In another simulation containing the trailing edge of the body and thus the region of absolute instability we observed self-sustained disturbance development in the wake. Here, the wake is drastically changed by the presence of three-dimensional disturbances.

NUMERICAL MODEL

Three different integration domains for the Navier-Stokes calculations were used for this study, as shown in Fig. 1. In the chosen cylindrical coordinate system z, r, φ with the origin located at the centerline at the trailing edge of the body, the two-dimensional (φ=0) domain I extended from z=-1.5 to z=10.5 in the streamwise direction, and from r=0 to r=3 in the radial direction, respectively. Lengths are nondimensionalized using the body diameter D. The corresponding dimensions for domain II are $2.5 \le z \le 26.5$, $0 \le r \le 3$, $0 \le \varphi \le 2\pi$, and $-1.5 \le z \le 22.5$, $0 \le r \le 4$, $0 \le \varphi \le 2\pi$ for domain III.

In domain I the steady axisymmetric base flow field was determined from Navier-Stokes calculations without the azimuthal velocity component. For these base flow calculations, we solved the Navier-Stokes equations in vorticity-stream-function formulation using a time-dependent approach. For this, a finite-difference method of $O(\Delta z^2, \Delta r^2, \Delta t^3)$ was employed. The Reynolds number based on the displacement thickness of the inflow velocity profile was 120. Then, the velocity profiles obtained from the base flow calculations were used to investigate the local stability characteristics by applying linear (quasiparallel) stability theory analysis. For the linear stability calculations, a shooting method with normalization and orthogonalization (to preserve linear independence of the different solutions) based on the same set of equations as derived by Morris [1] was employed. We found a region of absolute instability for the first helical mode, starting at the trailing edge of the body and extending almost to the end of the recirculation region (z=2.1).

For the simulation of spatially evolving three-dimensional disturbances, we developed a numerical method for the solution of the Navier-Stokes equations in vorticity - vector potential formulation. The basic approach is similar to the one used by Fasel et al. [2] in a spatial model for the simulation of boundary-layer transition. The numerical method is a combination of finite-difference approximations (in the streamwise and radial directions), and spectral approximations in azimuthal direction. New in the present work is the use of a vector potential. We used boundary conditions for the vector potential as proposed by Hirasaki and Hellums [3], and by Tutty [4]. At the outflow boundary we used a very effective technique to avoid upstream influence, that we developed earlier for spatial simulations of boundary-layer transition [5]. Within a so-called "relaminarization zone" (RZ) the disturbance vorticity is artificially suppressed by means of a weighting function leading to practically complete dissipation of any fluctuating component so that the "relaminarized" wake flow can then pass through the outflow boundary without causing reflections. Most importantly, this technique causes practically no upstream effect.

CODE VALIDATION

Test calculations for validation of our Navier-Stokes code were performed in the convectively unstable region of the wake, i.e. in domain II (see Fig. 1). Starting from a steady axisymmetric flow field as initial condition, the flow was periodically excited at the inflow boundary by introducing disturbances of small amplitudes using amplitude and phase distributions from linear stability results. The streamwise development of these disturbances, i.e. amplitude and phase distributions as well as amplification rates were then compared with linear stability results. The agreement between simulation and linear theory was found to be good, as can be seen from Fig. 2 for a Mode-1 disturbance. The small differences observed in these comparisons are likely due to nonparallel effects.

For the simulations with domain III, where a region of absolute instability is present, any reflections caused by the outflow boundary would strongly influence the global behaviour of the flow. Test calculations with different downstream boundary positions were performed to make sure that the RZ-technique is causing no disturbing upstream effects even for nonlinear saturated waves.

RESULTS

For simulations with the complete integration domain III, axisymmetric solutions of the Navier-Stokes equations were used as a base flow which was then initially excited in the recirculation area, for example, by a very small Mode-1 vorticity disturbance. This impulsively imposed initial disturbance triggers a strong three-dimensional development of the wake. After a period of linear growth, nonlinear saturation of the disturbances was observed. The temporal behavior in the saturated state (Fig. 3) suggests the onset of deterministic chaos.

The wake is characterized by clockwise and counter-clockwise rotating instability waves of varying strength. Fig. 4 visualizes a typical situation where both, the clockwise and counter-clockwise rotating components, have almost the same amplitude. With rotating and counter-rotating components of exactly the same amplitude the wake would have a plane of symmetry, because, by simple superposition, velocity components normal to this plane cancel each other. The orientation of this plane is completely random. For visualization, particles were released continuously along a concentric ring directly behind the trailing edge of the body, and their positions were computed by numerical integration using the flow field obtained from the Navier-Stokes solution. Views from directions orthogonal to each other are shown. Although in this simulation only four fully complex spectral modes were used in azimuthal direction, the resemblance to structures visualized by ink in a water-channel experiment (Fig. 5) under identical conditions is quite good. It should be noted, however, that Fig. 5 shows structures at different time levels, as we were unable to simultaneously take photographs from directions orthogonal to each other.

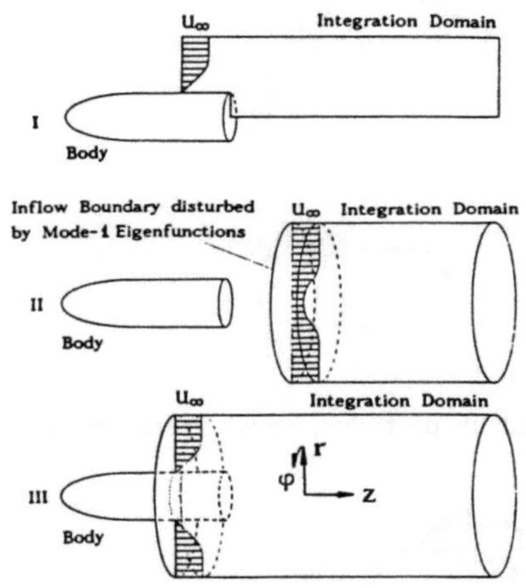

Fig. 1 Integration domains

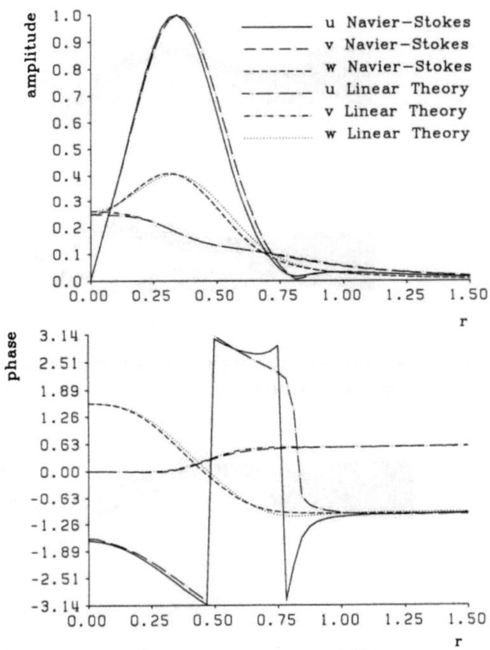

Fig. 2a

Fig. 2b

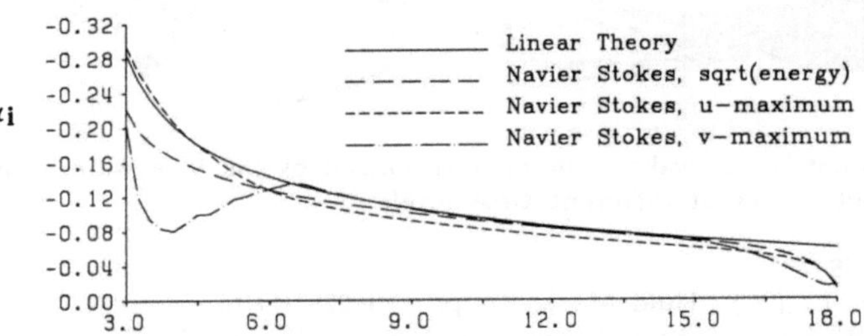

Fig. 2 Comparison between numerical simulation and linear stability theory for a Mode-1 disturbance in domain II.

a) Amplitude and phase distributions for velocity components at z=14.5

b) Spatial amplification rates, determined for various criteria

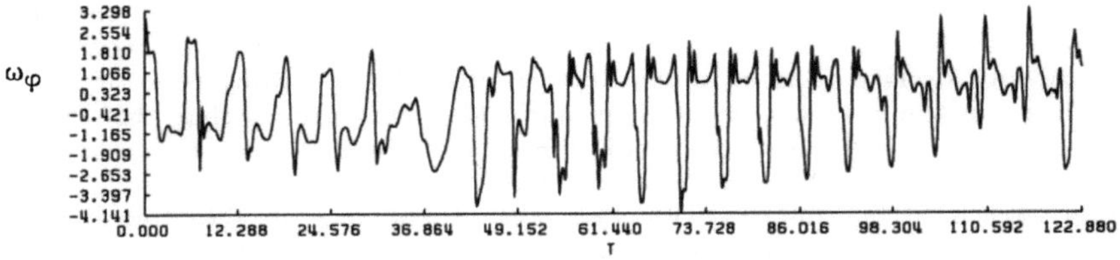

Fig. 3 Temporal behaviour of the vorticity component ω_φ at a position z=2.5, r=0 in the nonlinear saturated state

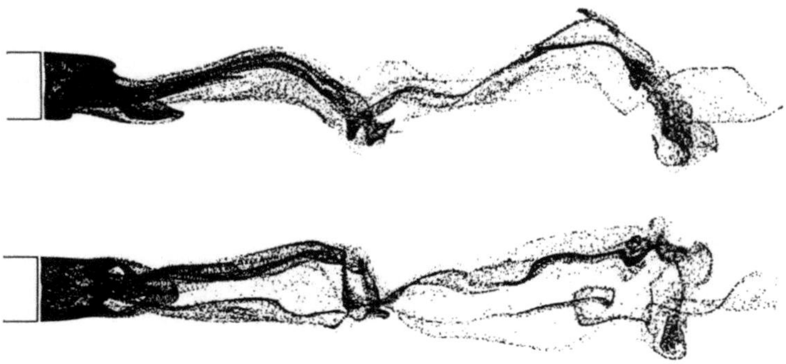

Fig. 4 Instantaneous wake structures visualized by particles, views from orthogonal directions

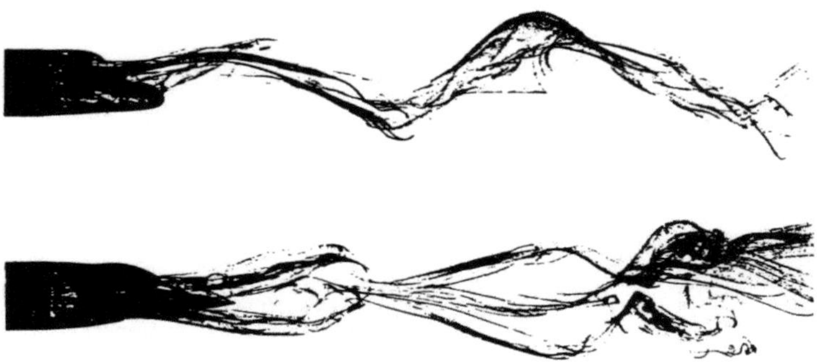

Fig. 5 Instantaneous wake structures visualized by ink in a water-channel experiment, views at different time levels

REFERENCES

1. Morris, P. J.: J. Fluid Mech. 77, pp. 511-529 (1976)
2. Fasel, H., Rist, U., Konzelmann, U.: AIAA J. 28, pp. 29-37 (1990)
3. Hirasaki, G. J., Hellums, J. D.: Quart. Appl. Math. 26, pp. 331-342 (1968)
4. Tutty, O. R.: J. Comput. Phys. 64, pp. 368-379 (1986)
5. Kloker, M., Konzelmann, U., Fasel, H.: Outflow boundary conditions for spatial Navier-Stokes simulations of laminar-turbulent transition in boundary layers. To appear in AIAA Journal.

Secondary Growth of 3-D Modes in Far-Wakes Through Tuned and Detuned Subharmonic Resonance

Thomas C. Corke and Mahmood Ghassemi

Fluid Dynamics Research Center
Mechanical and Aerospace Engineering Department
Illinois Institute of Technology
Chicago, Illinois
USA

Summary

The simultaneous generation of phase-coupled plane TS waves and pairs of oblique waves with equal but opposite wave angles are used to study the first growth of three-dimensional modes in the far-wake region of a 2-D body. This was accomplished using a computer controlled array of elements which introduced time-periodic, spanwise-phase-varying v-component velocity perturbations. To quickly achieve far-wake conditions, a thin symmetric airfoil at a 0° angle of attack was used as the wake generating body. The disturbances were introduced on the surface of the airfoil using electro-magnetic actuators. The experiment was conducted at a wake Reynolds number, $Re_w = 560$. The results document the resonant growth of the subharmonic 3-D mode. Comparisons to theory include mode amplification and eigenfunction modulus and phase. The effect of the nonlinear mode dynamics on the structure of the wake is also presented through velocity-tracer particle visualization. Contrasted to these results are the effects on the secondary mode growth when the initial modes are not harmonically related. Comparisons point to the role of so-called detuned resonance in this flow geometry.

1 Introduction

The focus of this work deals with documenting the conditions of a secondary instability growth of three dimensional modes in the far-wake region of two-dimensional bodies. By far wake we mean downstream of the point of flow reversals occurring close to bluff bodies. In this region, the mean velocity profile is Gaussian, and disturbances that are amplified within the wake are attributable to an instability of that basic state. The interest in this topic is largely motivated by the flow visualization records of Cimbala (1984) and Cimbala, Nagib and Roshko (1988). They had investigated the flow in a wake far downstream of a circular cylinder at a Reynolds numbers, $Re_d = U_0 d/\nu = 140$. Using both velocity sensors and smoke-wire flow visualization, they had documented the downstream decay of the initial Karman vortex street and the emergence of another mode with a different streamwise wave length. This mode first appeared at $x/D = 100$ and was determined to be due to a primary instability of the wake mean profile. In some of the cases, a three-dimensional structure with twice the wave length of the primary 2-D mode then also quickly emerged. In these cases, flow visualization records in the spanwise plane showed features which bore a striking resemblance to subharmonic 3-D secondary instability modes found in transitional boundary layers. Our interest was if a similar mechanism as in boundary layers was also in play in the wake, and what impact that had on the streamwise wake development.

Figure 1: Schematic of wake producing NACA 0008 airfoil, coordinate system, and sample phase-averaged smoke-wire flow visualization viewed in xz-plane.

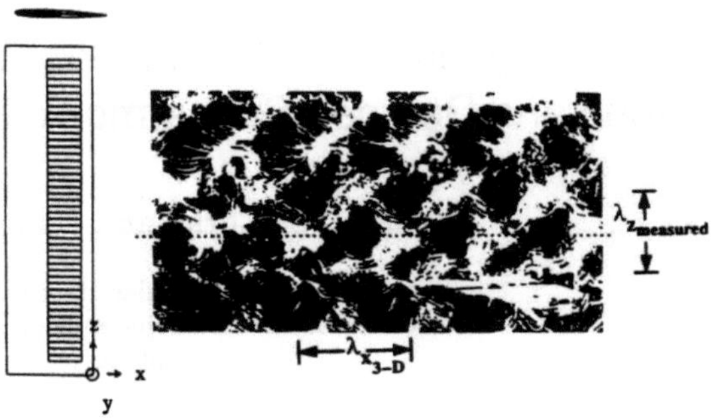

2 Experimental Procedure and Conditions

As with our previous work (Krull (1989) and Corke et al. (1991)), our approach was to introduce controlled time periodic disturbances which would seed pairs of oblique waves with wave angles of equal value but opposite sign, along with a plane wave. The initial amplitudes, frequencies, wave angles were all independently controllable. The disturbances were introduced by an array of v-component producing elements on the top and bottom surfaces of the wake producing body. An aerodynamic shape consisting of a NACA 0008 airfoil at a 0° angle of attack was chosen over a bluff body so that the characteristics of the far wake could be reached in a shorter streamwise distance.

The results are based on sampling UV-velocity time series at different points in 3-D space. The measurements fell into two types. The first type consisted of profiles taken across the wake, at different streamwise positions. In this case the spanwise location corresponded to the a point where the amplitude of the subharmonic 3-D mode was a maximum. This also corresponds to the location of oblique mode crossings. This is illustrated in Figure 1, which shows a schematic of the airfoil and coordinate system along with a sample flow visualization record in the spanwise plane (not to the scale of the airfoil). We use this to illustrate the different spanwise locations of the velocity measurements. The dotted line marks the spanwise location for measurements taken only in the xy-plane. The spanwise extent of the measurements corresponded to one spanwise wave length of the subharmonic mode, $\lambda_{z_{3-D}}$. This is shown at the right edge of Figure 1.

2.1 Experimental conditions

The free-stream velocity used in the experiment was 5.0m/s. The wake Reynolds number, based on an average of the mean profiles at $x = 3$ and 10mm, was 560. In that case the physical frequency of the natural (unforced) plane mode for the wake was 242Hz. The value predicted from linear theory was approximately 237.

For the "tuned" conditions we introduced disturbances which would produce a plane (2-D) mode at a physical frequency of 230Hz, and pairs of oblique waves at the subharmonic frequency (115Hz). The initial wave angle of oblique waves was chosen to be 75°. This gave a spanwise wave number which gave the largest secondary growth based on the analysis of Flemming (1986).

For the detuned conditions, the characteristics of the oblique modes were kept the same, and only the frequency of the plane mode was lowered to a different discrete value. The largest deviation from the tuned condition that we examined was 9%, which corresponded to a physical frequency of 215Hz. Over this range of frequencies, the

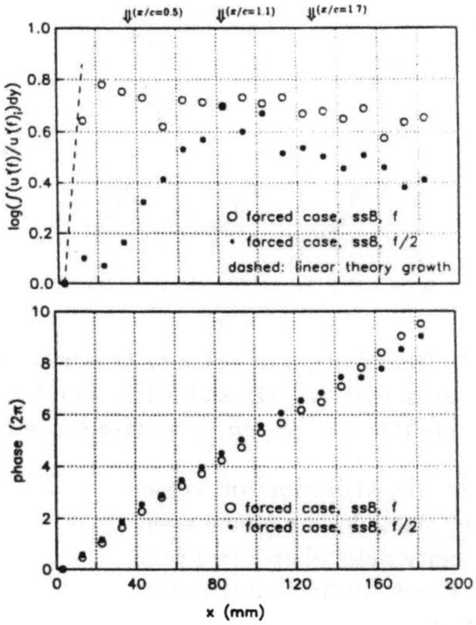

Figure 2: Streamwise development of amplitude and phase for the u-component fluctuations in the fundamental (f) and subharmonic ($f/2$) modes for forced condition. Subharmonic mode phase has been multiplied by a factor of two.

predicted linear amplification rate of the plane mode changes by less than 2%, so the influence of that aspect on the results should be minimal.

All of the input modes were set to be symmetric across the wake centerline. We define symmetry with respect to the v-component of velocity, so that a symmetric mode has an eigenfunction phase with a 0° phase shift on opposite sides of the wake centerline. Because of their topological feature such symmetric modes are also referred to as "sinuous".

3 Results

3.1 Tuned Conditions

We determine the total energy in a mode as the integral area under the eigenfunction modulus. The streamwise development of the total u-component energy in the fundamental and subharmonic modes is presented in the top half of Figure 2. The fundamental mode is shown as the open symbols. The dashed line corresponds to the growth rate for this mode based on linear theory. The arrows at the top of this plot mark the locations of the spanwise measurements.

Shown in the bottom half of the figure is the streamwise phase development for these modes. At any x-location, the local slope of the phase development corresponds to the local streamwise wave number, α_r. For each of the modes, the local phase speed is determined as $C_r = 2\pi f/\alpha_r$. Resonance requires that the phase speeds of the two modes be the same. Therefore, the ratio of the local slopes should be the ratio of their frequencies, namely 2:1. To better show this, the actual phase slope for the subharmonic mode was doubled for the presentation in the bottom part of Figure 2. In this case we observe that throughout the measurement region the two modes are traveling at the same phase speed.

Our method of introducing disturbances makes it easy to reconstruct the unsteady velocity field in a phase average sense. To visualize the structure of the wake in the yx-plane, we followed the evolution of a sheet of velocity-tracer particles which were

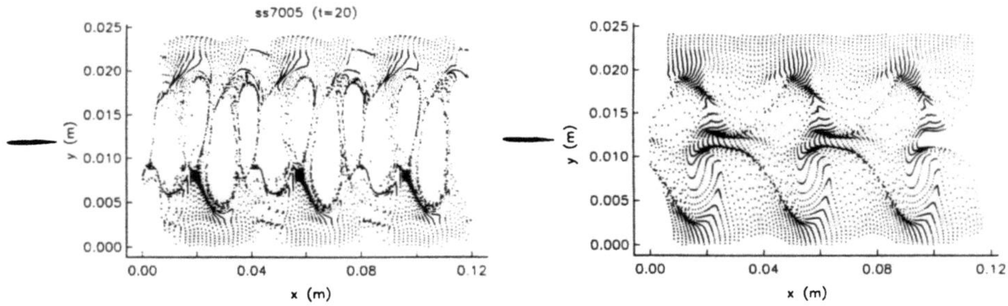

Figure 3: Distribution of velocity tracer particles in phase averaged velocity field containing all modes (*left*), and when band-pass filtered about subharmonic mode frequency (*right*): $x = 38mm$, spanwise centerline.

Figure 4: Distribution of velocity tracer particles in phase averaged velocity field when band-pass filtered about fundamental mode: $x = 38mm$.

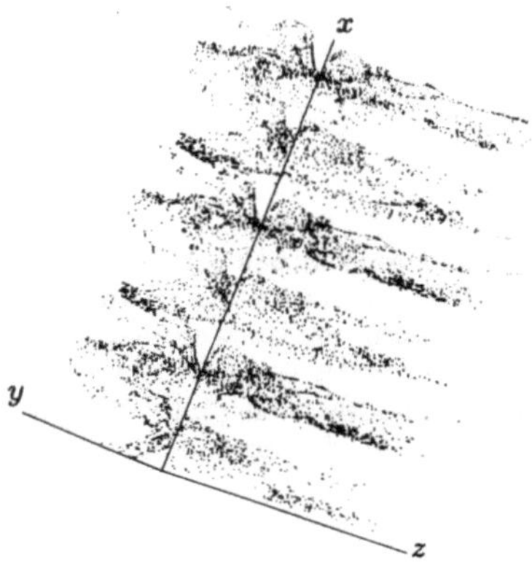

introduced into the measured, phase-averaged velocity field. Following this initial state, the particles are transported in the 2-D vector space, $\vec{x}$, to follow the local 2-D velocity vectors, $\vec{v}$, according to the relation $d\vec{x}/\vec{x} = d\vec{v}dt$. The velocity field includes the mean flow and is viewed from a moving frame traveling at the phase speed of the modes, derived from Figure 2 (bottom).

The velocity-tracer particle distribution corresponding to $x = 38mm$ is presented in Figure 3. At this stage of development, the structure of the wake is dominated by the fundamental mode. This is evident by the six cycles of the feature of the particle distribution seen at the left half of Figure 3. The fundamental mode was input as a symmetric mode so that topologically it appears "sinuous".

In order to view the structure of the subharmonic mode at this location, we have band-pass filtered the time series about the subharmonic mode frequency. The velocity-tracer particles were then inserted into that velocity field. The result is shown in the right part of Figure 3. In this case we now observe a feature with twice the wave length of the fundamental mode. As a condition for the secondary growth of the subharmonic 3-D mode, this mode is now antisymmetric. Topologically it appears "varicose".

We take similar uv-velocity measurements in the yx-plane, at the different the spanwise positions, to construct the velocity field in full 3-D space. For these we calculate the w-velocity component by applying conservation of mass within an element formed by sample points which are $\Delta x \cdot \Delta y \cdot \Delta z$ apart. The spanwise structure is visualized

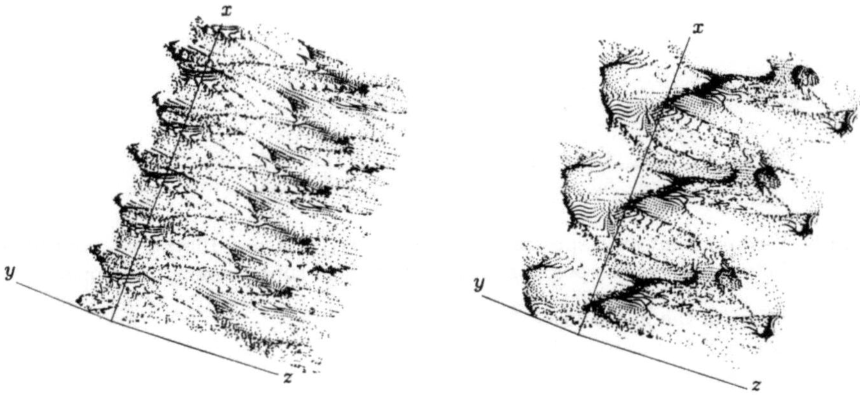

Figure 5: Distribution of velocity tracer particles in phase averaged velocity field when band-pass filtered about fundamental mode (*left*) and subharmonic mode (*right*) frequencies: $x = 83mm$.

by inserting an initially two-dimensional sheet of velocity tracer particles into the phase averaged velocity field. The sheet extends in the zx-plane, and is located at the y-location of the critical layer on one side of the wake centerline. The particles are now transported in 3-D vector space and again viewed in a moving frame of reference.

The velocity-tracer particle distributions at the upstream position ($x = 38mm$) for the fundamental mode is shown in Figure 4. At this location it is clearly 2-D. At the further downstream location, the amplitude of the subharmonic mode has increased to match that of the fundamental mode. The staggered 3-D structure associated with the subharmonic mode is clearly evident. This is shown in Figure 5 (right).

The spanwise structure of the mode at the fundamental frequency is however now quite different. The particle distributions show that it is now three-dimensional. The spanwise wave length of that mode is one-half that of the subharmonic 3-D mode. We find the origin of this mode as the result of a summing interaction of the subharmonic mode with itself. In wave number form this is $(\alpha_r/2, \pm\beta) + (\alpha_r/2, \pm\beta) = (\alpha_r, \pm 2\beta)$.

3.2 Detuned Conditions

In the previous results, the frequency of the 2-D primary mode was exactly the harmonic of frequency of the 3-D mode. However under natural (unforced) conditions, where the background disturbances fall in a broad band of frequencies, such a well tuned condition may not exist. Our intention is to simulate conditions like these, in order to document its effect on the mechanism for the secondary growth of 3-D modes, whereby an exact fundamental/subharmonic combination may not present.

Figure 6 shows auto-spectra and linear coherence of v-component fluctuations in the wake at an x-location within the linear region. In this case the frequencies of the input 2-D and 3-D modes are not harmonically related. The frequency and wave angles of the 3-D mode are the same as in the previous "tuned" cases. However, the frequency of the primary 2-D mode has been shifted by 9%, from the original 230Hz value to 215Hz. As a result of this the spectrum is filled with discrete peaks separated by 15Hz increments which correspond to the lowest difference mode frequency, for example by 230 − 215Hz.

A method to demonstrate that resonance occurs in this case under detuned conditions is depicted in Figure 7. For this the initial amplitude of the 3-D mode was fixed. The initial amplitude of the 2-D mode was slowly increased. The measure of the initial

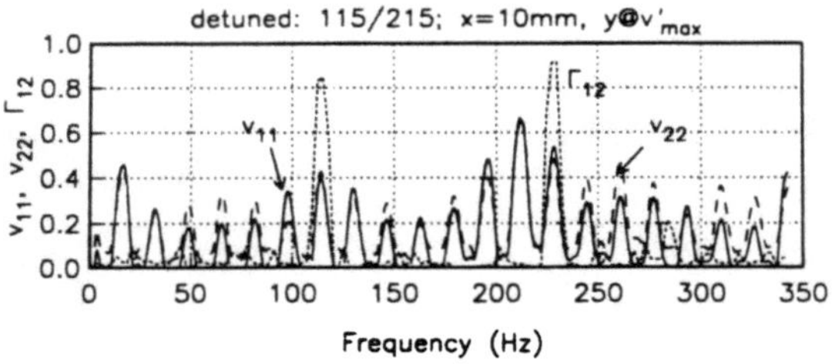

Figure 6: V-component auto-spectra at $x = 5mm$, v_{11}, and $x = 10mm$, v_{22}, and linear coherence between these two positions, Γ_{12}, for detuned subharmonic resonance.

Figure 7: Effect of the initial forcing amplitude of 2-D mode on the amplitude of velocity fluctuations in the 2-D mode (top) 3-D mode (bottom) at a fixed downstream location.

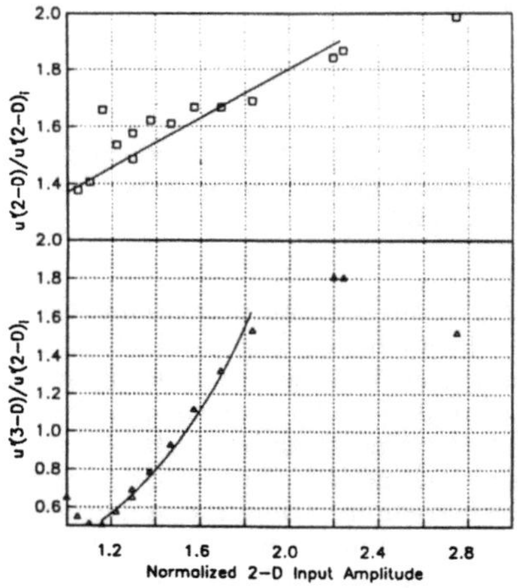

amplitude of the 2-D mode is shown on the x-axis. The response in terms of velocity fluctuations measured at a fixed distance downstream for the 2-D and 3-D modes is read from the vertical axis of the graphs shown respectively at the top and bottom parts of the figure. For the 2-D mode within the linear region, the output should vary linearly with the initial input amplitude. This is indeed the case. With fixed initial amplitudes, the growth of the 3-D mode is due to a resonant energy exchange from the 2-D mode. This is initiated when the amplitude of the 2-D mode exceeds a critical level. In this case, the change in amplitude of the 3-D mode is exponential, which agrees with an amplitude model for a weakly nonlinear three wave resonance.

References

Cimbala, J. M. 1984. Large structure in the far wakes of two-dimensional bluff bodies. PhD. thesis, California Inst. of Tech.; also Cimbala, J. M., Nagib H. M. & Roshko, A. 1988. *J. Fluid Mech.,* **190**, pp265-298.

Flemming, M. F. 1987. Secondary instability in the far wake. M.S. thesis, Illinois Inst. of Tech.

Krull, J. D. 1989. Three-dimensional mode resonance in far wakes. M.S. Thesis, Ill. Inst. of Tech.; also Corke, T. C., Krull, J. D. & Ghassemi, M. 1992. *J. Fluid Mech.,* **239**, pp99-132.

The Kelvin-Helmholtz Instability of the Separated Shear Layer from a Circular Cylinder

J.Sheridan[1], J.Soria[2], Wu Jie[1] and M.C. Welsh[2]

1 Department of Mechanical Engineering, Monash University, Clayton, Vic 3168, Australia
2 Division of Building Construction and Engineering, Commonwealth Scientific and Industrial Research Organization, Highett, Vic 3190, Australia

Summary

Experiments have been performed in which the nature of the Kelvin–Helmholtz instabilities in the separating shear layer from a cylinder were studied when perturbed by an applied sound field. The nature of their interaction with the Strouhal vortices was also examined.

Introduction

The secondary instabilities found in the near wakes of cylinders are important because of the role they play in the transition to turbulence. This transition is characterised by the interaction between the Strouhal vortices of the Karman–Vortex Street, the secondary convective instabilities in the separating shear layer, and the three dimensional streamwise vortices found in the braid region between the Strouhal vortices.

The work described here is part of an ongoing experimental and numerical study of these interactions. However, only experiments examining the Kelvin–Helmholtz instability, the resulting secondary vortices, and their interaction with the Strouhal vortices will be discussed.

Bloor [1] is credited with being the first to observe that small vortices form in the shear layer separating from a circular cylinder. She noted their similarity to boundary layer instability waves and suggested they might be important in the transition to a turbulent wake. Wei and Smith [2], using flow visualization and hot-wire anemometry, established the following power law relationship between the separated shear layer instability (f_{KH}) and the Strouhal shedding frequency (f_{St}).

$$\frac{f_{KH}}{f_{st}} \propto Re^{0.87}$$

They also suggested that if the vortices in the shear layer were perturbed in the spanwise direction they would experience three dimensional distortion which would result in the streamwise vortices. Kourta et al. [3] have also studied this phenomenon, in their case over the Re range 2000–60000, but found that their data was better fitted with an exponent of 0.5 in the above relationship; this corresponds to the value found by Bloor.

The work presented here investigates the Kelvin–Helmholtz instability of the separated shear layer from a cylinder over the Reynolds number range 2000–8000. The data presented was obtained by selectively exciting the instability modes of the separated shear layer using small acoustic perturbations and studying their characteristics.

Experimental Apparatus and Method

The majority of the experiments were conducted in the open–jet wind tunnel discussed in detail by Welsh et al. [4]. The operating range of the tunnel was 0–15m/s. The mean velocity

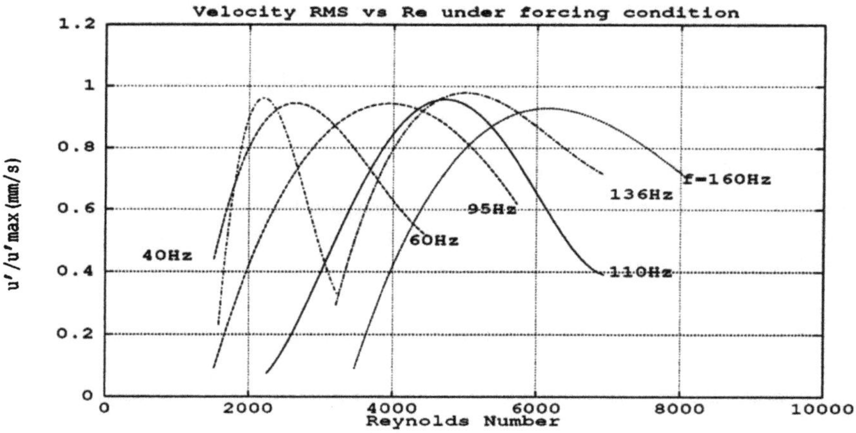

Figure 1: Variation of normalized peak rms velocity fluctuations with Re for different forcing frequencies at x/D = 0.52.

profile in the core of the jet was uniform within ±0.5%, while the longitudinal turbulence level was typically 0.3% when bandpass-filtered between 0.1 and 2 kHz.

A two dimensional velocity perturbation field was produced around a circular cylinder 25mm in diameter and 244mm long by two loud-speakers connected in anti-phase. These were positioned down stream of the jet and 350 mm apart; for the frequencies of the experiments described, the phase of the sound field was essentially uniform. Endplates with elliptic leading edges were used to minimize the end effects and to ensure the perturbation velocity was two dimensional. A TSI IFA-100 constant temperature hot-wire anemometer was used to measure the longitudinal velocity signal.

Flow visualization was performed in a water tunnel. Details of the water tunnel have been described by Welsh et al. [4]. It has a 244mm square working section, a maximum velocity in the working section of 400 mm/s, and was specially designed and built to have a pair of flexible walls in the test section. Thus, the walls can be oscillated in the cross flow direction imposing a perturbation velocity field analogous to that produced by the loud speakers in the open-jet wind tunnel. Hydrogen bubbles were used as seeding particles and a 4W Argon-Ion laser used to illuminate the near wake region behind the cylinder. Video records for a Reynolds number range of 2000-4000 were collected and could be played back at slow speed to observe the evolution of the flow structures.

Results

Figure 1 shows that when sound was applied at a particular Reynolds number there was a frequency at which the response, as indicated by the streamwise velocity fluctuation, was a maximum. There was, however, a range of frequencies over which a response was evident. This range tended to broaden with increasing Reynolds number and, as a result, flows at higher Reynolds number can be particularly responsive to disturbances over a wide band of frequencies.

The data points have not been included in this figure, however, it should be noted that there is more scatter in the data at higher Reynolds numbers. Figure 2 shows the relationship between the shear layer instability frequency, the Strouhal frequency and the Reynolds number. There is clearly good agreement with the 0.5 power law relationship proposed by Bloor [1] and Kourta et al. [3] at lower Reynolds number; at higher Reynolds numbers, however, there appears to be a deviation from this. The spanwise coherence of the separated shear layer was found to decay much more rapidly than that of the Strouhal vortices.

The spectrum of the longitudinal velocity showed evidence of velocity fluctuations at the frequencies of both the Strouhal and the secondary vortices. However, it also showed peaks at summations and differences of the two frequencies and their harmonics. This corroborates the findings of Kourta et al. [3] that there is nonlinear interaction between the two sets of vortices. It should be emphasized, however, that their work was for the natural shedding condition while in this work the Kelvin–Helmholtz instabilities are forced. When sound was applied at two distinct, incommensurate frequencies, both of which were within the band of frequencies to which the Kelvin–Helmholtz instabilities respond, the spectrum showed peaks corresponding to the Strouhal vortex and both the applied perturbations. There was clear evidence of nonlinear interaction between the Strouhal instability and both secondary instabilities. There was no evidence, however, of interaction between the two secondary instabilities; though it should be noted that this would be more difficult to detect due to the energy levels and frequencies involved.

Results from the flow visualization in the water tunnel in Fig.3 show how the shear layer develops in the near wake region for the unforced, Fig.3(a), and forced, Fig.3(b), conditions. On close examination of the video recordings it was concluded that under natural shedding conditions the secondary vortex develops with a higher degree of randomness both in terms of its frequency and evolving spatial location. When a very weak external transverse perturbation velocity field was applied the small scale vortices were seen to lock on to the perturbation. The level of the applied perturbation could be very small, in the experiments described they were typically less than 0.5% of free stream velocity.

Acknowledgements:

Dr. Sheridan and Wu Jie acknowledge the support of a grant from the Australian Research Council and a Monash–CSIRO Grant.

References

1. Bloor, M.S.: The transition to turbulence in the wake of a circular cylinder. J. Fluid Mech. 19 (1964) 290-304.

2. Wei, T. and Smith, C.R.: Secondary vortices in the wake of circular cylinders, J. Fluid Mech., 169, (1986) 513-533.

3. Kourta, A., Boisson, H.C., Chassing, P. and Haminh, H.: Nonlinear interaction and the transition to turbulence in the wake of a circular cylinder, J. Fluid Mech., 181,(1987) 141-161.

4. Welsh, M.C., Hourigan, K., Welch, L.W., Downie, R.J., Thompson, M.C. and Stokes, A.N.: Acoustics and Experimental Methods: The Influence of Sound on Flow and Heat Transfer, Experimental Thermal and Fluid Science, 3, (1991) 138-152.

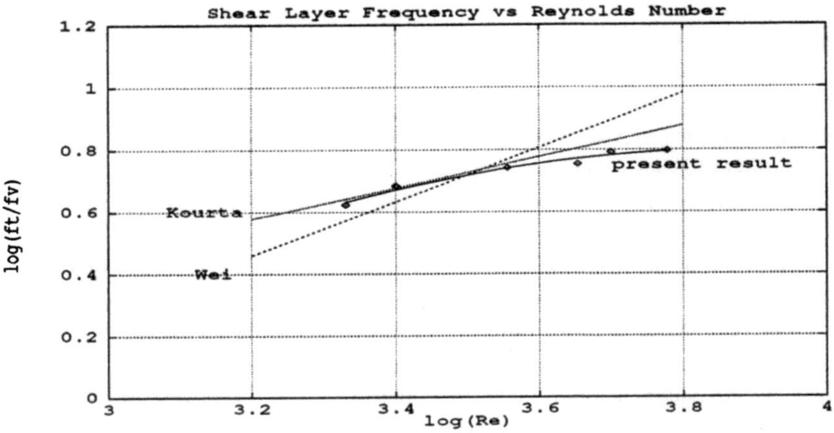

Figure 2: Variation of the most amplifiable shear layer frequency normalized by the Strouhal frequency with Re.

(a)

(b)

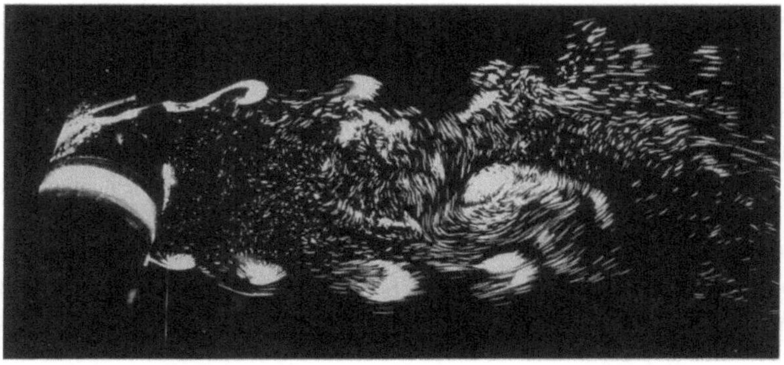

Figure 3: Flow visualization of the KH instability in the separated shear layer from a cylinder at Re=2700, (a) Unforced, (b) Forced.

Pressure Distributions on a Cylinder in Planar Oscillatory Flow at High Reynolds Numbers

JOHN R CHAPLIN

Department of Civil Engineering,
City University, London EC1V 0HB

Summary

Instantaneous pressure measurements on a cylinder in planar oscillatory flow are compared with results for steady flow within the Reynolds number range 2×10^5 to 8×10^5. It is shown that the Vorticity Dispersion Ratio follows the same trend in both cases, suggesting that with fine roughess, post-critical conditions in oscillatory flow may not occur at Reynolds numbers much less than 2×10^6

Introduction

Studies at large scale of the loading on a cylinder in waves and oscillatory flow have concentrated largely on the behaviour of the Morison drag and inertia coefficients. The nature of the underlying changes in the flow with Reynolds number, and with fine surface roughness, have received much less attention. There have in any case been very few measurements of forces in oscillatory flow in the range of Reynolds numbers normally associated (for steady flow) with critical conditions, and it is therefore not easy to identify where the post-critical regime (where Reynolds number is unimportant) begins. And it is not clear to what extent the important changes that occur at high Reynolds numbers are similar to the mechanisms that are well known in steady flow. When the motion is oscillatory, familiar concepts of separation may not be appropriate, but the roles of transition, wake re-encounter, and surface roughness are likely to be significant, and of changing importance though each cycle of oscillation. Post-critical conditions will occur only when the boundary layer is turbulent at all phases of the motion, from a region very close to the upstream stagnation point.

This paper is concerned with the pressure distributions observed on a cylinder with fine surface roughness, executing planar oscillatory flow at Reynolds numbers between 2×10^5 and 8×10^5. The drag and inertia coefficients derived from these measurements have been presented elsewhere [1], and have been shown to be compatible with earlier data. Here we are concerned with the alterations in the pressure distributions that cause the integrated loading to undergo significant change over the range of Reynolds number tested.

Experimental arrangements

Full details of the experimental arrangements are given by Chaplin [1]. The cylinder used in these tests was one section of that built for wave loading measurements in the Delta Flume, and had the same surface roughness ($k/d = 1.2\times10^{-4}$) as in one of the series of experiments in waves [2,3]. The cylinder's diameter is 0.5m, and in the present experiments a 1.36m long section was mounted vertically between 1.2m diameter endplates, on the carriage of a 2-dimensional random planar motion system. The whole cylinder was submerged, with the upper endplate 150mm beneath the water surface, and the lower endplate 25mm clear of the tank floor. At its mid-depth cross-section, the cylinder was provided with a ring of 14 equally spaced flush mounted pressure transducers, and signals from these instruments and from the position transducers of the carriage system were sampled 512 times in every cycle.

It can be shown that 14 pressure transducers are sufficient to provide integrated forces to a sampling error of around 2% [1], though clearly more are required to reveal detailed features of instantanous pressure distributions.

Experimental Results

Results are presented in terms of the Reynolds number $R_e = U_{max}d/\nu$, and Keulegan Carpenter number $K_c = U_{max}T/d$, where U_{max} and T are the amplitude and period of the ambient incident velocity, and d the cylinder's diameter.

At Keulegan Carpenter numbers between 10 and 25, the present measurements indicated significant increases in the drag coefficient over the Reynolds number range 2×10^5 to 8×10^5 [1]. In steady flow, such a change is to be expected; for oscillatory flow much less evidence is available, and it may be argued that the effect of the cylinder encountering its turbulent wake would render the drag coefficient less sensitive to the Reynolds number after the critical regime. This is the implication of measurements with a smooth cylinder, in waves [2], and in planar oscilatory flow [4]. But measurements in waves with the same degree of fine surface roughness as tested here were in this respect less conclusive [3].

Experimental results presented below include the Vorticity Dispersion Ratio V_{dr}, an index which in steady flow provides an indication of the nature of boundary layer separation. Pearcey *et al* [5] define the Vorticity Dispersion Ratio as the reduction in the vorticity flux in the boundary layer through the region of adverse pressure gradient, divided by the vorticity flux at the point of minimum pressure. In terms of the surface pressure coefficients (defined in the usual way in terms of the maximum ambient velocity), $V_{dr} = (C_{p2} - C_{p1})/(1 - C_{p1})$ where point 1 is where the pressure reaches a minimum, and 2 is at separation. Pearcey *et al* obtained the variation of V_{dr} for steady flow through the critical Reynolds number region for various surface roughnesses, and showed that typically it increases from about 0.1 for laminar separation, to about 0.6 at critical Reynolds numbers, falling to around 0.4 in the post-critical regime. The values of V_{dr} presented below (like the pressure coefficients also shown) refer to times of maximum velocity.

In oscillatory flow the pressure coefficient at the upstream stagnation point at a time of maximum velocity is in general not unity, because of the influence of the cylinder's wake generated during the previous half cycle. Results given in figure 1 show an increase with Reynolds number, with maximum values (corresponding to $K_c = 10$) around 1.5. Conversely, base pressure coefficients shown in figure 2 are in general lower than those observed in steady flow, but show little systematic variation with Reynolds number. The results for steady flow in figure 2 refer to surface roughnesses which bracket that of the cylinder used in the present tests.

Corresponding results for the minimum pressure coefficient are shown in figure 3. These may be influenced by the sampling error associated with the use of only 14 pressure transducers, but if so, closer pressure measurements would give rise to still greater departures from the results for steady flow. In all respects, $K_c = 10$ seems to give extreme results, probably associated with the corresponding strength of vortex shedding.

Figure 4 compares results for V_{dr} with those (for surface roughnesses that bracket the present conditions) from steady flow. The agreement is excellent, suggesting that similar changes with Reynolds number are taking place in both flows; on this basis one may speculate that for the surface roughness used here ($k/d = 1.2 \times 10^{-4}$) post-critical conditions in oscillatory flow will be achieved first at a Reynolds number similar to that observed for the corresponding steady flow, namely about 2×10^6.

Conclusions

The shortage of data on pressure distributions on a cylinder in oscillatory flow is a hindrance to the understanding of the importance of Reynolds number, and to the development of numerical models. This paper makes some simple comparisons between pressure coefficients observed in oscillatory flow at times of maximum velocity, with results of nominally similar measurements in steady flow. It is shown that the Vorticity Dispersion Ratio [5] exhibits similar behaviour in both cases, suggesting that for a cylinder with fine roughness, post-critical conditions have not yet been achieved in oscillatory flow.

References

1. Chaplin, J R Planar oscillatory flow forces at high Reynolds numbers' ASME J of OMAE, in the press, 1992.
2. Bearman P W, Chaplin J R, Graham J M R, Kostense J K, Hall P F and Klopman G, 'The loading on a cylinder in post-critical flow beneath periodic and random waves', BOSS, 213-225, 1985.
3. Bearman P W, 'Wave loading experiments on circular cylinders at large scale' BOSS, 471-487, 1988.
4. Sarpkaya T, 'In-line and transverse forces on smooth and rough cylinders in oscillatory flow at high Reynolds numbers', Tech Rpt NPS-69-86-003, Naval Postgraduate School, Monterey, Calif. 1986.
5. Pearcey H H, Singh S, Cash R F and Matten R B, 'Fluid loading on roughened cylindrical members of circular cross section', Department of Energy Report OT-O-8411, 1985.

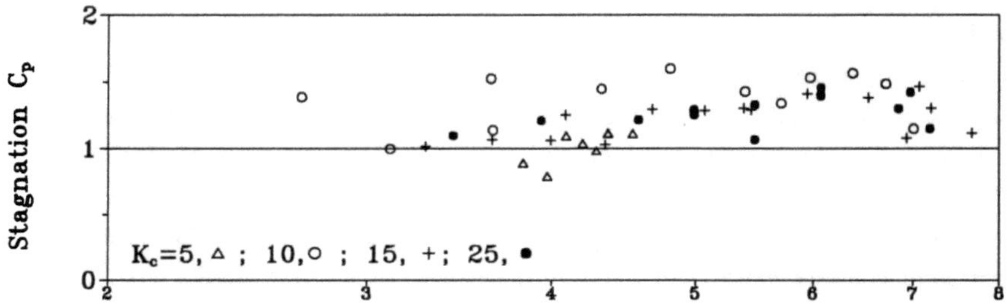

Figure 1. Stagnation pressure coefficients for planar oscillatory flow

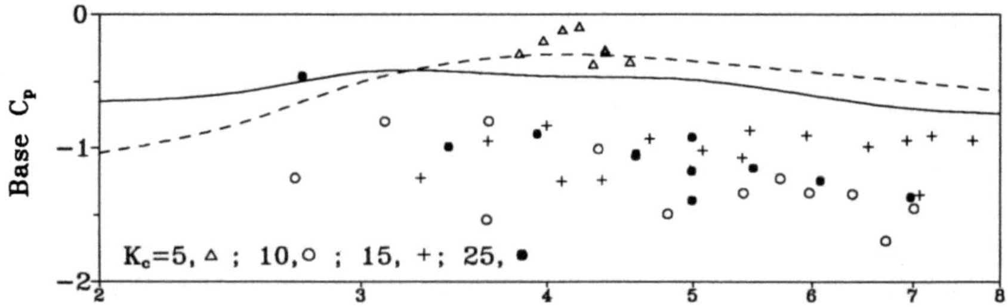

Figure 2. Base pressure coefficients, present results and steady flow data from [5]: solid line, $k/d=4x10^{-4}$; broken line, $4.3x10^{-5}$

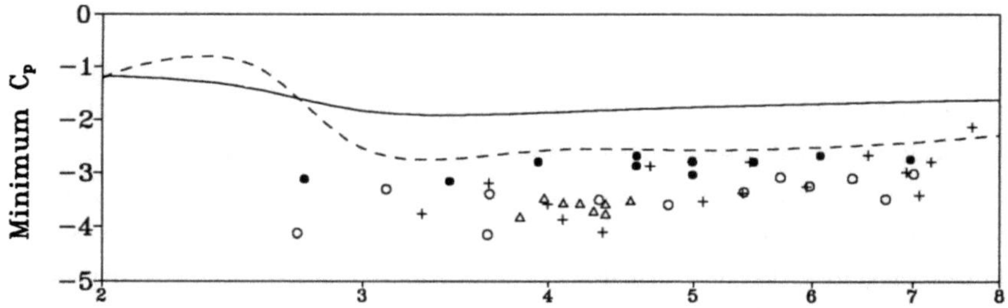

Figure 3. Minimum pressure coefficient. Legend as in figure 2.

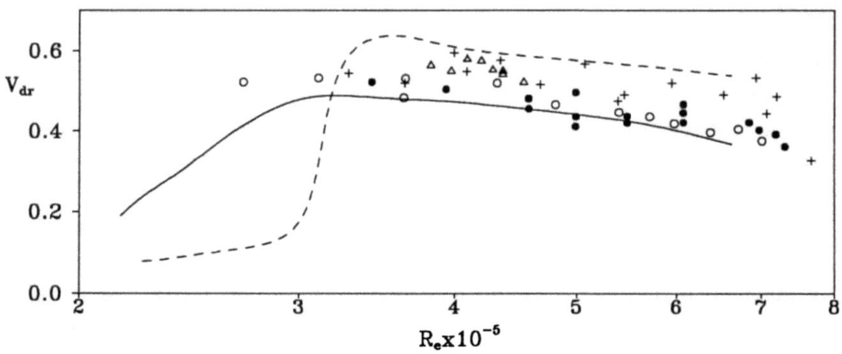

Figure 4. Vorticity Dispersion Ratio. Legend as in figure 2.

Vortex Shedding in Oscillatory Flow: Effects of Free-Stream Turbulence

KHALFAN AL-ASMI & IAN P CASTRO

Department of Mechanical Engineering
University of Surrey

Abstract

A comprehensive study of vortex shedding from bluff bodies in oscillatory flow has been undertaken (Al-Asmi [1], Al-Asmi & Castro [2]). This paper discusses a sub-set of the results, concentrating on the effects caused by the addition of free-stream turbulence. In particular, it is shown that, in the case of a triangular-shaped body, turbulence substantially increases the range of oscillatory driving frequencies (N) over which lock-in conditions occur, provided N is greater than twice the natural shedding frequency, n_o. The effects depend on the ratio of turbulence integral scale to body size, L_x/d, and the free-stream turbulence intensity, u'/U_o. Increasing either (or both) first extends the lock-in range but eventually leads to 'saturation', with further increases in u'/U_o causing a *reduction* in the extent of lock-in.

Introduction

It is well known that if a bluff body is allowed (or forced) to vibrate the vortex shedding process can be strongly modified. Features such as 'lock-in', when the vortex shedding frequency no longer follows changes in the free stream velocity but is dependent solely on the vibrational frequency of the body, have been well-documented and are fairly well understood. This is particularly true in the case of cross-stream oscillations. However, in the context of vortex meters used as devices for measuring the volumetric flow rate of a fluid in a pipe, it is the (axial) oscillations in the flow - which often occur in industrial pipework systems - that can lead to lock-in and consequent metering errors. Even without lock-in, if the shedding frequency is changed even marginally significant measurement errors can arise.

There have been a number of studies of the effect of in-line oscillations (of nominally sinusoidal form) on vortex shedding from bluff bodies (e.g. Hatfield & Morkovin [3], Griffin & Ramberg [4], Barnes & Grant [5] and Barbi et al [6]). In general it has been found that the vortex shedding is most readily synchronised when its frequency, n, corresponds to one half of the flow oscillation frequency, N. The velocity range over which synchronisation occurs increases with the oscillation amplitude. Most of these experiments have been in wind tunnels with low free-stream turbulence levels and have concentrated on bodies of cylindrical cross-section. However, in the context of pipe flows, the flow upstream of the vortex shedding obstacle is turbulent and the obstacle is usually sharp-edged. The effects of flow oscillations are therefore likely to be somewhat different from those which occur in the wind tunnel cases but they have not been studied in any detail until the present work, which is fully described by Al-Asmi [1]. Al-Asmi & Castro [2] discuss the way in which geometrical parameters (body shape, solid blockage and spanwise aspect ratio) affect the response to in-line flow oscillations; in this paper attention is concentrated on the effects of free-stream turbulence.

Experimental Details

The general experimental arrangement is fully described in [1] and [2]. A number of different body shapes were used and were mounted (with end plates) in the 0.9x0.6x4.5m working section of a low speed wind tunnel, which had a specially designed entrance section (upstream of the centrifugal fan) allowing accurately sinusoidal free-stream oscillations to be produced in the working section, without affecting the quality of the flow in any other way (Al-Asmi & Castro [7]). The total harmonic distortion, defined as the inverse ratio of the energy in the fluctuating velocity field around the oscillation frequency to the total energy across all other frequencies, was usually less than 0.4% and oscillation amplitudes (peak-to-peak) could be varied continuously up to around 40% of the mean velocity. Reynolds numbers were in the range 0.5×10^4 - 5×10^4. Free-stream turbulence was produced in the usual way via a square-mesh bi-planar grid mounted at various positions upstream of the bluff body. In the absence of the latter, the longitudinal integral scale and turbulence energy varied according to $L_x/M = 0.19(x/M - 2)^{0.4}$ and $(\overline{u'^2}/U_o{}^2)^{-0.8} = 6.62(x/M-2)$, respectively, where M is the mesh size. These results were not changed by the presence of the imposed oscillations in U_o, which are defined by $U=U_o[1 + \alpha \mathrm{Sin}(2\pi Nt)]$, with $\alpha=\Delta U/U$; likewise, the imposed oscillation had only a small effect on the turbulence field [1].

Major results and Conclusions

Figure 1 is a typical set of measurements of shedding frequency and base pressure as a function of reduced velocity, $U_r = U_o/Nd$, where U_o is the mean upstream velocity, for the triangular-shaped body. In this case the spanwise aspect ratio was 8, the blockage ratio was 0.17 and N=8.55Hz. Upstream turbulence levels were below 0.2%. The data have features similar to those obtained by previous workers. Note the increasing range of lock-in (at the upper end) as the amplitude of the oscillation (α) increases. The fact that similar changes do not occur at the lower end of the lock-in range is a feature of this particular body shape. Al-Asmi & Castro [2] show that lock-in occurs for all four bodies tested, provided the reduced oscillation amplitude (defined by $\varepsilon = \Delta U/2\pi Nd$ where ΔU is the amplitude of the upstream velocity oscillation) exceeds about 0.02 and N is close to twice the natural shedding frequency. The range of N (or, equivalently, the reduced velocity) over which such lock-in persists is strongly dependent on the body shape and ε. The lock-in ranges are significantly larger than those induced by cross-flow body oscillations and it was concluded that triangular or tee-shaped obstacles are more susceptible to flow pulsations than bodies of other shapes.

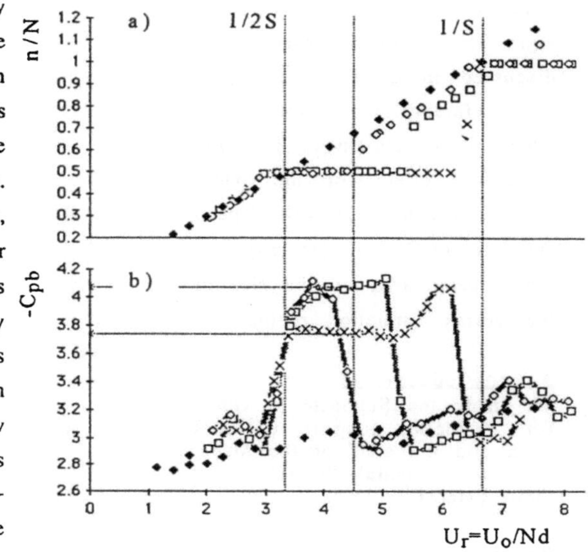

Fig.1. Variation of (a) shedding frequency and (b) base pressure with imposed oscillation frequency. α =:◆, 0 (steady flow); ◇ , 0.08; □ , 0.10; × , 0.15. S is Strouhal number, $n_o d/U$, at $\alpha=0$.

Figure 2 shows the behaviour of the Strouhal number, $S=nd/U_o$, and the base pressure coefficient, C_{pb}, with u'/U_o in the case of the triangular body in steady flow. Note that minima in both S and C_{pb} occur for u'/U_o around 8% - independent of L_x/d (in the range covered). In oscillatory flow the largest effects were also found for intensities around 8%. Figure 3 shows the critical displacement amplitude when free-stream turbulence is present. The u'/U_o and L_x/d ranges bracket those appropriate in the case of a vortex flow meter installed in fully developed pipe flow. For $N/n_o>2$ there are two distinct regimes in which increasing u'/U_o has an opposite effect on the synchronous range. For intensities up to about 8% the range is increased whereas further increases in intensity lead to a reduction in the synchronous range. In both regimes increasing L_x/d leads to increases in the synchronous range. For $N/n_o<2$ the results follow the corresponding smooth flow data quite closely - small differences between data in this range for the two body sizes are attributable to blockage effects [1,2]. In oscillatory flow it seems that frequency modulations allow an earlier coupling between the Strouhal and the free-stream fluctuations; this leads also to a shift in the resonant point (the minimum critical displacement amplitude) to lower reduced velocities - higher N, fig. 3. These results are discussed in detail in [1] (and with reference to earlier work), where it is concluded that the turbulence effects on the extent of the lock-in range will depend strongly on body geometry or, more specifically, on the size of the vortex formation region.

Acknowledgements
We thank the Sultanate of Oman for financial support of this work and the staff of the Mech. Eng. Departmental Workshop for their extensive help.

References

1. Al-Asmi, K: Vortex shedding in oscillatory flow. PhD Thesis, Univ.of Surrey, 1992
2. Al-Asmi, K.; Castro, I.P.: Vortex shedding in oscillatory flow: Geometrical effects. Flow Meas. and Inst. (in press), 1992.
3. Hatfield, H.M.; Morkovin, M.V.: Effect of an oscillating free stream on the unsteady pressure on a circular cylinder. J.Fluids Eng. (1973) 249-252.
4. Griffin, O.M.; Ramberg, S.E.: Vortex shedding from a cylinder vibrating in line with an incident uniform flow. J.Fluid Mech. 75 (1976) 257-271
5. Barnes, F.H.; Grant, I.: Vortex shedding in unsteady flow. J.Wind Eng.Ind.Aero. 11 (1983) 335-344.
6. Barbi, C.; Favier, D.P.; Maresca, C.A.; Telionis, D.P.: Vortex shedding and lock-on of a circular cylinder in oscillatory flow. J.Fluid Mech. 170 (1986) 527-544.
7. Al-Asmi, K.; Castro, I.P.: The production of oscillatory flow in wind tunnels. Submitted to Expt.in Fluids.

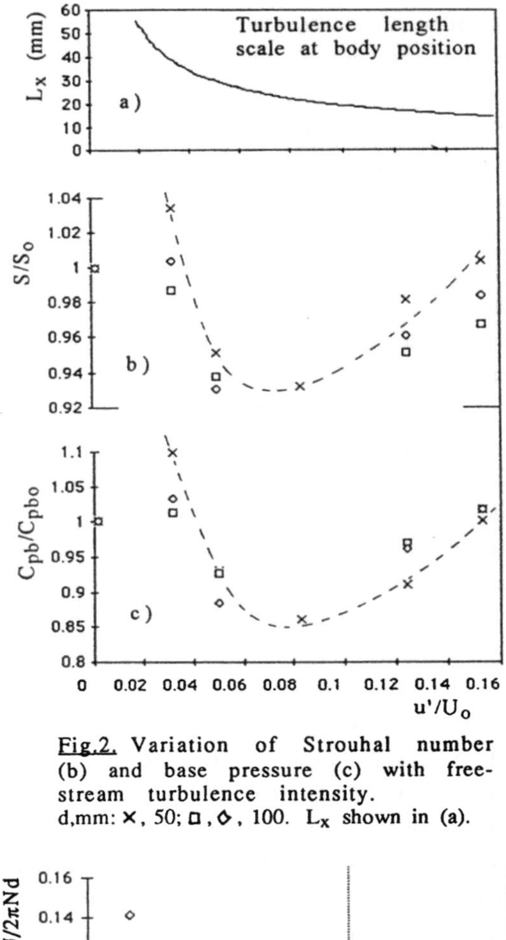

Fig.2. Variation of Strouhal number (b) and base pressure (c) with free-stream turbulence intensity.
d,mm: ×, 50; □, ◇, 100. L_x shown in (a).

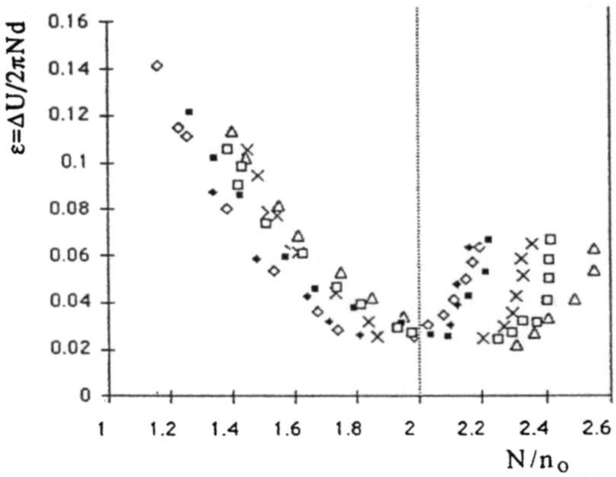

Fig.3. Effect of turbulence on the critical displacement amplitude.
■, smooth flow. u'/U_o, L_x/d: □, 0.05,0.59; △,0.08, 0.45; ×,0.15, 0.29; ◇, 0.05, 0.30; ◆, 0.12, 0.17. ◇, ◆, d=100; otherwise 50mm

Numerical Analysis of the Instabilities in the Near Wake of a Circular Cylinder at Low Reynolds Numbers

J.H. Gerrard

Department of Engineering, University of Manchester, UK

Summary

Numerical experiments are described which followed the experimental observation of the production of an oscillatory wake by the insertion of a control rod on the centre line of the near wake at Reynolds numbers, Re, (based on diameter, d) less than the critical Re. The computed values of C_L grew exponentially for a range of positions of the control on the wake axis.

Results are presented which show that the driving force at Re < Re_{crit} is small fluctuations in the near wake which convect downstream. Their interaction with the control rod produces oscillations at the control rod which feed back to the cylinder to initiate the exponential growth of C_L. At Re > Re_{crit} the exponential growth of C_L takes place without a control in the wake. The growth rate as a function of Re is determined and compared with other work.

Introduction

In experiments it was found by Al-Khafaji [1] that the normally steady wake at Re = 29 could be made to oscillate by the inclusion of a small diameter (d/8) rod on the centre line of the near wake at radial distance r = 3.8 to 6 radii from the cylinder centre. These results were directly opposed to the findings of Strykowski & Sreenivasan [2] and so it was thought that there may have been some effects in the towing tank, such as cylinder vibration, which were producing the oscillation: also the flow in our work started impulsively from rest. Consequently it was decided to use a numerical scheme to repeat the experiments numerically. In an existing code [3] vortices were introduced at the cylinder surface and from there were convected, and diffused and redistributed on separate meshes. The code was adapted to include a control rod on the axis by placing a vortex pair disposed across the wake axis. The strength of these oppositely signed vortices was k (the vortex strength factor) times the radial velocity, VR, one mesh length of the convection mesh from the wake axis at the same r. (Control vortex strength = CVS = k.VR). These vortices were convected and diffused in the same manner as those introduced at the cylinder surface. There was no effect if this was not done.

Results of numerical experiments

Figures 1 to 5 show the basic results. A control vortex asymmetry, CVA, the sum of the two control vortex strengths, develops and grows exponentially; figure 1. The CVS, figure 2, reflects the development of the flow. The recirculating near wake 'bubble' grows with time and the radial velocity at the control drops to zero when the bubble ends at its position. C_L grows exponentially out of the background which is also shown in figure 3. These $k = 0$ oscillations are due to numerical inexactness. C_D is shown in figure 4 descending to its steady level. The aim is to examine the origin and mechanism of the growth. When $Re > Re_{crit}$ computations show that the disturbances with $k = 0$ grow to eventually produce the saturated state which is the Karman vortex street. The noise shown in figure 3 was obtained with a single precision computation. A double precision computation (taking twice the computing time) shows growth from much smaller beginnings. At $T = 40U/a$ (a = cylinder radius) C_L, CVA and the near wake velocity asymmetry, VA, are 10^{-10} smaller with double precision whilst C_D and CVS are the same.

If figure 5 a sample of VA at four points is shown, VA is the difference in circumferential velocity at equal and opposite values of y. This has been calculated for many pairs of points at y values close to 1, 0.25 and 0. The form of the curves in figure 5 is similar with and without control present. At small times we are seeing numerical disturbances which later become organised. The near wake, except at the should of the cylinder, flaps from side to side. The waves of VA at these and later times convect downstream with a speed of 0.86 U which is the same as the speed of particles released from $x = 0$ at the outer edge of the boundary layer. This is the case for control present and absent except for the first peak of VA with control present. In this case the maximum of VA appears over the near wake instantaneously. CVA leads C_L by 0.25 to 0.38 of a period: VA leads C_L by the smaller time of 0.2 to 0.3 period, for the control factor $k = 0.04$ with the control at $r = 5a$.

The time variation of C_L and CVA was found to be exponential. The amplitudes of both were of the form $c \exp(mT)$. Most results were obtained with $k = 0.04$. No oscillations were observed for $r < 4a$ in times upto $T = 60$. As r increased m reached a broad maximum of ca. 0.075 at $r \approx 6a$. $r = 6.35a$ is the critical position below which splitter plates affect the shedding at higher Re but at $Re = 30$ even at $r = 14a$ the value of m was still 0.05. For $0.03 < k < 0.12$ the value of m was found to fit the approximate relation $m = 0.026 \exp(25k)$.

Recent work has concentrated on variation with Re and results are presented for $30 < Re < 60$. At low Re, growth was followed up to $T = 50$ with $k = 0.03$ at $r = 5$: the control was then switched off and the decay followed. At $Re \geq 45$ growth initiated by numerical noise was studied. At $Re = 45$ and 53 the waveform was irregular and scarcely left the background level by $T = 115$ and so growth and decay, as at lower Re, was investigated also. At $Re = 53$ the decay exponent plotted in figure 6 ended at $T = 120$ and thereafter the amplitude was constant indicating that $Re_{crit} < 53$. At Re = 45, 53 and 60 and Strouhal number, S, agreed with the universal curve of Williamson [4]. At Re = 30, 35 and 40, $0.1 < S < 0.13$ and above the extrapolation of the universal curve. In growth and decay S decreased with T to a constant level. When the growth was from small disturbances S increased to the constant level. Comparison with, σ_r, the bifurcation parameter in the Landau truncation of the stability equation determined by Provansal et al. [5] shows fair agreement in figure 6.

Conclusions

Exponential growth under the action of a control body in the wake has been found. This action is triggered by disturbances convected downstream from the near wake and takes place for a range of positions of the control on the wake axis provided that the control is more than four radii from the centre of the cylinder. Some findings remain unexplained and further investigation continues.

References

1. Al-Khafaji, A.A.: Instabilities in the near wake of a circular cylinder at low Reynolds number. Ph.D. thesis, (1989) University of Manchester.
2. Strykowski, P.J.; Sreenivasan, K.R.: On the formation and suppression of vortex 'shedding' at low Reynolds number. J. Fluid Mech. 218 (1990) 71-107.
3. Benson, M.G.; Bellamy-Knights, P.G.; Gerrard, J.H.; Gladwell, I.: A viscous splitting algorithm applied to low Reynolds number flows. J. Fluids & Struc. 3 (1989) 439-479.
4. Williamson, C.H.K.: Defining a universal and continuous Strouhal-Reynolds number relationship for laminar vortex shedding of a circular cylinder. Phys. of Fluids 31 (1988) 2742-2744.
5. Provansal, M.; Mathis, C.; Boyer, L.: Bernard-von Karman instability: transient and forced remiges. J. Fluid Mech. 182 (1987) 1-22.

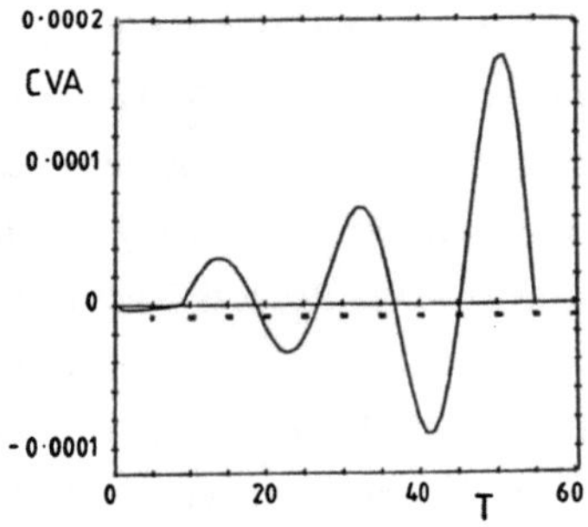

Figure 1 CVA, k = 0.04, R = 5.0

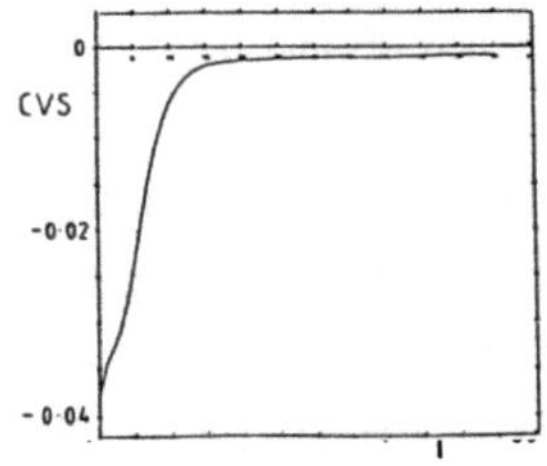

Figure 2 CVS, k = 0.04, R = 5.0

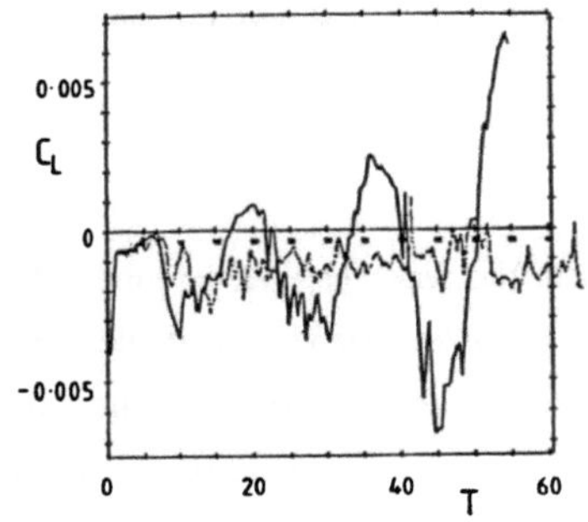

Figure 3 C_L development, k = 0.04, R = 5.0.
Broken curve, k = 0.

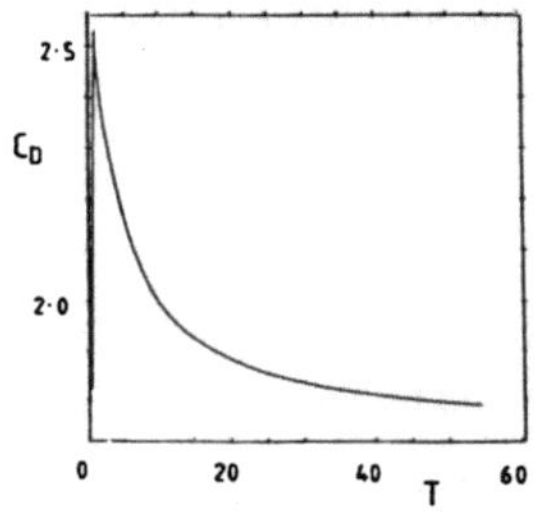

Figure 4 C_D, k = 0.04, R = 5.0

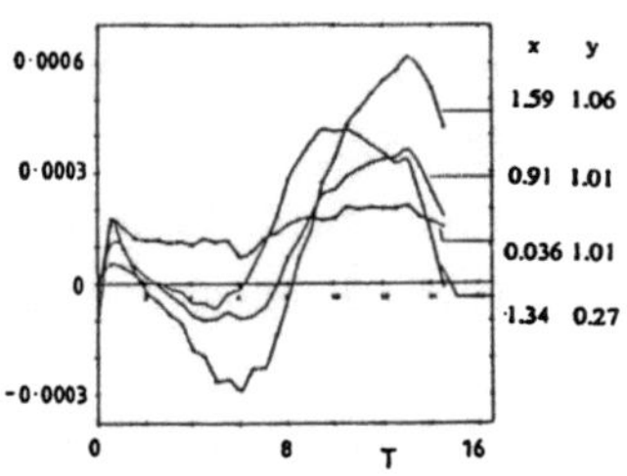

Figure 5 Near wake velocity asymmetry
k = 0.05, R = 5.0.

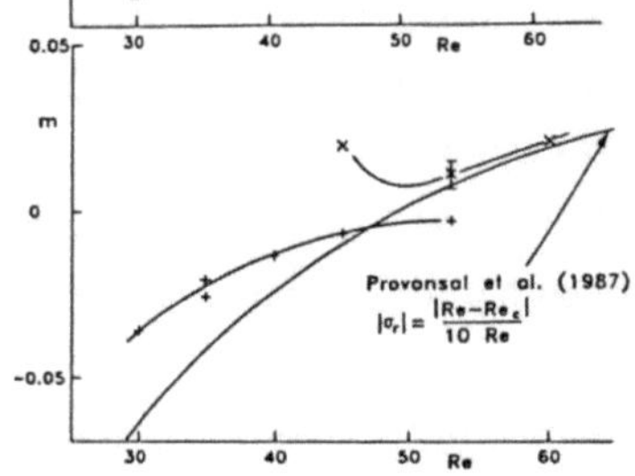

Figure 6 Rate of growth and decay as a function of Re

The Effect of Mean Flow Unstediness on the Dynamics of a Plane Wake

R. W. Miksad, M. R. Hajj*, R. S. Solis, and E. J. Powers

College of Engineering
The University of Texas at Austin
Austin TX 78712

ABSTRACT

The generation of low-frequency velocity components and their influence on the dynamics of a plane wake are examined. The results indicate that the influence and role of low-frequency modes on wake transition and spectral energy transfer depends somewhat on the mechanism by which these modes are generated. In both steady and unsteady wakes, low frequency modes play an important role in establishing modulations and enhancing sideband interactions. Time and spectral-domain measurements indicate that low-frequency mean flow variations introduce very strong and coherent modulations of the velocity fluctuations. Nonlinear couplings and modulations measurements show that low-frequency modes arising from spectral interactions in steady wakes are primarily limited to spectral regions involving the primary instability mode, its harmonics and near sidebands. In contrast, the unsteady wake is characterized by a broader and richer array of couplings through out the spectrum.

I. INTRODUCTION

In many shear flows, low-frequency fluctuations play an important role in the transition to turbulence via their interactions with higher frequency fluctuations to enhance spectral energy redistribution. Miksad et al. (1982 & 1983) showed that spectrally generated low-frequency modes indirectly modulate the instability components through generation of sidebands. Solis (1986) extended the argument of Shen (1961) to show that low frequency mean flow unsteadiness will directly modulate the amplitude and phase of the instability modes of a plane wake. The basic result is that the role main of low frequency modes, whether spectrally generated or introduced by mean flow unsteadiness, is to induce modulations of the fluctuation velocity field.

The experiments were conducted in an open-return type wind tunnel with a downstream sonic throat. In one set of experiments the natural transition of a steady wake was examined as a base reference case. In these steady experiments the dominant low frequency fluctuations were centered at 33Hz and were generated by nonlinear interactions between growing instability modes. In a second set of experiments, the cross-sectional area of the sonic throat was varied periodically to produce an unsteady mean flow of the form $u(t) = U_0\,(1+\varepsilon\, e^{i2\pi f_m t})$, where f_m the frequency of mean flow variation was set equal to 33 Hz. This value was chosen to allow a

---

* Current Address: Department of Engineering Science and Mechanics, Virginia Polytechnic Institute and State University, Blacksburg, VA 24061.

comparison of the modulation mechanisms introduced by mean flow unsteadiness, and those induced by spectrally generated low-frequency modes. f_m is an order of magnitude smaller than the frequency $f_o = 587$ Hz of the naturally most unstable instability mode of the wake. ε, which represents the normalized amplitude of the mean flow unsteadiness with respect to the mean velocity U_0, was set equal to 0.015.

II. EXPERIMENTAL RESULTS

The effects of low-frequency variations in the mean flow on the dynamics of a wake were examined by comparing time and spectral domain measurements of unsteady and steady wakes. Sample traces of the normalized maximum velocity fluctuations, u'_{rms}/U_0 measured under natural excitation conditions are shown in figure 1. Note that the velocity fluctuations in the unsteady wake are more strongly modulated at all locations. These modulations diminish with downstream distance, but can still be noticed in the time trace at x = 40.0 cm.

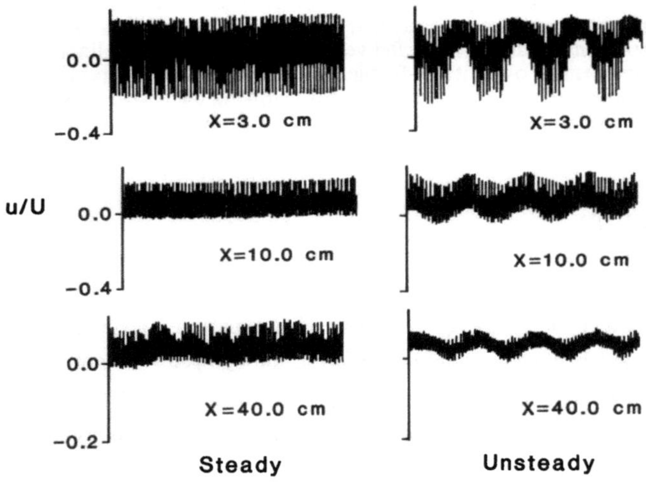

Figure 1: Normalized maximum u'_{rms} at x = 3.0, 10.0 and 40.0 cm under steady and unsteady conditions. Both flows excited by natural background noise.

Spectral energy distribution of the velocity field in the steady and unsteady wakes are shown in figure 2. A comparison of the spectra brings out two significant features. First, the sideband structure of velocity components at $(nf_o \pm mf_m)$ surrounding the fundamental frequency f_o, and its harmonics nf_o, is broader in the unsteady wake than in the steady wake. Second, the spectral valleys between the peaks fill in more rapidly in the unsteady wake. The difference in spectral characteristics between steady and unsteady wakes seems to arise from the degree of coupling involved in the interactions between the different components of the flow.

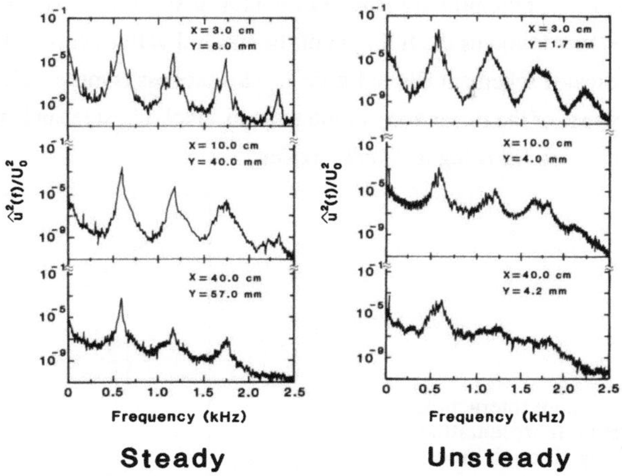

Figure 2: Spectral distribution of the velocity fluctuations at x = 3.0, 10.0, and 40.0 cm under steady and unsteady conditions. Both flows excited by natural background noise.

Auto-bicoherence measurements of the velocity fluctuations presented in figure 3 show that in the steady wake high levels of coupling are measured for interactions which reinforce the fundamental mode f_0, and the harmonic band structures centered at, nf_0. In the unsteady wake, high levels of spectral coupling are measured for interactions which feed energy to the spectral valleys between the harmonic band structures. These interactions involve sideband fluctuations (surrounding the harmonic modes) which are introduced by modulations set up by the unsteady mean flow.

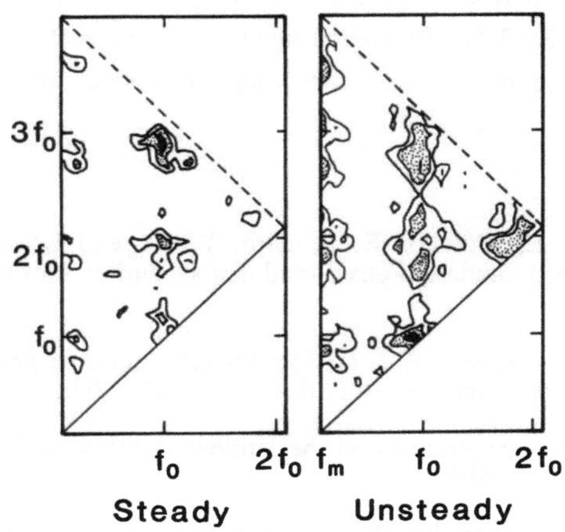

Figure 3: Auto-bicoherency measurements of the velocity fluctuations at x = 5.0 cm under steady and unsteady conditions. Both flows excited by natural background noise.

The results indicate that the primary feature of nonlinear spectral energy transfer in both steady and unsteady wakes involves the filling in of the spectral valleys between the harmonic band structures, as indicated schematically in figure 4. This process is more efficient and rapid in unsteady wakes because of the richer array of interactions involving sidebands introduced via mean flow modulations of the growing instability modes.

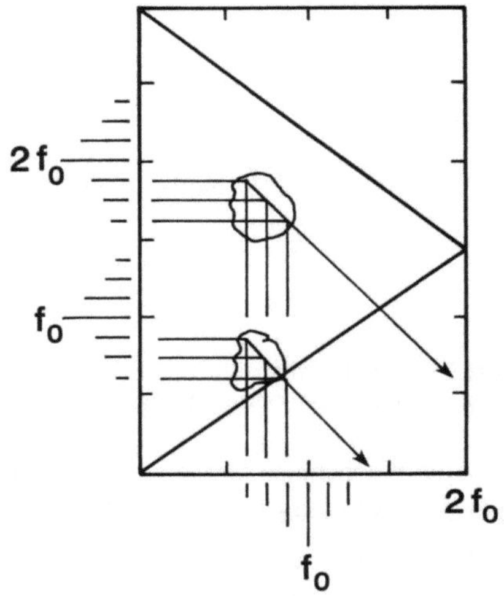

Figure 4: Schematic of sum interaction scheme and bicoherence representation of spectral valley filling through cross-sideband interactions

ACKNOWLEDGMENTS

This paper is based in part upon work supported by the Texas Advanced Technology Program under Grant No. ARP 3280. The digital signal processing techniques were developed under Air Force Office of Scientific Research (AFOSR) Contract No. F49620-89-C-0044.

REFERENCES

Miksad R. W., Jones, F. L., Powers, E. J., Kim, Y. C., and Khadra L. W. (1982): Experiments on the role of amplitude and phase modulations during transition to turbulence. J. Fluid Mech., v123, 1-29.

Miksad R. W., Jones, F. L., Powers, E. J. (1983): Measurements of nonlinear interactions during transition of a symmetric wake. Phys. Fluids, v26, 1402-1409.

Shen, S. F. (1961) : Some considerations on the laminar stability of time-dependent basic flows, J. Aerosp. Sci., v28, 397-404.

Solis, R. S. (1986) : Experiments on the effects of mean flow unsteadiness on the transition to turbulence of a laminar wake, Ph. D. Dissertation, The University of Texas at Austin, Austin, Tx, 78712.

Numerical Investigation of Wake Instabilities

Marek Morzyński
Hermann Föttinger–Institut, Technische Universität, Berlin
Frank Thiele
Abteilung Turbulenzforschung, DLR, Berlin

Summary
The application of non-parallel flow stability analysis to the wake problem is shown. Stream function, curvilinear FDM formulation and solution of the eigenvalue problem is described in details. The results consist of critical values, determined for the circular cylinder flow and the attempt to analyze the wake instability mechanism.

Introduction

For external flow around cylinders at very small Reynolds numbers the separation is not present. For higher ones the separation appears and the wake is growing. For a certain value of the Reynolds number, called critical the unsteady motion develops. Further increase in the Reynolds number results in laminar-turbulent transition, developing in different parts of the flow.

The wake instability is an object of intensive investigation. The task of the research is to determine the mechanism of instability (which potentially gives the possibility to control it) and to predict as accurate as possible the critical values. These parameters are the critical Reynolds number, the frequency of the unsteady motion and the growth-rate of the infinitesimal disturbance.

The linear stability theory bases on splitting of a flow to a steady part and disturbance. It has been successfully applied to parallel flows. This enhanced its generalization to slightly non-parallel flows on expense of more complicated formulation. Even fully 2D flows are analyzed using this method under the assumption that average velocities carry the stability characteristics of the flow. The extensive bibliography of the problem can be found in [1, 2, 3]

The parallel flow assumption was forced by the absence of technical means to solve the general non-parallel problem. Recent development of computers and computational methods enable the formulation and solution of linear stability problem, at least for 2D [4-7] and hopefully for 3D in the future.

Formulation

The non-parallel flow stability problem is solved in stream function formulation. The incompressible Navier-Stokes equation takes the form:

$$\left[\frac{\partial}{\partial t} + (\nabla \times \vec{\psi}) \cdot \nabla - \frac{1}{Re}\Delta\right]\Delta\vec{\psi} = 0 \tag{1}$$

This expression is general for 2D and 3D. The disadvantage of the formulation is the high order derivatives in the equation. The important advantage of the pure stream function formulation is the use of only one unknown. This is reflected in reasonable size of the eigenvalue matrices resulting from 2D considerations

The procedure of linear stability analysis requires splitting of the unsteady quantity to steady part and disturbance and assumes periodic behavior of the latter one:

$$\vec{\psi}(x,y,t) = \bar{\psi}(x,y) + \vec{\psi}'(x,y,t) \qquad ; \qquad \vec{\psi}'(x,y,t) = \vec{\varphi}(x,y)e^{-i\lambda t} \tag{2}$$

Limiting the considerations to linear analysis and small value of the disturbance allows writing the disturbance equation in the form:

$$i\lambda\Delta\vec{\varphi} - (\nabla \times \bar{\psi}) \cdot \nabla\Delta\vec{\varphi} - (\nabla \times \vec{\varphi}) \cdot \nabla\Delta\bar{\psi} + \frac{1}{Re}\Delta^2\vec{\varphi} = 0 \tag{3}$$

The discretization of the equation (3) is accomplished with the second order accurate difference formulas based on the 13-point stencil.

The stability problem is written in curvilinear body-fitted orthogonal coordinates:

$$g^{ii}\left[i\lambda\varphi\Big|_{ii} - e^{mn3}(\psi\Big|_n\varphi\Big|_{iim} + \varphi\Big|_n\psi\Big|_{iim} + \frac{1}{Re}g^{jj}\varphi\Big|_{iijj}\right] = 0 \tag{4}$$

The metric coefficients g and its derivatives are calculated exactly by the algebraic manipulation package. The assumed form of the metric coefficients limits the amount of all coefficients necessary to define the transformation to $g(\xi,\eta)$ and its first order derivatives $g_{,\xi}$ and $g_{,\eta}$. The remaining derivatives are vanishing simplifying in this way the computations.

The orthogonal mesh necessary for the computations is generated by the conformal mapping. For the airfoil calculations the Kármán-Trefftz transformation is used.

To the discretized system of linear equations the boundary conditions have to be introduced. In the simplest case the homogeneous boundary conditions can be applied. The convective boundary conditions used in the Navier-Stokes solver are also adapted to the disturbance equation. Forcing the homogeneous boundary conditions means that the disturbance has to cross the zero axis on the boundary (Fig.7). It can influence the number of waves in the computational domain and by this the frequency. The convective boundary conditions are applied only to the outflow boundary. The practical computations show however, that the influence of different types of boundary conditions is smaller than it could be expected.

Finally the discretized equations lead to a generalized, complex eigenvalue problem. The number of unknowns used in the computations is of order of 20 000. For this reason none of the known packages of the eigenvalue solvers could be used.

Solution procedure

The eigenvalue problem is solved iteratively via inverse iteration method. The solution is sought in the neighborhood of the guess value of the eigenvalue. It is easy to determine the guess value of the frequency for the wake instability if the reference length is taken as the dimension of the cylinder perpendicular to the main flow direction. The initial growth-rate value is always taken to be zero.

The Newton-Raphson method applied to the eigenvalue problem gives:

$$(A - \lambda^{(n)}B)\eta^{(n+1)} = B\varphi^{(n)} \tag{5}$$

The eigenvalue λ and eigenvector φ are determined simultaneously. For the assumed value of λ the solution procedure involves finding η from the linear equation:

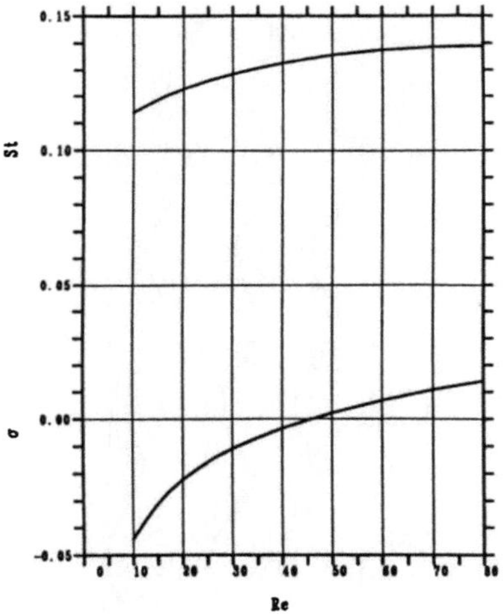

Figure 1: The growth-rate and the Strouhal number for the circular cylinder flow.

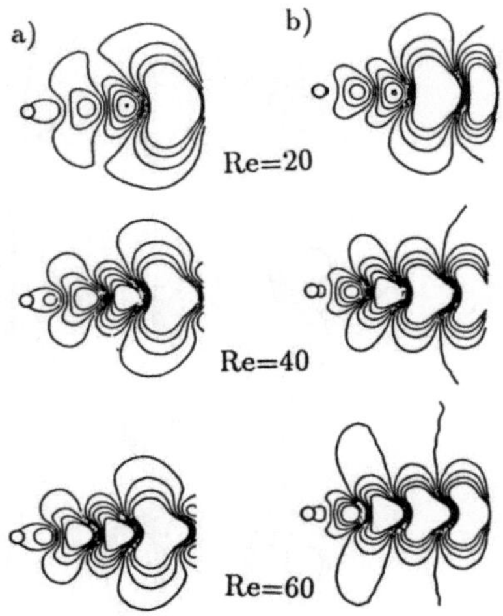

Figure 2: Real (a) and imaginary (b) part of the eigenvector.

$$i\lambda\Delta\vec{\eta} - (\nabla \times \bar{\psi}) \cdot \nabla\Delta\vec{\eta} - (\nabla \times \vec{\eta}) \cdot \nabla\Delta\bar{\psi} + \frac{1}{Re}\Delta^2\vec{\eta} = \Delta\vec{\varphi} \tag{6}$$

and the correction of the eigenvalue from:

$$d\lambda^{(n+1)} = \frac{1}{(e^r)^T\eta^{(n+1)}} \quad ; \quad (e^r)_i = \delta_{ir} \tag{7}$$

The comparison of the equation (6) to the Navier-Stokes equation (1) shows the extensive similarities, so any solver can be used for this purpose after appropriate change of coefficients. In case of modified procedure also the Newton-chord steps can be applied:

$$(A - \lambda_o B)^{-1} B\varphi^{(n)} = \eta^{(n+1)} \tag{8}$$

This approach applying the elements of Navier-Stokes equation solver is producing the solution much faster. The similarity of the equations is also the reason of similar numerical difficulties resulting from non adequate cell Reynolds number, poor conditioning of the matrix for the convection-dominated flows etc. The instability of the wake for the bluff bodies takes place for relatively low Reynolds numbers so there are no severe difficulties with the solution of the equation (8) but extending of the method and generalizing of the non-parallel stability approach can meet some numerical problems. The precautions used in the Navier-Stokes solvers as upwinding, relaxation and smoothing can be applied here too. These remarks are also true for any other method of solution of the eigenvalue problem (8).

Numerical results

The task of the investigation is to determine the critical values characterizing the bifurcation of the wake and explanation of the instability phenomena.

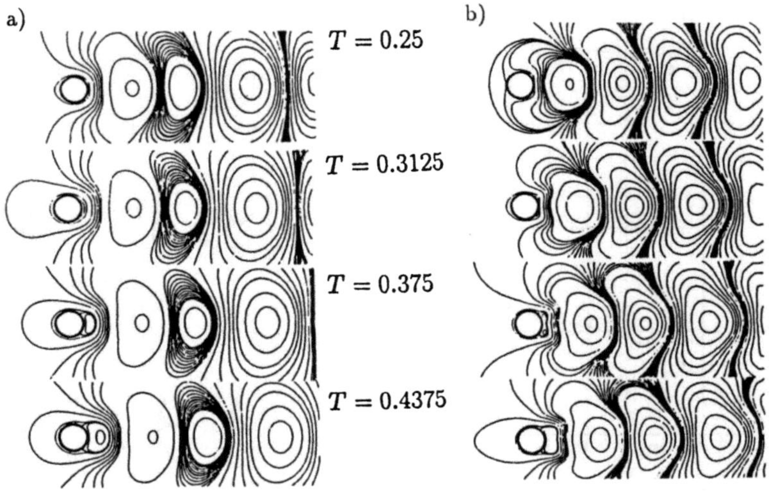

Figure 3: Temporal evolution of the disturbance a) Re=20 b) Re=60

Circular cylinder is a flow geometry extensively investigated, which can be used as reference for the numerical results. The flow phenomena in the circular cylinder wake are representative for any wake instability.

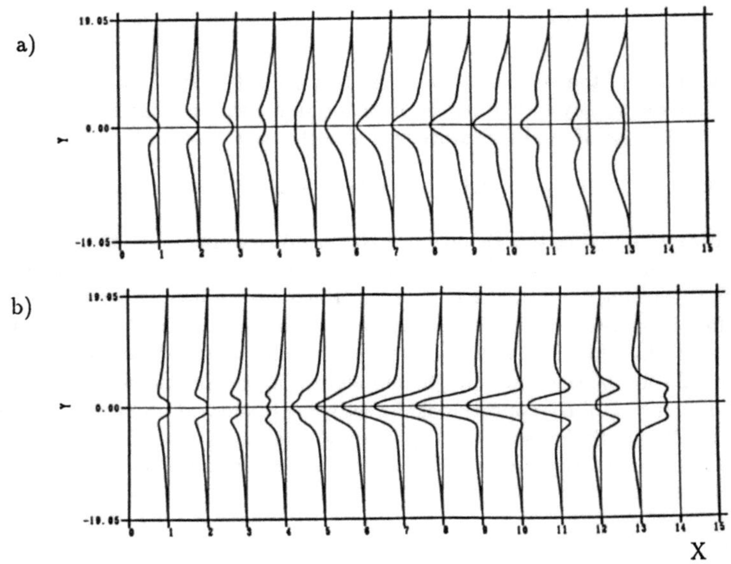

Figure 4: Disturbance behind the cylinder (T=0.375) a) Re=20 b) Re=60

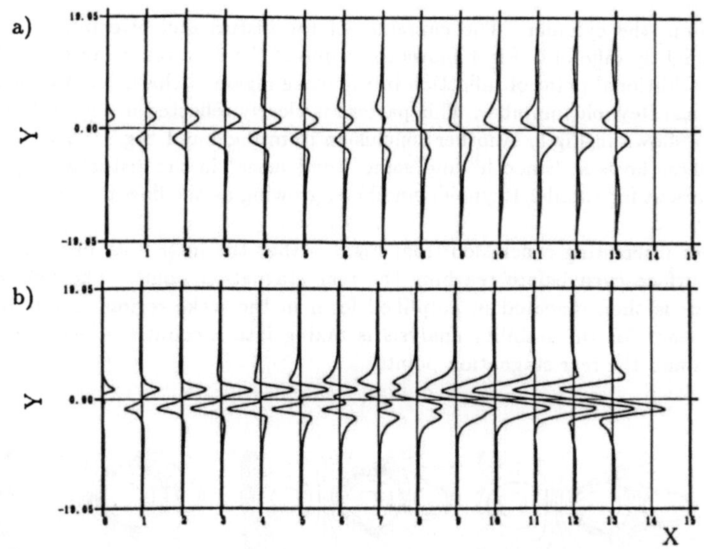

Figure 5: Velocities of the disturbance (T=0.375) a) Re=20 b) Re=60

The first step of the stability analysis is computation of the steady flow solution. It is known that for symmetrical cylinder flow one can obtain the steady state solution for Reynolds numbers much higher than the critical one. There is no change in the equations, which "informs" the system that the steady solution is not physical. Even more interesting is that this is also true for unsymmetrical wakes. For both cases the numerical difficulties appear if the wake approaches assumed "infinity" distance because none of the convective boundary conditions is really fully "non-reflective" for the flow with recirculation.

The steady solution is used as the input data for the eigenvalue calculations. Repeating this procedure for different Reynolds numbers one obtains the complex eigenvalue and the associated eigenvector. Eigenvalues for the flow around the circular cylinder are shown in Fig.1. The results agree with [6] and [9].

In Fig.2 the real and imaginary eigenvector patterns are shown. The disturbance is given by:

$$\vec{\psi}'(x,y,t) = e^{\lambda_{Im}}[\vec{\varphi}_{Re}cos(\lambda_{Re}t) + \vec{\varphi}_{Im}sin(\lambda_{Re}t)]$$

Because there is no distinguished "t=0" the real and imaginary eigenvector fields can differ even for the same geometry and Reynolds number. Only the full temporal evolution of the disturbance exhibits the patterns characteristic for different Reynolds numbers.

In Fig.3 the disturbance is shown for the time interval T=0.25-0.4375. Below the critical Reynolds number the "separation" of the disturbance streamlines starts building at the rear of the cylinder and travels downstream rising in time.

Above the critical Reynolds number this evolution is different. There are two "separations" growing symmetrically from the rear part of the cylinder. They vanish in close distance from the cylinder. The character of the disturbance streamlines in the wake region is slightly different. Fig.4 shows the value of the disturbance in the wake's cross-sections. Additional point of inflection in the wake region is characteristic for the curves above critical Reynolds number. This pattern is clearly reflected in the disturbance velocity curves shown in Fig.5. Another conclusion from Fig.4 and Fig.5 is that the "switch" between these flows is "smooth" and some "tendencies" in the disturbance patterns are already present for smaller Reynolds numbers, growing as the flow becomes closer to the instability.

Another interesting conclusion from Fig.3 is that the instability originates from the cylinder surface even before reaching the rear stagnation point. The behavior of the disturbance is then repeated in amplified form in the wake region, but it is not clear how important for the stability analysis is taking into account the region between the separation and the rear stagnation point.

Figure 6: Disturbance patterns for T=0-$\frac{5}{32}$, Re=40, axis ratio $\frac{3}{2}$, 1, $\frac{2}{3}$

In Fig.6 the disturbance patterns for different cross-sections of the cylinder are shown. For Re=40 and l/d=3/2 the flow is stable, for l/d=2/3 instability develops. For the same Reynolds number the instability is much more likely for the ellipsis with the long axis located perpendicular to the flow direction. Here the disturbance can "split" much easier.

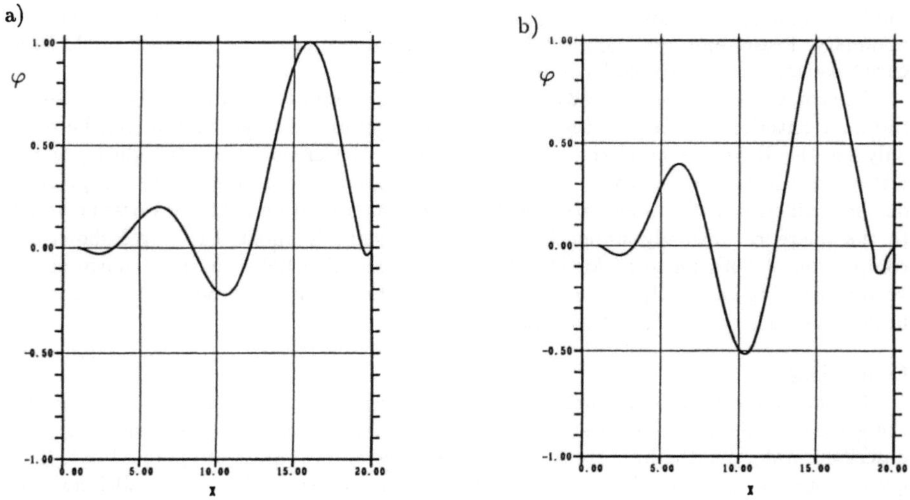

Figure 7: Disturbance value on the rear centerline of the flow a) Re=20 b) Re=60

Our earlier [8] results confirm this well known fact but the mechanism and disturbance development were not studied there.

It is characteristic for the non-parallel flow stability analysis that the disturbance value is growing in the direction of the "outflow". Fig.7 shows the value of the disturbance on the rear centerline of the flow. For sub-critical Reynolds number the value of the third half-period is even smaller than the second one. The maximum, normalized to one is reached near the outflow boundary. For Re=60 each half-period has the amplitude greater than the previous one. Jackson [5], using FEM, velocity-pressure formulation and non-parallel stability analysis obtained similar disturbance pattern, growing in the flow direction. The direct consequence of the growing value of the disturbance is that the superposition of the disturbance and the steady solution coincide with the unsteady simulation close to

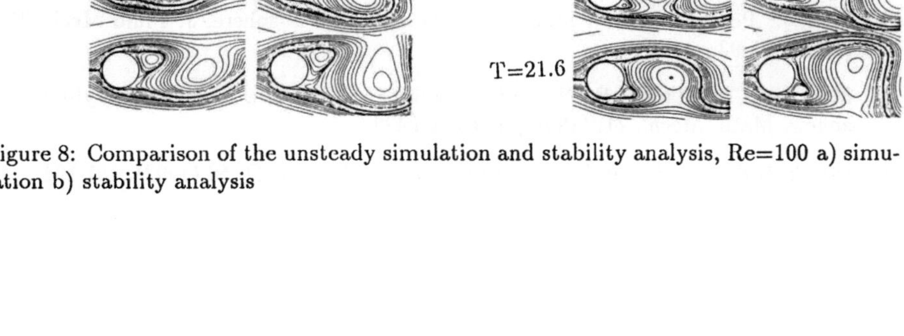

Figure 8: Comparison of the unsteady simulation and stability analysis, Re=100 a) simulation b) stability analysis

the body (Fig.8). The disturbance reaches always the maximum value near the outflow boundary. Prescribing the "infinity" boundary condition in very far distance would have cure this but it is impossible for numerical reasons.

The comparison of the unsteady simulation and the result of stability analysis is always the final criterion for judging the reality of the solution. The simulation can be done only for the Reynolds number (much) higher than the critical one. Weighting of the disturbance in the steady solution - disturbance superposition is arbitrary. Depending on the coefficient, determined intuitively the eigenvalue solution is fitting worse or better to the reference unsteady simulation. The usual value is about 0.1. Fig.8 shows the comparison of both patterns for the circular cylinder, Re=100. The mechanism of the vortex shedding especially in the neighborhood of the body is reproduced properly and both results are in good qualitative agreement.

Conclusions

The non-parallel stability analysis is able to determine almost exact critical values for the wake flow. Some characteristic processes, leading to the cylinder wake instability are described in details, the important for the phenomenon flow areas are pointed out. This gives deeper insight into the whole wake instability problem. The non-parallel stability analysis seems to be the only method able to resolve slight changes in the flow field near the critical Reynolds number. The advantages of the formulation and potential numerical problems are shown. The method, as the most general among the linear stability ones is able to solve a wide spectrum of stability problems.

References

[1] Kleiser, L.; Numerical simulations of transition in wall-bounded shear flows. Annu. Rev. Fluid Mech., 23, (1991), 495-537.

[2] Koch, W.; Local instability characteristics and frequency determination of self excited wake flows. J. of Sound and Vibration 99, (1985), 53-83.

[3] Huerre, P.; Monkevitz, P.A.; Local and global Instabilities in spatially developing flows. Annu. Rev. Fluid Mech., 22, (1990), 473-537.

[4] Morzyński, M.; Thiele, F.; Global non-parallel stability analysis of wake and boundary layer flows. Z. angew. Math. Mech., 72, (1992), 5, T403-T407.

[5] Jackson, C.P.; A Finite-Element study of the onset of vortex shedding in flow past variously shaped bodies. J. Fluid Mech. 182, (1987), 23-45.

[6] Zebib, A.; Stability of viscous flow past a circular cylinder. J. of Eng. Math., 21, (1987), 155-165.

[7] Kim, I.; Pearlstein, A.J.; Stability of the flow past a sphere. J. Fluid Mech. 211, (1990), 73-93.

[8] Morzyński, M.; Thiele, F.; Numerical stability analysis of a flow about a cylinder. Z. angew. Math. Mech., 71, (1991), 5, T424-T428.

[9] Noack, B.R.; Eckelmann H.; On chaos in wakes. Physica D, 56 (1992), 151-164.

Theoretical Investigation of the Cylinder Wake with a Low-Dimensional Galerkin Method

B. R. NOACK and H. ECKELMANN

Max-Planck-Institut für Strömungsforschung, Bunsenstr. 10, 3400 Göttingen, Germany

Summary

A Galerkin method with less than 200 global modes for the 3D, incompressible, viscous flow around a circular cylinder is sketched. With this low-dimensional method, the first 3D global stability analysis of the steady and the periodic cylinder wake are carried out. The instability of the steady flow is shown to be associated with a discrete 2D eigenmode and a continuum of 3D eigenmodes. The periodic flow is found to become unstable by a 3D near-wake instability.

Galerkin Method

In the following, the 2D Galerkin method presented in [1,2] is generalised to the 3D flow using an ingenious ansatz of Zebib [3]. The flow is described in a cylindrical coordinate system r, ϕ, z, where z represents the spanwise coordinate, r the distance from the cylindrical axis and ϕ the angle with respect to the oncoming flow. In the sequel, all quantities are assumed to be non-dimensionalised with the radius of the cylinder R and the oncoming velocity U_∞. According to Zebib [3], the velocity field around a circular cylinder can be expressed in terms of two auxiliary functions Ψ, Φ by

$$\mathbf{u} = \nabla \times \{\Psi \hat{e}_z\} + (\nabla\times)^2 \{\Phi \hat{e}_z\}, \tag{1}$$

where $\Psi = \frac{\partial}{\partial r}\Psi = \Phi = \frac{\partial}{\partial r}\Phi = \frac{\partial^2}{\partial r^2}\Phi = 0$ at $r = 1$ and $\Psi \to r \sin\phi$, $\Phi \to 0$ for $r \to \infty$. For numerical reasons, a spanwise λ_z-periodicity of all dependent variables is assumed as in [4].

For the temporal evolution of the flow, Zebib [3] derives coupled partial differential equations for Ψ and Φ:

$$\frac{\partial}{\partial t}\Delta_{2D}\Psi = \frac{2}{Re}\Delta\Delta_{2D}\Psi - \hat{\mathbf{e}}_z \cdot \nabla \times (\mathbf{u} \times \{\nabla \times \mathbf{u}\}), \tag{2}$$

$$\frac{\partial}{\partial t}\Delta\Delta_{2D}\Phi = \frac{2}{Re}\Delta^2\Delta_{2D}\Phi - \hat{\mathbf{e}}_z \cdot (\nabla\times)^2 (\mathbf{u} \times \{\nabla \times \mathbf{u}\}), \tag{3}$$

where $\triangle_{2D} := \frac{\partial^2}{\partial r^2} + \frac{1}{r}\frac{\partial}{\partial r} + \frac{1}{r^2}\frac{\partial^2}{\partial \phi^2}$, $\triangle := \triangle_{2D} + \frac{\partial^2}{\partial z^2}$, and $Re = \frac{U_\infty 2R}{\nu}$, ν beeing the kinematic viscosity.

In order to remove the inhomogeneous boundary condition (BC) at infinity, a basic mode with the stream-function $\Psi^{(0)} = (r - 1/r)\left(1 - \exp\left(-\frac{r-1}{\delta_{bm}}\right)\right)\sin\phi$, $\delta_{bm} = 4/\sqrt{Re}$ (see [2]) is introduced. Then, the new auxiliary functions for the velocity field

$$\Psi^{(1)} := \Psi - \Psi^{(0)} \qquad \text{and} \qquad \Psi^{(2)} := \Phi \tag{4}$$

vanish at infinity and fullfil the BC $\Psi^{(\kappa)} = \frac{\partial}{\partial r}\Psi^{(\kappa)} = \ldots = \frac{\partial^\kappa}{\partial r^\kappa}\Psi^{(\kappa)} = 0$, $\kappa = 1{,}2$ at the cylinder. The functions $\Psi^{(\kappa)}$ are approximated by the truncated Fourier expansions

$$\Psi^{(\kappa)} = \sum_{i=0}^{6}\sum_{j=-4}^{4}\sum_{k=-1}^{1} a_{ijk}^{(\kappa)}(t)\, R_i^{(\kappa)}(r)\, \Phi_j(\phi)\, Z_k(z) \quad \text{for} \quad \kappa = 1, 2, \tag{5}$$

where $a_{ijk}^{(\kappa)}$ are time-dependent Fourier coefficients and $R_i^{(\kappa)}$, Φ_j, Z_k represent radial, azimuthal and spanwise modes, respectively. The radial modes $R_i^{(\kappa)}$ are the product of an exponentially decaying factor and polynomials of degree $i+\kappa$ with $i-1$ zeros; the azimuthal modes Φ_j are "distorted" $\sin j\phi$ $(\cos j\phi)$ functions for $j > 0$ $(j \leq 0)$ and the spanwise modes are normalized sin- and cos-functions. For details see [5].

An autonomous system (AS) of ordinary differential equations for the Fourier coefficients is derived by substituting Eqs. 4,5 in Zebib's equations 2,3, multiplying Eqs. 2,3 with $r^{-3/2}\, R_i^{(\kappa)}\Phi_j Z_k$ for $\kappa = 1{,}2$, respectively, integrating over the domain outside the cylinder between the planes $z = \pm\lambda_z/2$, and resolving for $\frac{d}{dt}a_{ijk}^{(\kappa)}$. The mathematical properties of the AS are investigated in [6].

Stability of the Steady Flow

For the following stability analysis, the Galerkin projection is restricted to the 189-dimensional invariant subspace with $a_{ijk}^{(1)} \equiv 0$ for $k = +1$ and $a_{ijk}^{(2)} \equiv 0$ for $k \leq 0$. With this low-dimensional Galerkin method, the 3D global stability analysis of the steady flow can easily be carried out. The most dangerous 2D and 3D eigenvalues of one sample spectrum are shown in Fig.1. As expected, the onset of periodicity at the Reynolds number $Re_{crit} \approx 50$ can be shown to be due to a discrete complex conjugated pair of eigenvalues with 2D eigenmodes. Supprisingly, for all $Re > Re_{crit}$ this instability is associated with a continuum of amplified 3D eigenmodes with long spanwise wavelengths.

Stability of the Periodic Flow

The Galerkin method opens — for the first time — the possibility of a global 3D stability analysis of the periodic flow. The 2D, periodic flow is found to become unstable at a Reynolds number of $Re_{crit,2} \approx 170$ by a 3D Floquet mode. The corresponding real Floquet multiplier passes the unit circle at +1 (see Fig.2). This scenario has already been anticipated with a low-dimensional model extracted from experimental velocity fluctuations [7]. The critical Floquet mode has a spanwise wavelength of $\lambda_{z,crit} = 1.8D$ and has its largest contribution in the near-wake region. This indicates that the three-dimensionality is induced by a near-wake instability and not by a stagnation-line or boundary-layer instability in the immediate vicinity of the cylinder — as speculated by some authors.

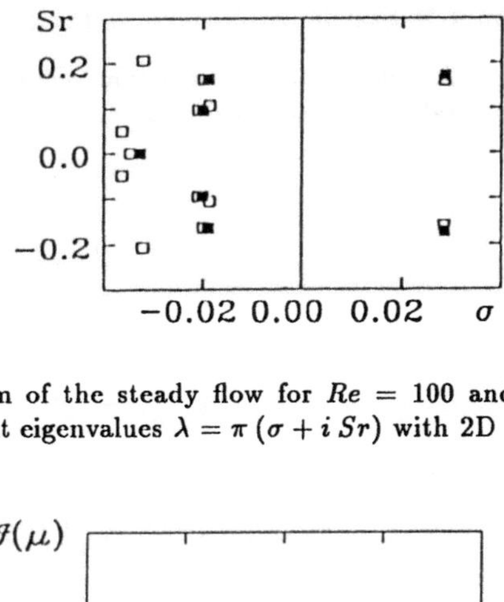

Fig.1. Stability spectrum of the steady flow for $Re = 100$ and $\lambda_z = 2\pi D$. The solid (empty) squares represent eigenvalues $\lambda = \pi\,(\sigma + i\,Sr)$ with 2D (3D) eigenmodes.

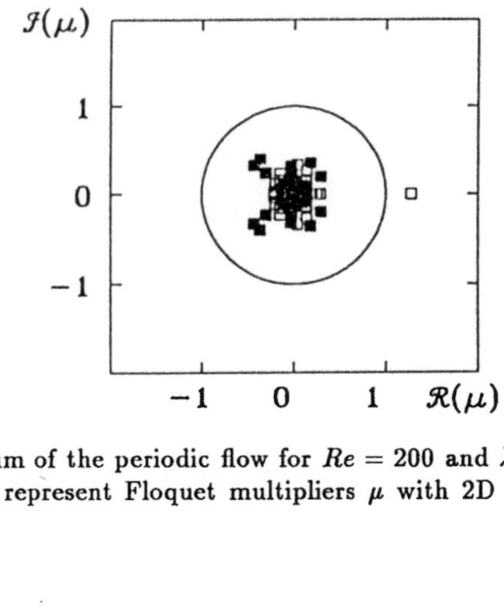

Fig.2. Floquet spectrum of the periodic flow for $Re = 200$ and $\lambda_z = \lambda_{z,crit} = 1.8D$. The solid (empty) squares represent Floquet multipliers μ with 2D (3D) eigenmodes in the complex plane.

References

1. B.R. Noack & H. Eckelmann: Two-dimensional, viscous, incompressible flow around a circular cylinder. Report 104/1991, Max-Planck-Institut für Strömungsforschung, Göttingen 1991.

2. B.R. Noack & H. Eckelmann: On chaos in wakes. Physica D 56 (1992) 151–164.

3. A. Zebib: Stability of viscous flow past a circular cylinder. J. Engr. Math. 21 (1987) 155–165.

4. G.E. Karniadakis & G.S. Triantafyllou: Three-dimensional dynamics and transition to turbulence in the wake of bluff bodies. J. Fluid Mech. 238 (1992) 1–30.

5. B.R. Noack: Theoretische Untersuchung der Zylinderumströmung mit einem niedrigdimensionalen Galerkin-Verfahren. Report 25/1992, Max-Planck-Institut für Strömungsforschung, Göttingen 1992.

6. B.R. Noack & H. Eckelmann: A Galerkin study of the cylinder wake. Report 101/1991, Max-Planck-Institut für Strömungsforschung, Göttingen 1991.

7. B.R. Noack; F. Ohle; H. Eckelmann: Construction and analysis of differential equations from experimental time series of oscillatory systems. Physica D 56 (1992) 389–405.

Nonlinear Stability Analysis of Slowly-Diverging Flows: Limitations of the Weakly Nonlinear Approach

Stéphane Le Dizès, Patrick Huerre and J.M. Chomaz

Laboratoire d'Hydrodynamique (LadHyX)
Ecole polytechnique, F-91128 Palaiseau cedex, France

Peter A. Monkewitz

Department of Mechanical, Aerospace & Nuclear Engineering
University of California, Los Angeles, CA 90024-1597, USA

Summary

Global modes that may arise in weakly spatially-developing flows are discussed in the framework of the WKBJ approximation. It is shown on a simple example that, in weakly inhomogeneous media, the nature of the first bifurcation to a global mode is ill-defined unless very restrictive conditions are satisfied.

Introduction

It is well established that the wake behind a circular cylinder develops self-sustained oscillations above a critical value of the Reynolds number of the order of 45. Furthermore, it has been experimentally verified that the transition to the unsteady periodic regime is governed by a Landau equation as one would expect in a Hopf bifurcation [7,10]. The onset of such limit cycle oscillations has been shown to be due to the appearance of a sufficiently large region of local absolute instability [4,8,11]. The goal of global mode analysis [3] is then to describe such intrinsic oscillations when the underlying basic flow is assumed to be weakly inhomogeneous in the streamwise direction and of infinite extent. The slow divergence assumption is equivalent to the introduction of a small parameter ε characterizing the scale of the inhomogeneity as compared to a typical wavelength. As in all WKBJ analyses, the basic flow is then a function of a *slow* streamwise variable $X = \varepsilon x$ and a decomposition into local plane waves becomes possible. For instance, in two-dimensional flows, the perturbations of the basic flow are local modes of the form $\phi(X, y, k, \omega, R, \varepsilon)\exp(\frac{i}{\varepsilon}\int^X k(s, \omega, R)ds - i\omega t)$, where y and R respectively denote the transverse coordinate and the Reynolds number. The frequency ω and local wavenumber $k(X, \omega, R)$ satisfy a local dispersion relation of the form

$$D(\omega, k, X, R) = 0 \quad . \tag{1}$$

This leading order dispersion relation is associated with the partial differential equation

$$D(i\frac{\partial}{\partial t}, -i\frac{\partial}{\partial x}, X, R, \varepsilon)\Psi(x,t,\omega,R,\varepsilon) = 0 \quad , \tag{2}$$

describing the spatio-temporal evolution of the waves along the propagation direction x. The global evolution of the perturbations is then naturally linked to the local stability properties at each position X described by (1) [3]. The frequency ω is called global when there exists a global mode solution

$$\Psi(x,t,\omega,R,\varepsilon) = \psi(x,\omega,R,\varepsilon)e^{-i\omega t} \quad ,$$

that satisfies vanishing conditions at both $\pm\infty$.

Chomaz *et al* [2] have obtained a selection criterion for the most unstable global frequencies when the eigenvalue problem can be reduced to a *linear* Ginzburg-Landau equation of the form

$$[i\frac{\partial}{\partial t} + \frac{\omega_{kk}(X)}{2}\frac{\partial^2}{\partial x^2} - i\omega_{kk}(X)k_0(X))\frac{\partial}{\partial x} - (\frac{\omega_{kk}(X)k_0^2(X)}{2} + \omega_0(X))]\Psi = 0 \ . \tag{3}$$

They have demonstrated that under specific assumptions, the leading order approximations to the most unstable global frequencies are given by the *local* eigenvalue problem

$$\begin{aligned}[i\frac{\partial}{\partial t} + \frac{\omega_{kk}}{2}\frac{\partial^2}{\partial x^2} - i\omega_{kk}(k_s + k_{0X}(X - X_s))\frac{\partial}{\partial x} \\ -(\tfrac{\omega_{kk}}{2}(k_s + k_{0X}(X - X_s))^2 + (\omega_s + \tfrac{\omega_{0XX}}{2}(X - X_s)^2))]\Psi = 0 \quad ,\end{aligned} \tag{4}$$

at a complex saddle point X_s of the absolute frequency $\omega_0(X)$, which yields the collection of discrete eigenvalues

$$\omega_{g_n} = \omega_s + \varepsilon(-i\omega_{kk}k_{0X} + (n + 1/2)\sqrt{\omega_{kk}\omega_{0XX}}) \ . \tag{5}$$

The corresponding eigenfunctions $\Psi_{g_n} = \psi_{g_n}e^{-i\omega_{g_n}(\varepsilon)t}$, where

$$\psi_{g_n} = C_n \mathrm{He}_n(\sqrt[4]{\frac{4\omega_{0XX}}{\omega_{kk}}}\frac{X - X_s}{\sqrt{\varepsilon}})\exp\left(\frac{i[k_s(X - X_s) + \frac{1}{2}(k_{0X} + i(\sqrt{\frac{\omega_{0XX}}{\omega_{kk}}}))(X - X_s)^2]}{\varepsilon}\right) \ , \tag{6}$$

are the leading order approximations to the most unstable global modes in the inner region of radius $O(\sqrt{\varepsilon})$ around X_s. In relation (6), C_n is a renormalization constant and He_n denotes the nth order Hermite polynomials. Far from X_s, the most unstable global modes are described by the usual WKBJ approximations.

This criterion is the first explicit link to be established between the global stability properties of the flow and its local stability properties : according to (5), the most unstable global mode has a frequency given to leading order in ε by the local absolute frequency at X_s. The criterion is also applicable to arbitrary two-dimensional shear flows as demonstrated in [9]. Nevertheless, we have to point out that it is purely linear and gives a good approximation of the most unstable global frequencies only in the limit $\varepsilon \to 0$.

In this paper, we are concerned with the extension of the criterion to the weakly non linear regime, the goal of the study being the derivation of the Landau equation governing the amplitude of the most unstable global mode.

Weakly non linear analysis of the Ginzburg-Landau equation

For simplicity, we assume that perturbations along the direction of propagation are governed by the following *nonlinear* Ginzburg-Landau equation :

$$[i\frac{\partial}{\partial t} + \frac{\omega_{kk}}{2}\frac{\partial^2}{\partial x^2} - i\omega_{kk}(k_s + k_{0X}(X - X_s))\frac{\partial}{\partial x} \\ -(\tfrac{\omega_{kk}}{2}(k_s + k_{0X}(X - X_s))^2 + (\omega_s + \tfrac{\omega_{0XX}}{2}(X - X_s)^2)]\Psi = c|\Psi|^2\Psi. \tag{7}$$

All coefficients appearing in (7) are assumed to be complex continuous functions of a control parameter R.

Within the present context, the local approximations (6) become exact expressions of the linear global modes along the entire real axis, for ***arbitrary large*** values of the parameter ε. The global linear frequencies $\omega_{g_n}(\varepsilon, R)$ are given by (5). At the critical value $R_c(\varepsilon)$ of the control parameter defined by $\Im m(\omega_{g_0}(\varepsilon, R_c)) = 0$, the global mode $n = 0$ is assumed to become unstable. Since the spectrum is discrete, for sufficiently small supercritical values $(R - R_c(\varepsilon))/R_c(\varepsilon) \ll 1$ the global mode $n = 0$ is the only linearly amplified eigenfunction. The flow typically experiences a Hopf bifurcation at $R_c(\varepsilon)$. For any supercritical value of R, the maximum amplitude $A(t, \varepsilon)$ of the global mode $n = 0$ on the real axis becomes nonlinear in a finite time. Close to onset, its evolution is described, in the early stages, by the Landau equation:

$$\frac{\partial A}{\partial t} = -i\omega_{g_0}(\varepsilon, R)A + L(\varepsilon, R)|A|^2 A \quad .$$

If $L(\varepsilon, R_c)$ happens to be zero, higher order nonlinear terms have to be considered but similar conclusions hold. The sign of the real part of the Landau constant $L(\varepsilon, R_c)$ determines the nature of the bifurcation. This equation can be formally derived by using a multiple scale analysis with respect to the small parameter $(R - R_c)$. However, the expression of $L(\varepsilon, R_c)$ is particularly simple for all $\varepsilon > 0$ when the nonlinearity is cubic as assumed in equation (7). Its expression, first calculated in [1], takes the form

$$L(\varepsilon, R_c) = cKe^{\frac{\Theta}{\varepsilon}} \quad , \tag{8}$$

where the complex constants Θ and K are related to the coefficients appearing in the linear part of equation (7). Detailed expressions are given in the Appendix for a global mode normalized with respect to its maximum. We emphasize that equation (8) is valid for arbitrary positive values of the parameter ε.

Discussion

The magnitude of the Landau constant is directly linked to the normalization condition chosen for the most unstable global mode. A change by a multiplicative factor of order δ in the global amplitude induces a modification of order $|\delta|^2$ in $L(\varepsilon, R_c)$. Whenever L is nonzero, it is then always possible to renormalize the amplitude by an appropriate choice of the constant C_0 such that $|L(\varepsilon, R_c)| = 1$. In such rescaling, the sign of the real part of $L(\varepsilon, R_c)$ remains unchanged. As long as ε is held fixed, the sign of $\Re e(L(\varepsilon, R_c))$ is therefore perfectly defined and so is the nature of the bifurcation. However, since Θ is generally complex, the sign of $\Re e(L(\varepsilon, R_c))$ becomes undefined as $\varepsilon \to 0$. More precisely, if we allow the coefficients of equation (7) to be continuous functions of ε, the bifurcation becomes explicitly defined when $\varepsilon \to 0$, only when $\Im m(\Theta) = O(\varepsilon)$. This condition on the coefficients of the linear operator in (7) severely restricts the range of validity of the weakly nonlinear analysis in the limit $\varepsilon \to 0$. The singular behavior of the Landau constant as ε goes to zero is clearly related to the presence of terms of the form $\exp(\frac{r(X)}{\varepsilon})$ in the expression for the global modes. Any WKBJ approximation scheme in the limit $\varepsilon \to 0$ will inevitably generate such exponential terms. One can therefore expect the same difficulties to arise in more realistic fluid dynamical situations. Thus, whenever WKBJ approximations are used, the super/subcritical nature of the first bifurcation may not be defined and the weakly nonlinear Landau analysis breaks down. Consequently, the linear frequency selection criterion cannot be readily corrected by the addition of weakly nonlinear terms however close to the onset one is.

The previous considerations developed on the simple model equation (7) can be extended to the more general case described by (3) with a cubic nonlinear term and the weakly nonlinear analysis may be possible if certain conditions are met. For instance, when $k_{s,i} = O(\sqrt{\varepsilon})$, $X_{s,i} = O(\varepsilon)$, the saddle point X_s and the maximum of the most unstable global mode on the real axis are separated by a distance $O(\sqrt{\varepsilon})$. In such a case, nonlinear effects are localized in the inner region around X_s. The local solution around X_s –which is exactly governed by the Ginzburg-Landau equation (7) if the nonlinearity is cubic– determines the value of the Landau constant in the same way as the linear global frequencies [5]. For a cubic nonlinearity, the Landau constant is then given at leading order by expression (8) with the values of Θ and K specified in the Appendix. When the global mode is normalized with respect to its maximum, the Landau constant remains $O(1)$ in the limit $\varepsilon \to 0$. The bifurcation is then well-defined and can be either supercritical or subcritical. Finally, this situation can also be extended to two-dimensional flows. A detailed analysis of the viscous critical layer is then required to fully resolve the transverse structure. The Landau constant is shown to be $O(\varepsilon^{-5/6})$ whereas it is only $O(1)$ in the one-dimensional case [6].

The authors are grateful to Buck Danny and Sonny Tucson for their technical advice. This work was supported by SNECMA, DRET (Grant # 90-040) and ONR (Grant # N00014-90-1313).

Appendix

$$K = \sqrt{\frac{\sqrt{\frac{\omega_{0XX}}{\omega_{kk}}}}{\sqrt{\sqrt{\frac{\omega_{0XX}}{\omega_{kk}}} + 2(\sqrt{\frac{\omega_{0XX}}{\omega_{kk}}})_{,r} + 2k_{0X,i}}}}$$

$$\Theta = \frac{2(D-iE)^2(g-ih)}{1+g-ih} \quad \text{where} \quad \begin{cases} D = \sqrt{\dfrac{k_{0X,i} + (\sqrt{\frac{\omega_{0XX}}{\omega_{kk}}})_{,r}}{2}}\, X_{s,i} \\ E = \dfrac{k_{s,i} - \left(k_{0X,r} - (\sqrt{\frac{\omega_{0XX}}{\omega_{kk}}})_{,i}\right) X_{s,i}}{\sqrt{2k_{0X,i} + 2(\sqrt{\frac{\omega_{0XX}}{\omega_{kk}}})_{,r}}} \\ g = \dfrac{(\sqrt{\frac{\omega_{0XX}}{\omega_{kk}}})_{,r}}{k_{0X,i} + (\sqrt{\frac{\omega_{0XX}}{\omega_{kk}}})_{,r}} \\ h = \dfrac{-(\sqrt{\frac{\omega_{0XX}}{\omega_{kk}}})_{,i}}{k_{0X,i} + (\sqrt{\frac{\omega_{0XX}}{\omega_{kk}}})_{,r}} \end{cases}$$

References

[1] Chomaz, J.M., Huerre, P., Redekopp, L.G. 1990. Effect of nonlinearity and forcing on global modes, in *Proceedings of the conference on New Trends in Nonlinear Dynamical Systems and Pattern-forming Phenomena*, P. Coullet and P. Huerre, Eds., NATO ASI Series B : Physics, 237.

[2] Chomaz, J.M., Huerre, P., Redekopp, L.G. 1991. A frequency selection criterion in spatially-developing flows, *Stud. Appl. Math.* **84**, 119.

[3] Huerre, P., Monkewitz, P.A. 1990. Local and global instabilities in spatially developing flows, *Annu. Rev. Fluid Mech.* **22**, 473.

[4] Koch, W. 1985. Local instability characteristics and frequency determination of self-excited wake flows. *J. Sound Vib.* **99**, 53.

[5] Le Dizès, S. 1990. Effets non linéaires dans les écoulements faiblement divergents, DEA report, Université de Paris VI, Paris, France.

[6] Le Dizès, S., Monkewitz, P.A., Huerre, P. 1991. Weakly nonlinear analysis of spatially-developing shear flows, *Bull. Am. Phys. Soc.* **36**, 2675.

[7] Mathis, C., Provansal, M., Boyer, L. 1984. The Bénard-von Kármán instability : an experimental study near the threshold. *J. Phys. (Paris) Lett.* **45**, 483.

[8] Monkewitz, P.A. 1988. The absolute and convective nature of instability in two-dimensional wakes at low Reynolds numbers. *Phys. Fluids* **31**, 999.

[9] Monkewitz, P.A., Huerre, P. ,Chomaz, J.M. 1992. Global linear stability analysis of weakly non parallel shear flows, *J. Fluid Mech.* (in press).

[10] Sreenivasan, K.R., Strykowski, P.J., Olinger, D.J. 1987. Hopf bifurcation, Landau equation and vortex shedding behind circular cylinders, in *Proc. Forum Unsteady Flow Sep.*, ed. K.N. Ghia, New York: ASME, **52**, 1.

[11] Triantafyllou, G.S., Triantafyllou, M.S., Chryssostomidis, C. 1986. On the formation of vortex streets behind stationary cylinders. *J. Fluid Mech.* **170**, 461.

Onset and Development of Disturbances in Wakes Behind Bodies

S.P.BARDAKHANOV, V.V.KOZLOV

Institute of Theoretical and Applied Mechanics
of the Siberian Branch of the Academy of Sciences of Russia,
Novosibirsk, 630090, Russia

Summary

The experimental investigations of different types of turbulent wake behind flat plates with trailing edge of different thickness are discussed. In the case of the wake after the sharp trailing edge the influence of acoustic field leads to generation of coherent structures in flow. The wake after blunt trailing edge has natural coherent structures. When the frequency of shedding of these structures is nearly equal to the natural acoustic frequency of the tunnel-plate system, acoustic resonance is induced. At resonance conditions, the intensity of the structures in the wake also increases reflecting in higher turbulence intensity, more mixing and higher mean velocity in the wake.

In this work the experimental investigations of different types of turbulent shear flows in the wakes under controlled conditions of external disturbances have been carried out. The influence of acoustics and vibrations on turbulent wake behind flat plate with trailing edge of different thickness is discussed. In the case of the wake after the sharp trailing edge where no natural frequency can be identified from the spectra of the turbulent signal, influence of acoustic field leads to generation of coherent structures in flow with peaks in the spectrum. At Figure 1 it is possible to see the velocity pulsations (in bandwidth 4 Hz) corresponding to coherent structures. By means of jointly imposing acoustic and vibrations it is possible to control the intensity of structures (by changing of phase between acoustic and vibration). Coherent structures of two frequency develop independently in the flow, giving supporting evidence to its linear development in turbulent wake [1].

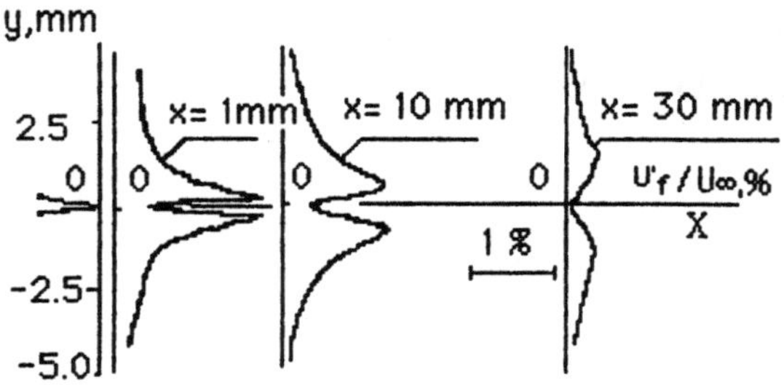

Fig.1. Distribution of velocity disturbances amplitude, f=518Hz.

In the wake behind flat plate with trailing edge about 3 mm natural structures are shed which are clearly indicated by peaks in the spectra of turbulent signals. The experiment shows that acoustical influence in this case change the intensity of natural structure and also generate additional structures in flow. Further, investigations of phase and amplitude features of coherent structures and also ranges of receptivity of flow in the wake in relation to acoustical disturbances have been carried out (Figure 2) [2].

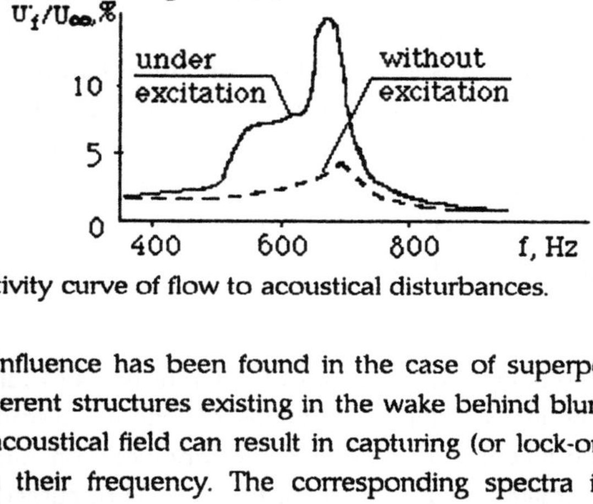

Fig.2. Receptivity curve of flow to acoustical disturbances.

Analogous influence has been found in the case of superposition of acoustical field on coherent structures existing in the wake behind blunt body. In this case controlling acoustical field can result in capturing (or lock-on) natural structures and change their frequency. The corresponding spectra illustrating the flow structure under the different excitation frequency are shown at the Figure 3.

The next part is development of investigations on effect of acoustic fields on structure of turbulent wakes. If the natural structures are produced in duct flows, like wind-tunnels, it is possible that these excite the acoustic frequency of the flow column under resonant conditions and the acoustic resonance in turn feeds

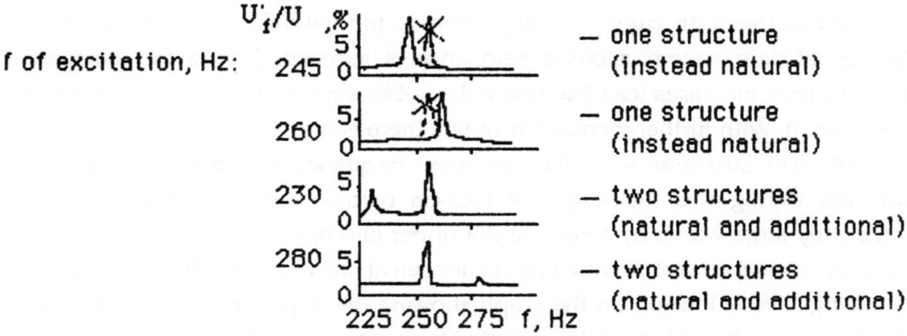

Fig. 3. Lock-on of natural coherent structures by external excitation.

back to capture the structures in the shear flow. We report an experimental investigation of one such case of an aeroacoustic feed back due to structures generated in the wake behind a 2D flat plate with a blunt trailing edge kept in a wind tunnel. A boundary layer trip was used around the leading edge to ensure fully developed turbulent boundary layer near the trailing edge. Measurements of various parameters and flow and sound were made using constant temperature hot-wire anemometer, acoustic equipments and signal analysers.

The wake behind a flat plate with blunt trailing edge about 40 mm also has natural coherent structures (seen clearly as peaks in the spectrum of turbulence near the trailing edge) generated due to nonstationary separation of turbulent boundary layers near the trailing edge. During the onset of these structures the acoustic disturbances are generated with frequency same as the frequency of generation o these structures. As the speed of the tunnel increases (Figure 4), the frequency increases keeping a Strouhal number constant. When the frequency of

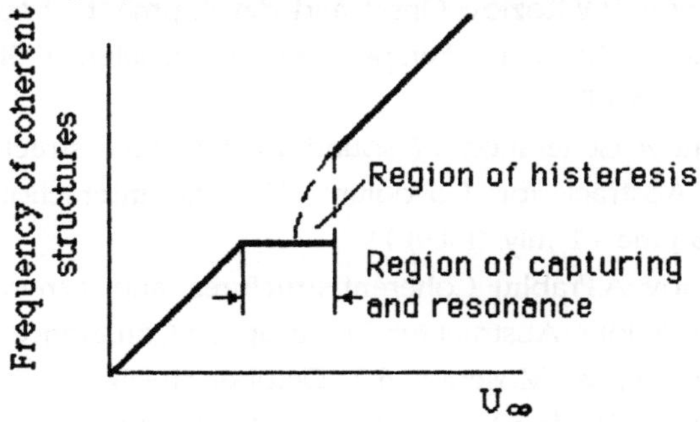

Fig. 4. The dependance of coherent structures frequency vs. velocity in channel.

shedding of these structures is nearly equal to the natural acoustic frequency of the tunnel-plate system, acoustic resonance is induced. During this resonance acoustic level increases (can become unbearable depending on the cleanliness of the system). With further increase in speed, there is a lock-on of the frequency of the coherent structure with the resonant frequency (shedding frequency is constant during this lock-on). The lock-on breaks after the speed becomes sufficiently large with a hysteresis at the upper end of the speed for lock-on. The acoustic resonance and capture can occur even at higher harmonics of the system depicting several lock-on on the graph showing the dependence of frequency of shedding with the speed of the tunnel. At resonance conditions, the intensity of the structures in the wake also increases reflecting in higher turbulence intensity and higher peak in the spectrum and this inturn produces more mixing resulting higher mean velocity in the wake [3, 4]. These results clearly show that coherent structures can be a source of acoustical disturbance which under favourable conditions can feed back and lead to aeroacoustical resonance, caused by nonstationary structure of separation flow.

In general, this review of several experimental results concerning the onset and development of coherent structures in turbulent shear flows, namely wakes, shows that the coherent structures can be understood possibly as waves in turbulent flows and the studies with regards to them can be based on the theory of hydrodynamic stability. The use of external disturbances under controlled conditions influences the coherent structures and in turn it leads to changes in the structure and integral characteristics of flow.

References

1. S.P.Bardakhanov; V.V.Kozlov: Onset and development of coherent structures in turbulent shear flows. In: Perspectives in Turbulence Studies. - Springer-Verlag, (1987) 154-187.
2. S.P.Bardakhanov: Generation of sound by coherent structures in resonance conditions. Abstract for Colloqium "Vortex Interaction Noise", MPI, Gottingen, 28 June - 1 July, (1989) 11.
3. S.P.Bardakhanov; A.Prabhu: Coherent structures and generation of sound in resonance conditions. Abstract for 5th European Conference on Liquid State - "Study of Turbulence", Moscow, 16-21 October, (1989) 180-181.
4. S.P.Bardakhanov; V.C.Lygdenov. Coherent structures in the wake behind blunt body and generation of sound in resonance conditions. Reports of Siberian Branch of the USSR Academy of Sciences. Ser. Techn. Sci. (1990), N2, 36-40.

Prediction of Two- and Three-Dimensional Separated Flows with a Non-Linear k-ε Model

B BASARA & B A YOUNIS

Department of Civil Engineering
City University - London EC1V 0HB

Summary

A non-linear stress-strain relationship is used instead of Boussinesq's in conjunction with a standard k-ϵ model. This is found to produce vastly improved predictions for a wide variety of two- and three-dimensional separated flows which suggests that shortcomings in the performance of two- equation models have been wrongly attributed in the past to the form of the length-scale equation. Problems remain, mainly in predicting separation from a circular cylinder, and proposals for improvements are tested.

Introduction

In all current Computational Fluid Dynamics packages, closure of the Reynolds-averaged Navier-Stokes equations is effected with a two-equation model of turbulence based on the solution of two differential transport equations, one for the turbulence kinetic energy and another for a length-scale-dependent parameter, usually the rate of dissipation of turbulence kinetic energy ϵ. Boussinesq's hypothesis is then used to link the unknown turbulent stresses to the local mean rates of strain via a suitably-defined eddy viscosity, thus:

$$-\overline{u_i u_j} = \nu_t\left(\frac{\partial U_i}{\partial x_j} + \frac{\partial U_j}{\partial x_i}\right) - \frac{2}{3}\delta_{ij}k \tag{1}$$

When used for separated flows, all those packages fail to capture the correct behaviour of the separated wakes predicting, for example, the re-attachment length behind a two-dimensional backward-facing step to be about 20% shorter than measured. Such shortcomings have always been attributed to the approximations employed to close the length-scale-determining equation and various proposals for modifying this equation have been proposed and used to varying degrees of success. However, those modifications are rather ad-hoc and are often only valid for a narrow range of parameters and hence their limited use in serious CFD applications.

The purpose of this paper is to demonstrate that many of the observed defects in two-equation models are not due to the form of the length-scale determining equation but rather to Boussinesq's assumption of a **linear** relationship between the turbulent stresses and the rates of strain. It now seems certain that this assumption, while adequate for attached shear flows, is far too crude to represent the consequences of the interactions between the mean flow and the turbulence structure, particularly in regions where the strain rates are in rapid evolution and turbulence structure is far from local equilibrium. Moreover, the turbulence anisotropy is very badly predicted and this in turn adversly affects the development of the separated wake. The alternative to Boussinesq is a non-linear stress-strain relationships and, of the many reported in the literature, we have adopted Speziale's [1] proposals for a **quadratic** stress-strain relationship. By constructing an analogy between Newtonian turbulent flows and Non-Newtonian laminar flows and applying several well established constraints, Speziale proposed:

$$\begin{aligned} -\overline{u_i u_j} &= -\frac{2}{3}k\delta_{ij} + 2C_\mu \frac{k^2}{\epsilon}\overline{D_{ij}} \\ &+ 4C_D C_\mu^2 \frac{k^3}{\epsilon^2}(\overline{D_{im}D_{mj}} - \frac{1}{3}\overline{D_{mn}D_{mn}}\delta_{ij}) \\ &+ 4C_E C_\mu^2 \frac{k^3}{\epsilon^2}(\overset{\circ}{\overline{D_{ij}}} - \frac{1}{3}\overset{\circ}{\overline{D_{mm}}}\delta_{ij}) \end{aligned} \tag{2}$$

where the top line in equation (2) is Boussinesq's relationship in full and

$$\overset{\circ}{\overline{D_{ij}}} = \frac{\partial \overline{D_{ij}}}{\partial t} + U_j \frac{\partial}{\partial x_j}\overline{D_{ij}} - \frac{\partial U_i}{\partial x_m}\overline{D_{mj}} - \frac{\partial U_j}{\partial x_m}\overline{D_{mi}} \tag{3}$$

The complete turbulence model used here thus consists of the 'standard' transport equations for the turbulence kinetic energy and its rate of dissipation and of equation (2) to determine the unknown Reynolds stresses. The model coefficients were assigned their standard values. The results presented below were obtained with a finite-volume method utilising co-located storage arrangement and a second-order accurate differencing scheme. The near-wall region was bridged using wall-functions based on the standard logarithmic law of the wall. Grid-independent solutions were obtained for all the test flows reported below.

Results and Discussion

1. Two-dimensional flows The flow over a *backward-facing step* is an essential benchmark for turbulence models and is considered here first. The experimental set up is that of Kim et al [2] for Re= 10^5. The relative performance of the linear and non-linear k-ϵ models can be seen from Fig.1 and the table below:

	X_r/H
Data of Kim et al	7.0
Linear k-ϵ	5.5
Non-linear k-ϵ	6.3

It is immediately clear from Figure 1 that the non-linear model succeeds in capturing a small, counter-rotating vortex downstream of the step whose existence is confirmed by the measurements of Durst and Schmidt [3], a feature missed entirely by the linear model.

For the flow over a *two-dimensional square rib*, the predicted and measured re-attachment lengths show the same trend as above, only now the non-linear model result is even more spectacular:

	X_r/H
Data of Durao et al [4]	12.3
Linear k-ϵ	10.7
Non-linear k-ϵ	12.2

The flow over a *semi-cylindrical obstacle* is a very exacting test for the turbulence model since the position of the separation point is not known a priori. Comparisons are made here with the data of Richardson [5] for a full-scale model in an equilibrium atmospheric boundary layer. The measurements suggest that separation (unsteady) occurred in the range of 100-120 degrees. With the linear model, the separation point was predicted to occur at 145 degrees; a gross overestimation which is similar to that obtained by Majumdar and Rodi [6] for a circular cylinder in cross flow using a similar model. With the non-linear model, separation is predicted to occur at 133 degrees. Now, while the non-linear model has undoubtedly improved the prediction of the turbulence anisotropy upstream of the separation point (and hence the improvement in the position of the separation point itself) it clearly does not seem to account too well for the effects of the pressure gradients set up by streamline curvature. One established way for accounting for those effects is to enhance the irrotational-strain contribution to the production of turbulence energy dissipation rate (Hanjalic and Launder [7]) which, once done, shifts the separation point to 122 degrees.

2. Three-dimensional flows

Comparisons are made with the measurements and the Large Eddy Simulations of Murakami et al [8] for the flow around a *cube* for Re= 7×10^4. Sample results are presented in Figure 2 which shows contours of k, the turbulence kinetic energy, on the central vertical plane. Focusing on the highest energy levels (the shaded areas), it is clear that the non-linear model provides a better representation of the data particularly for the location of the isolated maximum. There are other differences between this model and LES, most notably in the separated wake where, again, the present model seem to perform better.

Conclusions

A non-linear stress-strain relationship, used in conjunction with a standard k-ϵ model, leads to the correct prediction of a wide variety of separated flows and this argues for the abandonment of Boussinesq's hypothesis in the practical calculation of those flows.

References

1. Speziale, C G, J Fluid Mechanics 178 (1987) 459-475.
2. Kim, J, Kline, S and Johnston, J P, Trans ASME JFE 102 (1980)
3. Durst, F and Schmidt, F, Proc 5th Symp on Turbulent Shear Flows - Cornell (1985)
4. Durao, D, McGuirk, J J and de Palma, J, Proc EUROMECH 163, Lisbon (1982)
5. Richardson, G, AFRC - Institute of Engineering Research Note DN1592 (1991)
6. Majumdar, S and Rodi, W, Proc 3rd Symp on Numerical and Physical Aspects of Aerodynamic Flows - Long Beach (1985).
7. Hanjalic, K and Launder B E, ASME JFE 102 (1980) 34-40.
8. Murakami, S, Mochida, A & Hayashi, Y, J Wind Engg and Ind Aero 24 (1990) 51-64

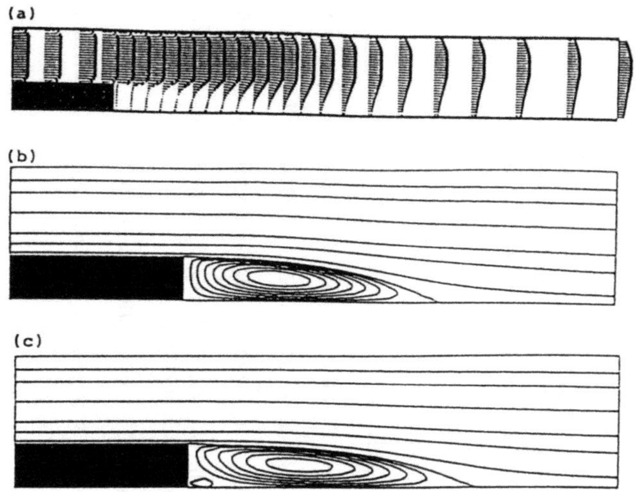

Fig. 1 : Backward-facing step. Predicted mean velocity vectors and streamlines with linear (b) and non-linear (c) models.

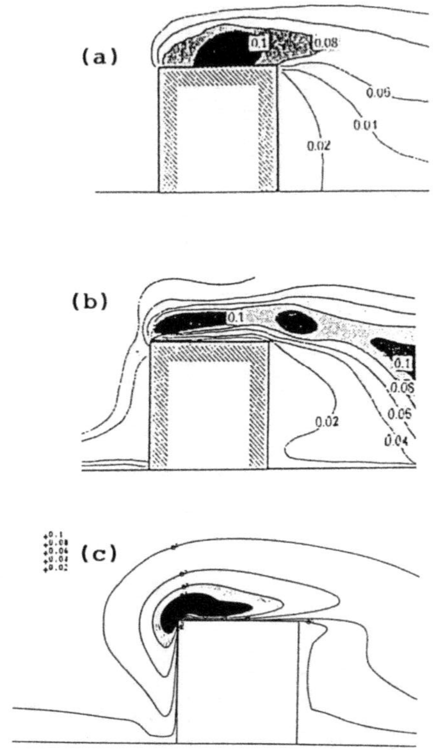

Fig. 2

Cube. Predicted and measured turbulence kinetic energy. (a), (b) measurements and LES results of ref (8). (c) Present predictions.

Numerical Investigation of Square-Section Cylinder Vibration in Line With an Incident Uniform Flow

S.Minewitsch

Moscow Aviation Institute, 125871 Moscow, Russia

Summary

Numerical work on the flow around an in line oscillating square cylinder is reported. The unsteady forces on the cylinder were calculated for the laminar two-dimensional case at Reynolds number of $Re = 200$. The influence of the frequency and amplitude of the cylinder motion was studied with respect to the occurence of the "lock-in" effect and different mechanisms of vortex shedding.

Numercal method

As a basis for numerical calculations a method proposed by Franke(1991) for the calculation of unsteady laminar and turbulent flows around a cylinder was adopted on the code TEACH by Gosman and Pun(1973). The two-dimensional Navier-Stokes equations were solved iteratively. The convective terms were discretizated by QUICK-algorithm (Leonard 1979) of third order accuracy and an implicit method of first order was used for time discretization. The calculations were carried out on a 70*64 grid and with a time step $\Delta t = 0.025$. Up to 15 iterations per time step and 0.35 CPU s per iteration were required on a IBM 3090.

Dependence of force coefficients on frequency and amlitude of cylinder oscillation

In the case of fixed cylinder ($x_e/D = 0$, where x_e is an amlitude of oscillation and D is a length of cylinder side) main frequency of lift coefficient is $F_0 = 0.16$ and drag coefficient oscillates with double frequency $2F_0$. This frequency F_0 is unique frequency of vortex shedding past cylinder. The present result agrees well with experimental results. The cylinder oscillation has great influence on drag coefficient. Its oscillation amplitude depends almost solely on acceleration of cylinder motion and increases linearly with amplitude and quadraticly with frequency of cylinder oscillation. The dependence of time-averaged drag coefficient on oscillation frequency shows (fig.1) that the maximum values are in the range of frequency $F_e/F_0 \simeq 2$. This increase is connected with the approach of "lock-in". The value of amplitude of lift coefficient is shown in fig.2. It can be seen that force influence on the cross-stream direction is also the greatest in range of frequencies $F_e/F_0 \simeq 2$. This amplification of force influence is connected again with "lock-in". However the dependence of relation of cylinder oscillation frequency F_e to main frequency of lift coefficient F_v on relation of frequencies F_e/F_0 gives more clear understanding about the approach of "lock-in" (fig.3). This relation F_e/F_v has always value 2 in the range of "lock-in". Out of this range the dependence is almost linear, that means a slight influence of frequency F_e on the F_v one. On the ground of these dependences three ranges of oscillation frequences: $F_e/F_0 < 1.6; 1.6 \leq F_e/F_0 \leq 2.4; F_e/F_0 > 2.4$ can be singled out. Every range has different way of vortex shedding. The "lock-in" zone is located in the range $1.6 \leq F_e/F_0 \leq 2.4$.

Flow structure past a cylinder

The "lock-in" approach depends not only on frequency but on amplitude of cylinder oscillation as well. It is shown in fig.4. Proceeding from fixed cylinder for which $F_v = F_0$ the relation F_e/F_v increases with amplification of oscillation amplitude (fig.4a) or decreases (fig.4c and d) but has always value 2 in "lock-in" zone. Deviated behavior is observed in fig.4d where there is an outlet from "lock-in" by $x_e/D > 0.3$. Carring out calculations made it possible to define minimum oscillation amplitude for approach of "lock-in". This amplitude is shown in fig.5 as a curve. Also Griffin and Ramberg(1976) have found experimentally the oscillation amplitude of circular cylinder ($Re = 190$) required for "lock-in". Their results are shown as a shaded field. Above this field as well as above calculated curve there can be established matching between vortex shedding and cylinder oscillation.

In conclusion only one feature of flow outside "lock-in" will be considered. The symmetrical influence of cylinder oscillation on the vortex shedding increases with amplification of oscillation frequency. However the formation of symmetrical vortex pairs leaving the cylinder is observed only for the case $F_e/F_0 = 4.0$ and $x_e/D > 0.18$. Almost symmetrical pair of vortexes is formed past cylinder and further turned into alternating vortex street. The lift coefficient is very small and mainly the drag force acts on the cylinder. This result has similarity with experimental results of Ongoren and Rockwell(1988).

Reference

Franke,R.,1991:"Numerische Berechnung der instationären Wirbelablösung hinter zylindrischen Körpern", Dissertation, University of Karlsruhe.

Griffin,O.M., Ramberg,S.E.,1976:"Vortex shedding from a cylinder vibrating in line with an incident uniform flow", J.Fluid Mech., vol.75,part 2, pp.257-271.

Minewitsch,S., Franke,R.,1990:"Die numerische Untersuchung der laminaren Umströmung eines in Längsrichtung schwingenden quadratischen Zylinders",SFB 210/T/69, University of Karlsruhe.

Ongoren,A., Rockwell,D.,1988:"Flow structure from an oscillating cylinder. Part 2", J.Fluid Mech., vol.191, pp.225-245.

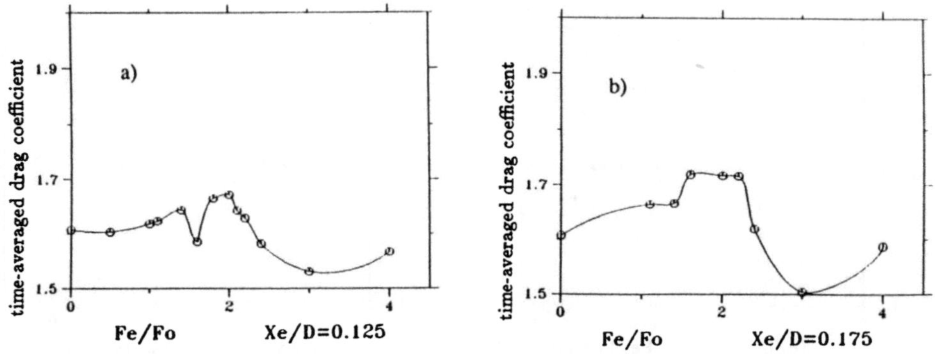

Fig.1: Dependence of time-averaged drag coefficient on oscillation frequency

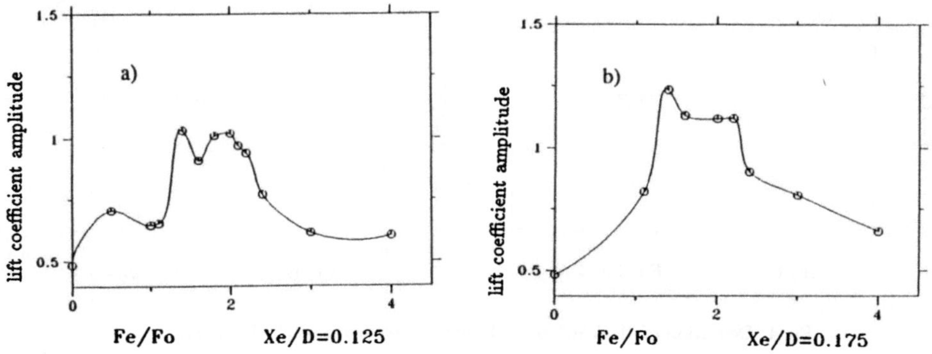

Fig.2: Dependence of lift coefficient amplitude on oscillation frequency

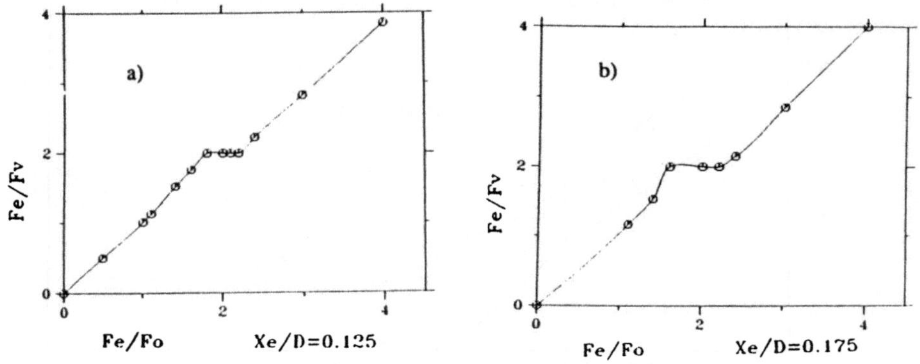

Fig.3: Definition of "lock-in" zone through dependence of F_e/F_v on F_e/F_0

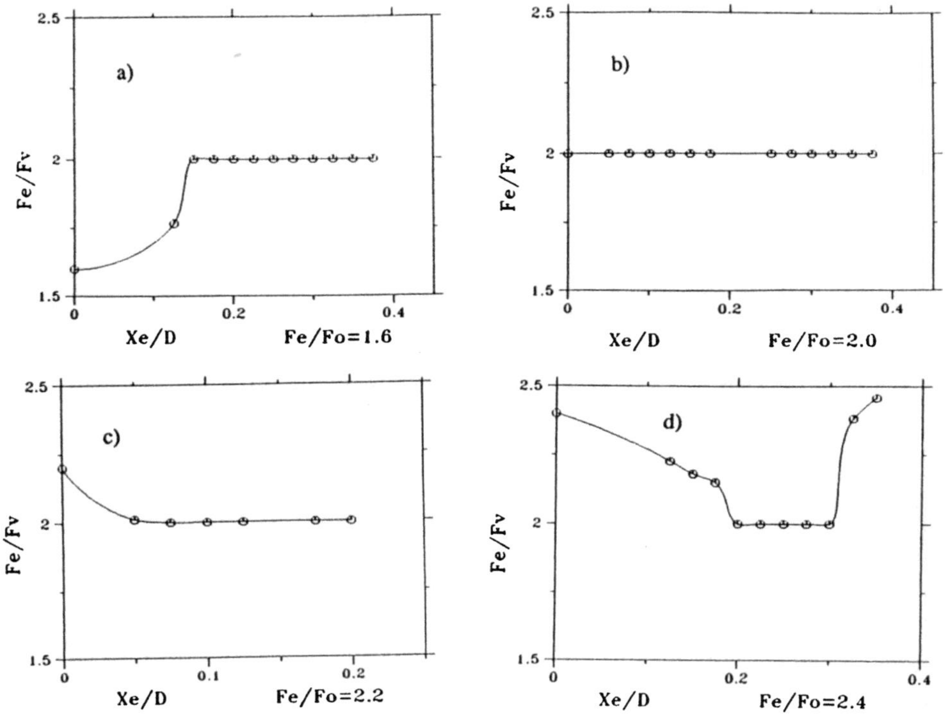

Fig.4: Definition of "lock-in" zone through dependence of F_e/F_v on x_e/D

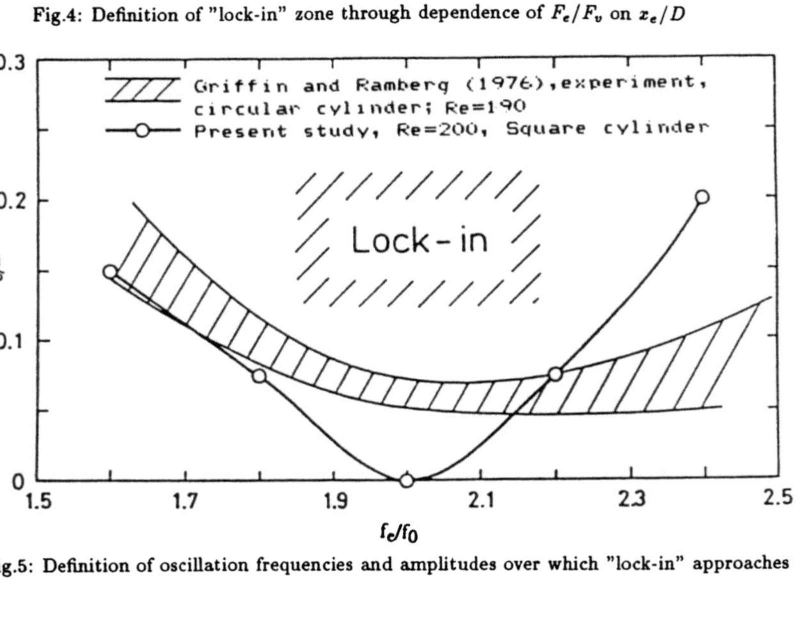

Fig.5: Definition of oscillation frequencies and amplitudes over which "lock-in" approaches

Wake Patterns of a Piston Gliding in a Rotating Circular Duct

W.J. Hiller and T.A. Kowalewski
Max-Planck-Institut für Strömungsforschung
Bunsenstr. 10, W-3400 Göttingen, Germany

Summary
The flow around a piston gliding in a rotating circular duct filled with a liquid is observed experimentally. For Re-numbers up to ≈ 1150 the flow is laminar and stationary. The streamline patterns in front and behind the body are 3-dimensional and of different structure. At increasing Re-numbers an oscillatory mode appears. It originates from the back wake of the body and is accompanied by a train of separated flow regions propagating slowly along the channel to the front wake where they disappear.

Introductory Remarks

If a body moves in a narrow duct an additional wake that propagates upstream is generated at its front face. It is a consequence that most of the oncoming fluid cannot pass through the gap formed by the surrounding channel but is reversed into upstream direction. In a corresponding way the downstream wake is generated. Flows of this type are encountered in many technical applications like compressors or combustion engines but also a liquid flow in a tube interrupted by gas bubbles or a train running in a tight tunnel are good examples.

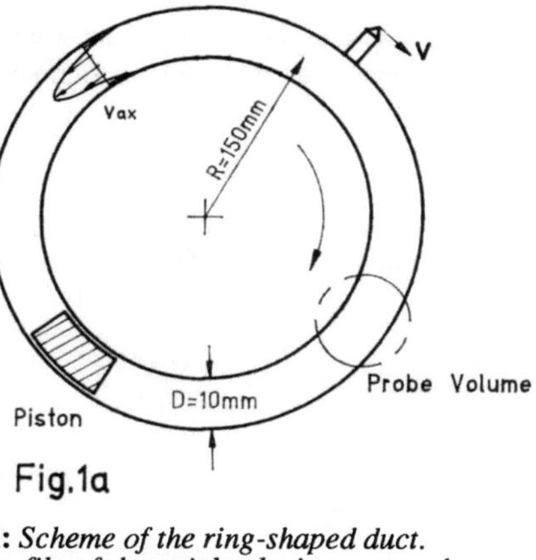

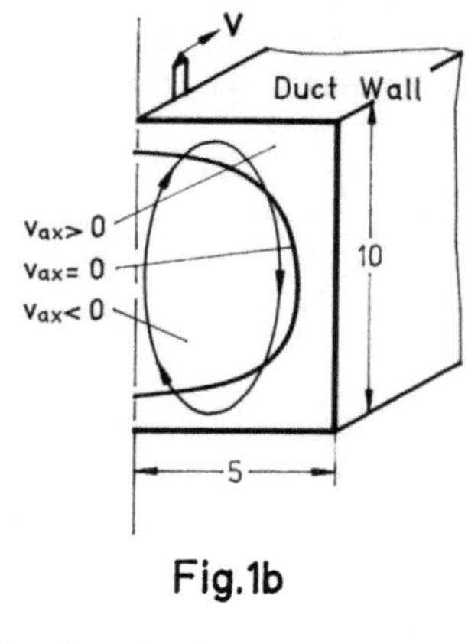

Fig. 1: *Scheme of the ring-shaped duct.*
1a: *Profile of the axial velocity* v_{ax} *and contour of the piston in the centre symmetry plane. If the duct is rotated at moderate velocities the piston is lifted to a stable equilibrium position.* **v** *is the velocity of the duct at its mean radius R=15cm. The direction of* **v** *is marked in all figures by an arrow fixed to a pointer outside the duct. All velocities are observed from the lab system.*
1b: *Regions of different axial velocity directions and principal shape of a streamline due to secondary motion in an azimuthal cross section of the duct. Only half of section up to the centre symmetry plane is displayed.*

Experimental

The experiments to be presented here are performed in a closed transparent circular duct of a mean radius of curvature of R=15cm and a square cross section of D=1cm side length. The cross section of the piston is also of square shape and fills 80% of that one of the duct. To reduce the friction between the duct walls and the piston and to guide the latter one symmetrically inside the duct, teflon spacers are mounted on the piston. As a flow medium glycerol-water mixtures are used in which small tracer particles of about 150µm are suspended. By rotating the duct slowly (≈1 rps) in a vertical plane around its axis of symmetry the fluid is convected by viscous forces and the metal piston is lifted till it reaches a stable equilibrium position. There, the piston is at rest and one has -at least for low Re-numbers- a stationary flow that can be easily observed. Conventional optical methods like PIV and a special visualization technique developed by Bartels et al. [1] that allows for on-line discrimination between "slow" and "fast" traces are applied. Also of importance for the investigation of the topological structure of the flow is the visual inspection of the tracer paths for which a stereo-microscope is used. Since the flow system is closed the wake generated at the front of the body is directly connected with that one emerging at its rear side. In rough approximation the flow structure can be described by an outer zone that is convected with the moving walls and a core region that moves into opposite direction. At the end faces of the piston these two regions are reconnected in a complex way. A schematic view of the duct and a typical velocity profile at some distance of the piston are displayed in Fig. 1.

Results

For low Re-numbers (Re= V·D/ν, with V mean velocity of the duct relative to the piston, ν kinematic viscosity of the liquid) the flow is laminar and stationary. At the back side of the piston where the streamlines change from the core to the wall region a torus shaped separated flow regime appears. As the flow is 3-D due to the curvature of the duct the dead-water is not completely closed. Some of the more outer core streamlines enter it at the inner perimeter of

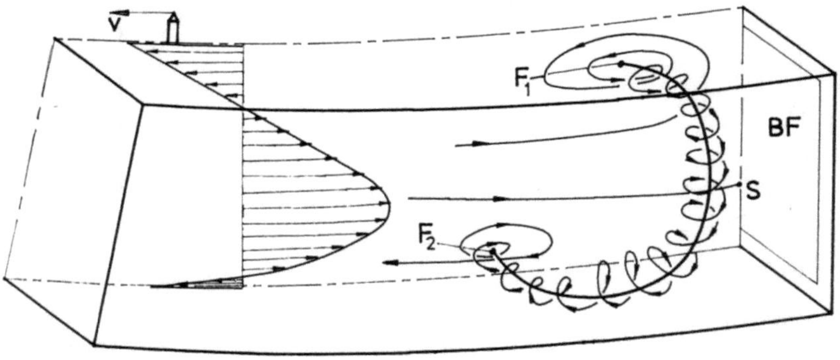

Fig. 2: *Flow structures behind the back face* **BF** *of the piston. For low velocities streamlines spiral around the torus shaped vortex core from focus* $\mathbf{F_1}$ *to focus* $\mathbf{F_2}$. *Foci and stagnation point* **S** *lie on the centre symmetry plane marked by a dashed line. Centre downstream axial velocity profile is not to scale. Only one half of the duct section is displayed.*

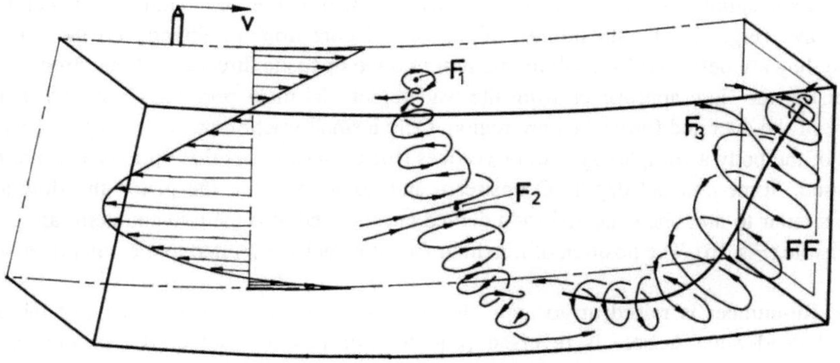

Fig. 3: *Flow structures ahead of the front face* **FF**. *Focal point* $\mathbf{F_1}$ *lies on the center symmetry plane,* $\mathbf{F_2}$ *and* $\mathbf{F_3}$ *are shifted towards the side wall of the duct. Upstream centre axial velocity profile not to scale.*

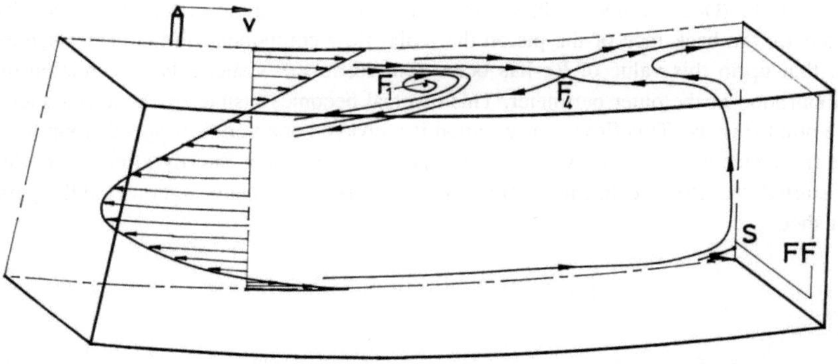

Fig. 4: *Flow structures on the centre symmetry plane ahead of the front face.* $\mathbf{F_4}$ *is a saddle point*

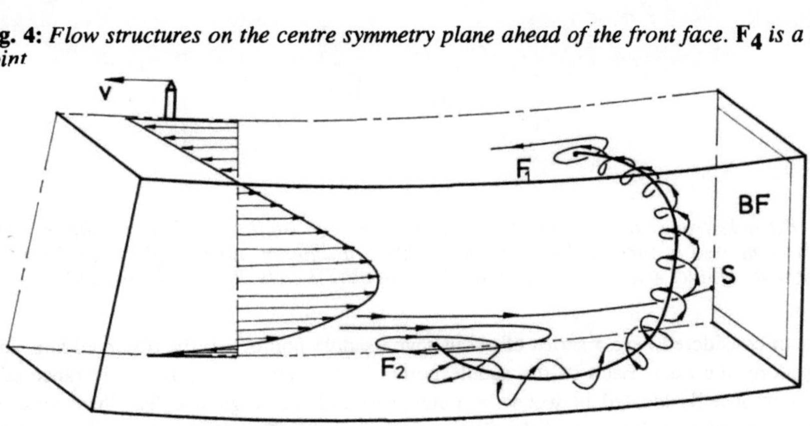

Fig. 5: *Flow structures behind the back face for Re > 340. Streamlines spiral around from focus* $\mathbf{F_2}$ *to focus* $\mathbf{F_1}$. *The stagnation point* **S** *is shifted closer to the outer wall of the duct.*

the duct and spiral around the axis of the torus into the direction of the outer duct wall where they leave (Fig. 2). The streamlines of the central core impinge directly on the rear face where they are deflected by 180° in one turn into the opposite direction. On the front side of the piston the flow approaches from the wall region. Its main portion is reversed into the centre of the duct and forms the core region. Only a small fraction passes through the gap. In front of the body a complex system of vortices (Figs. 3,4) appears that up to now is not fully analysed. Some channel depths D upstream and downstream of the piston the flow looks quite similar to that one expected for a curved channel and also the two counterrotating cells are visible (Fig.1b). The position of maximum axial velocity is shifted to the outer duct wall.

If the Re-number is raised up to appr. 340 the flow direction along the axis of the torus shaped dead-water bubble is reversed (Fig. 5). For this Re-number the location of the maximum of axial velocity and consequently also the position of the stagnation point are now so close to the outer wall of the duct, that the direction of the radial velocity components changes. The structures in front of the piston seem to remain unaffected. The transition between these two states shows hysteresis.
At Re $\approx$ 1140 (Dean-number $\approx$ 208) the flow becomes oscillatory. Some channel depths downstream the back face of the piston the radial flow component on the centre symmetry plane, that up to this value of Re has been positive, reverses suddenly its direction in the neighbourhood of the outer perimeter. This reversal becomes visible by the appearance of a saddle point (Fig. 6). This flow configuration is conveyed at a very low speed along the duct to the front face of the piston where it disappears in the wake. Shortly after this cycle has been started the flow reattaches again and after some moments a new saddle point is established.

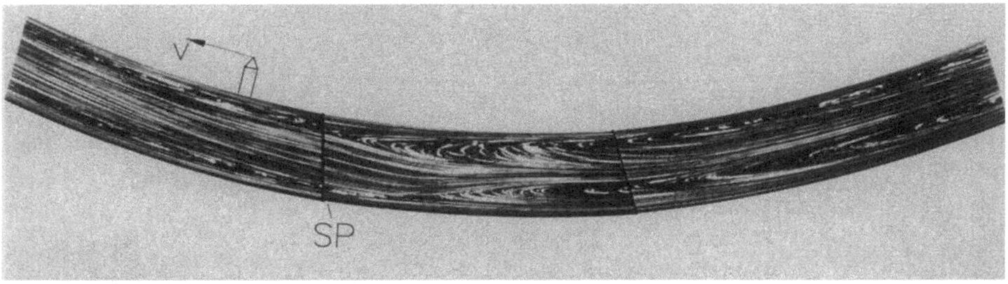

Fig. 6: *Particle traces in the centre symmetry plane for oscillatory flow conditions. The image is composed of three individual photos taken at different times at the same location. Mean velocity of duct wall 75.3cm/s, of saddle point* **SP** *7.2 cm/s. Re=1350, De=247.*

In the duct considered here a series of about seven saddle points may be observed at a time moving along the duct. Each of the saddle points is accompanied by the appearance of a second pair of cells located between the outer wall and the original cells. Their sense of rotation is opposite to that one of the first pair displayed in Fig. 1b. On the inner wall the velocity direction remains unchanged.

For the laminar flow through a curved duct at rest the stationary and unstationary four cell mode is well known and has been investigated experimentally by e.g. Hille et al. [2] and Ravi Sankar et al. [3] and theoretically e.g. by Winters [4]. In a rotating duct only the unsteady case relative to the piston has been observed. A further increase of Re results into a more and more stochastic generation of saddle points and finally ends up in disordered motion. Whether there are additional periodic or even stationary states is not known at the moment. At present, we are planning a more detailed investigation of the 3-D structure of the oscillatory mode and a survey of the higher Re-number regime.

References

1. Bartels-Lehnhoff, H.-H.; Hiller, W.J.; Kowalewski, T.A.: Computer-aided discrimination of slow and fast tracer paths. Experiments in Fluids **13**, (1992) 239-248.

2. Hille, P.; Vehrenkamp, R.; Schulz-Dubois, E.O.: The development and structure of primary and secondary flow in a curved square duct. J. Fluid Mech. **151**, (1984) 219-241.

3. Ravi Sankar, S.; Nandakumar, K.; Masliyah, J.H.: Oscillatory flows in coiled square ducts. Phys. Fluids **31** (6), (1988) 1348-1359.

4. Winters, K.H.; A bifurcation study of laminar flow in a curved tube of rectangular cross-section. J. Fluid Mech. **180**, (1987) 343-369.

Flow Measurements on an Axially Symmetric Body at Large Angle of Attack

Patrick Leehey and Yuksel Gur

Department of Mechanical Engineering
Massachusetts Institute of Technology, Cambridge, MA 02139, U.S.A.

Abstract

Computations have meager success in predicting the forces and moments acting on an axially symmetric body at large angles of attack because of error in separation locations. Work was done to provide experimental data for use in improving computational algorithms. Fluid flow measurements were performed at 15° and 10° angles of attack on an axially symmetric body of length L=25.9" and a ratio of length to diameter of L/d = 9.5. In our tests, we used three tunnel speeds, corresponding to length Reynolds Numbers, $Re_L = 0.7 \times 10^6$, 1.3×10^6, and 1.8×10^6.

Experimental Techniques and Results

We used surface fence gauges to determine both axial and cross-flow mean wall shear stresses at two different measurement locations. We also performed wall pressure measurements on the nose and at the shear measurement locations. Flow visualization experiments were performed to determine surface streamline patterns using a TiO_2, kerosene and oleic acid mixture. We also studied the effect of a boundary layer trip, and optimized trip location and its geometry for successful turbulent boundary layer tripping. A new technique was used to make fluctuating axial and cross-flow shear stress measurements by using two phase-matched microphones below the pressure slots on either side of the surface fence.

For the untripped flow, the axial and azimuthal mean shear stresses are even and odd respectively, in azimuthal angle, Φ (see Figure 1). This is true down to very fine scales of measurement, indicating both gauges are capable of very accurate measurements even near points of flow reversal. For the untripped flow, there is a complex separated flow region on the lee side of the body involving large shear stress variations. We believe these large shear stress variations are created by longitudinal vortices with cores close to the body surface. The peaks of large cross-flow mean shear are the azimuthal locations of the longitudinal vortices and the direction of large cross-flow shear indicates the rotation direction of each vortex.

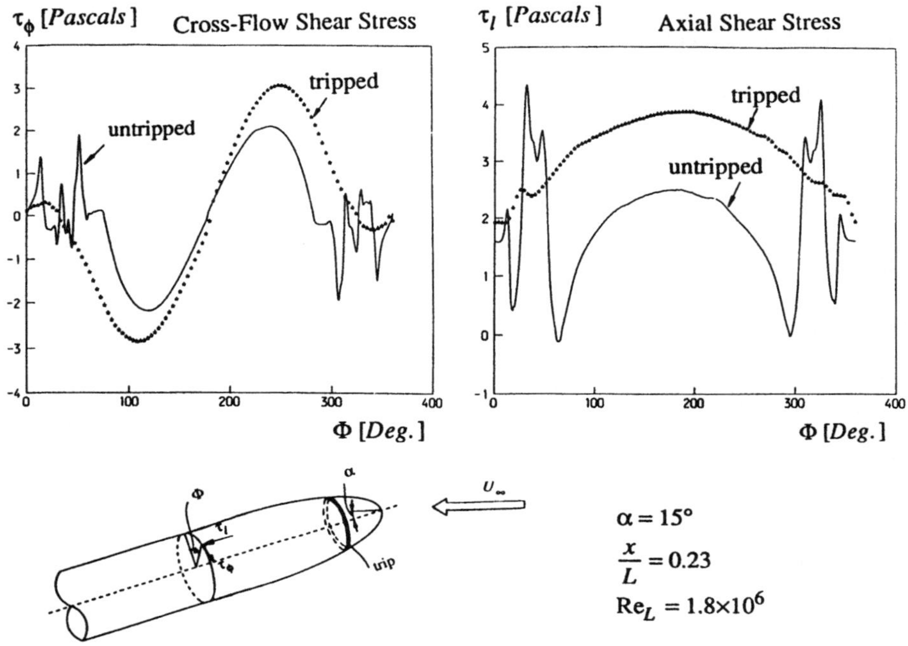

Figure 1. Cross-flow and axial shear stress distributions for both tripped and untripped flows

Figure 2 shows the fluctuating axial wall shear stress spectra at $\Phi = 0°$ with and without a boundary layer trip. Levels of the low frequency components both in the tripped and untripped cases are nearly the same suggesting that these are due to wake unsteadiness. In the tripped case, the fluctuating wall shear spectrum contains additionally a broad band high frequency component typical of fully turbulent boundary layer flow. Fluctuating axial shear stress spectra in the presence of a boundary layer trip at different azimuthal positions have the same shape, indicating that flow is turbulent on the leeward side as well as windward side.

Cross-spectral measurements between two shear stress gauges on opposite sides of the body demonstrate that the low frequency spectra (in the order of 30 Hz in our experiments) correlated completely around the test body (see Figure 3).

By measurement we have determined that this is not vibration effect, but is presumably due to the pressure field induced by the unsteady wake.

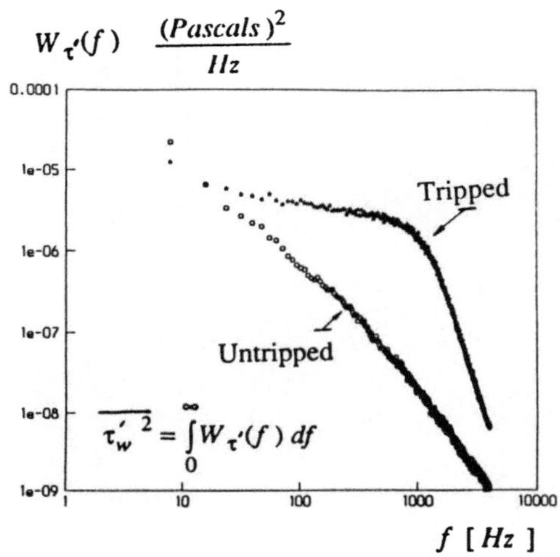

Figure 2. Fluctuating axial shear stress spectra at $\Phi = 0°$

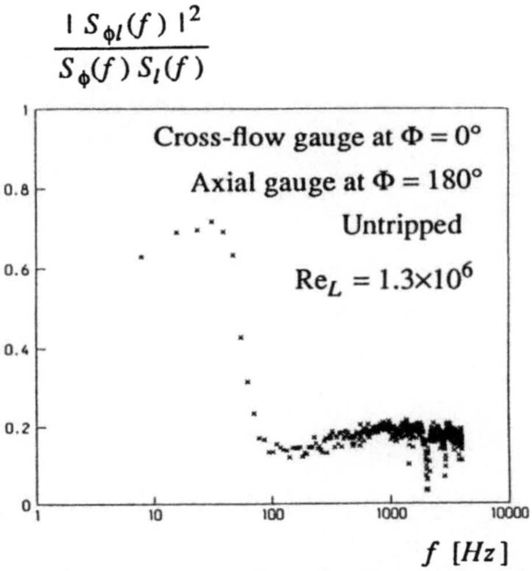

Figure 3. Coherence between axial and cross-flow shear

With one exception the flow visualization photographs and the mean surface shear stress measurements agree even to fine detail in the untripped condition at the forward measurement location. The one exception is that the main cross separation occurs for negative mean cross-flow shear stress. The visualization appears to show a separation on a line where there is a balance of cross-flow shear, gravity, and azimuthal pressure gradient. The latter two are dependent upon film thickness. Figure 4 shows one of our flow visualization photographs with mean cross-flow shear stress data superimposed. We determined that kerosene is not evaporating, but is convecting downstream during flow visualization development.

The fine structures, in the untripped mean wall shear stress measurements, are obliterated by tripping leaving only primary and secondary pairs of cross-flow separations, all of which are the open separation type. With exception of the nodal point of attachment at the nose, no first order surface shear singularities were observed in flow visualization for either tripped or untripped flows. This is somewhat surprising, in light of the results of Bippes and Turk [1]. They performed flow visualization studies on a hemisphere-cylinder body at incidence and found first order singularities such as saddle points and nodes of separation and attachment.

It is possible that unsteadiness of our flow obscured these singularities in the surface flow visualization.

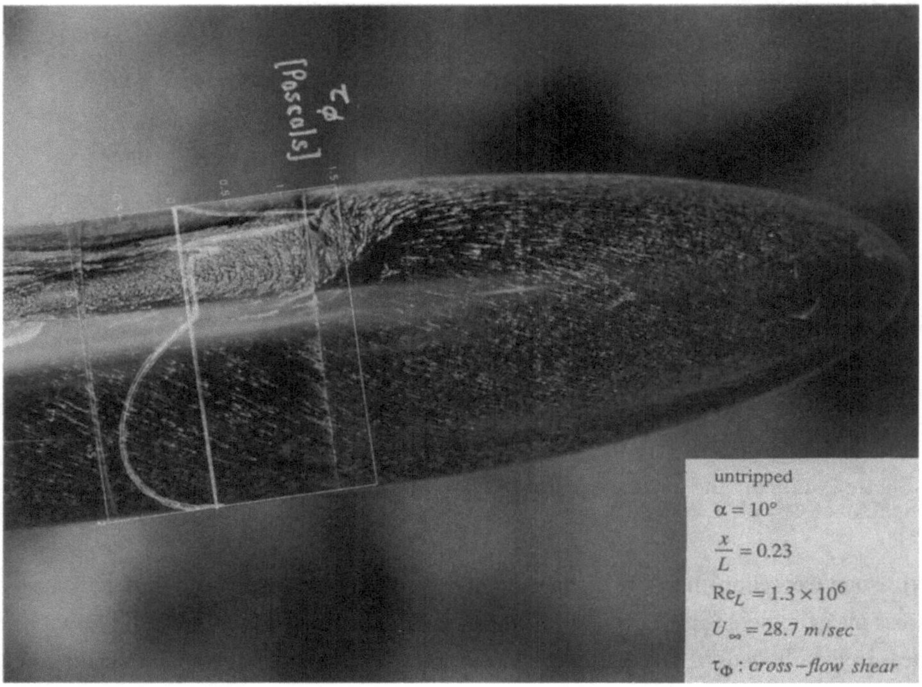

Figure 4. Flow visualization picture with cross-flow shear stress data superimposed

Acknowledgement

Work was supported by the Jason Associates Corporation, the prime contractor to the Defense Advanced Research Agency, and the Electric Boat Division of General Dynamics Corporation.

References

1. Bippes, H. and Turk, M.: Oil Flow Patterns of Separated Flow on a Hemisphere Cylinder at Incidence, DFVLR-FB84-20, Aerodynamicshe Versuchsanstalt Gottingen, 1983.

Numerical Study of Three Dimensional Wake Structure Behind a Yawed Circular Cylinder

Tetuya KAWAMURA and Tsutomu HAYASHI*

NASA Ames Research Center, Moffett Field, CA 94035, USA.
Present address : Department of Computer and Information Science, Chiba University, Yayoi 1-33, Inage-ku, Chiba 263, Japan.

* Department of Mechanical Engineering, University of Nottingham, University park, Nottingham, NG7 2RD, UK. Present address: Department of Mechanical Engineering, Tottori University, Minami 4-101, Koyama, Tottori 680, Japan.

Summery

Incompressible flows around a yawed circular cylinder at Reynolds number 2000 have been calculated by the finite difference method based on the third-order upwind scheme for approximating the nonlinear terms of Navier-Stokes equations. Flows around a perpendicular cylinder are also calculated for comparison. The calculated pressure distributions around the cylinder are compared with experimental data and the agreement is satisfactory. The results of this computation show the details of the flow especially in the wake region.

Introduction

Study of flow around a yawed circular cylinder is very important not only from the academic but also from the practical point of view. Flow around the structures in the ocean and flow in the heat exchangers are the typical examples of such flows. Because of this importance, we can find a number of experimental studies which are concerned with the flow around a yawed circular cylinder. On the other hand, there are few numerical investigations of such flows due to the numerical difficulty caused by the three dimensionality of the flow although numerical studies are very desirable for the better understanding of the flow.

The object of the present study is to numerically simulate the flow around the yawed circular cylinder and to make clear the detail flow structure especially in the wake region where the experimental studies are difficult to carry out in many cases. Special attention is paid to the question whether the two dimensionality is valid even in the wake region, in other words, whether the flows are similar to each other in the cross section parallel to the main flow. If this is valid, it is possible to generalize this flow.

Numerical method

The numerical method used in this study is essentially the same as the well-known MAC method. However, since the Reynolds number of the flow studied here is not low (= 2000), we paid special attention to the nonlinear terms of viscous incompressible Navier-Stokes equations. We used the third order upwind scheme which is successfully used to simulate high Reynolds number flow to approximate nonlinear terms.

We did not use any turbulence model in this study because we could not find suitable models applicable to this complex flow although we are now trying to apply some k-ε like models at higher Reynolds Number.

Since our object is not to study the effect of boundary layer generated by the end-plate, we impose free-slip boundary condition on the end-plate although we impose non-slip condition on the surface of the cylinder. Then we changed yaw angle ϕ of the cylinder to see the effect of the angle on the flow. The distance between two end-plates are 9 times as long as the diameter of the cylinder. Grid number in typical computations are 41 x 54 x 80 in radial, axial and circumferential directions respectively. Fig. 1 shows the typical grids.

Results and Discussion

In order to verify our numerical computations, first, we compare calculated pressure to our measurement. Fig. 2 is one of such

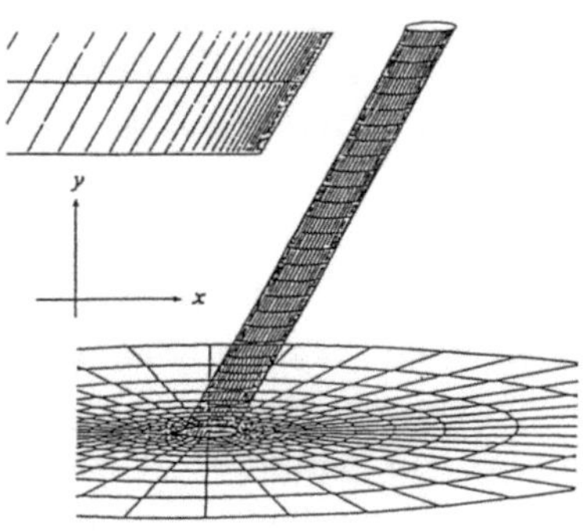

Fig.1 Typical grid system

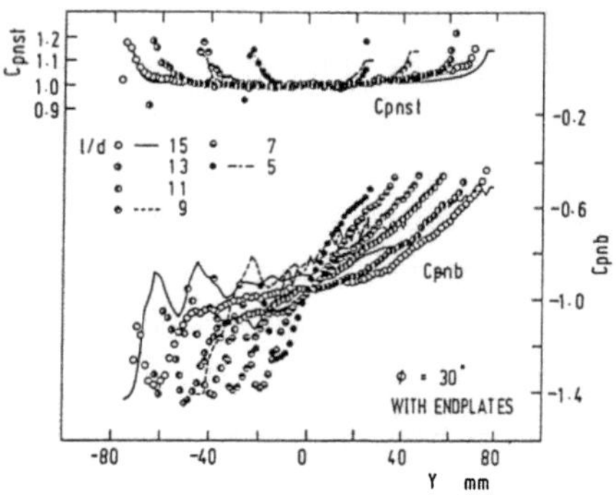

Fig.2 Computed (lines) and measured (symbols) pressure coefficient at $\theta = 0°$ and $180°$ plane.

comparisons showing the pressure of various length cylinder along the axis at $\theta = 0°$ and $180°$ where θ is the angle from the front stagnation point. Lines and symbols are the results of computations and experiments respectively. We can see the agreement is satisfactory. Fig.3 is the time averaged streamlines and Fig.4 is the time averaged velocity vectors in $\theta = 0°$ and $180°$ plane. We can see weak three dimensionality (i.e. weak periodicity in axial direction) even in this infinite length case.

We can see strong three dimensionality for the finite length cylinder. Although the flow is quite complicated for yawed cylinder, we can roughly understand the three dimensional wake

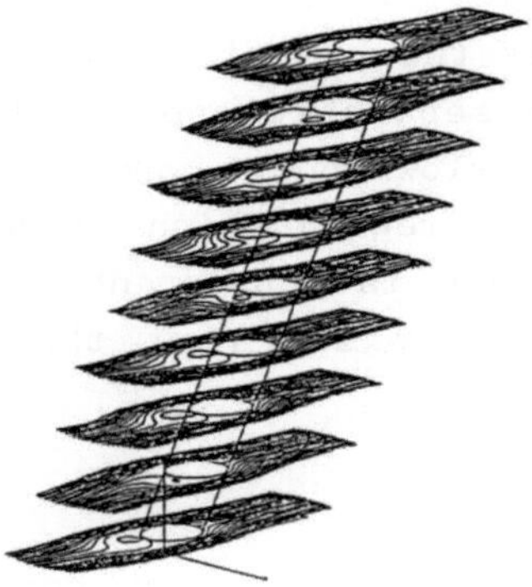

Fig.3 Time-averaged streamlines in various cross sections.

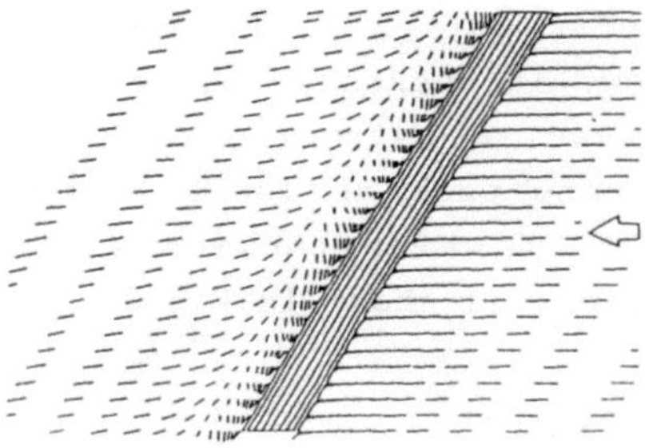

Fig.4 Time averaged velocity at $\theta = 0°$ and $180°$ plane (infinite length cylinder)

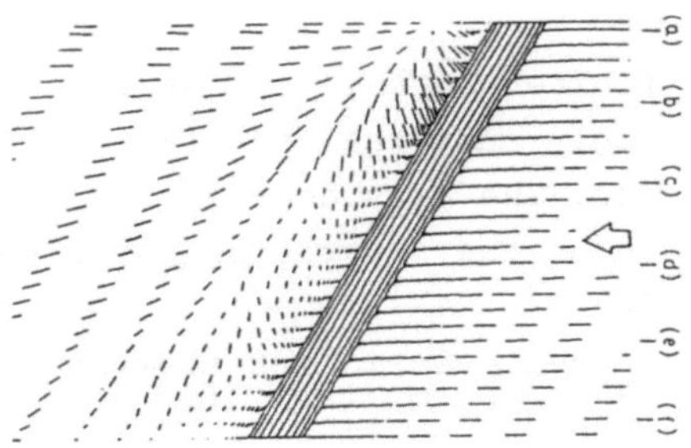

Fig.5 Time-average velocity at $\theta = 0°$ and $180°$ plane (finite length cylinder)

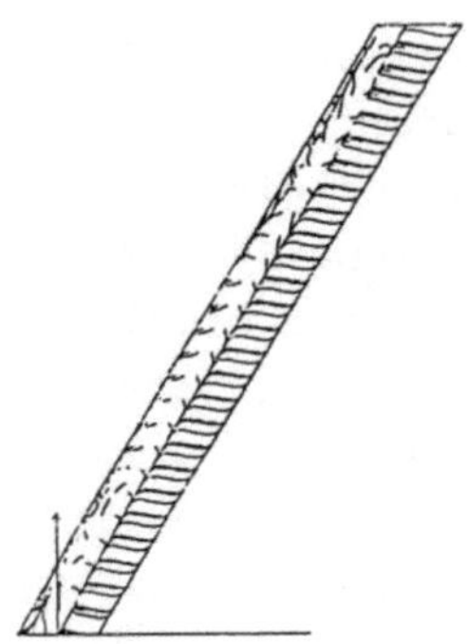

Fig.6 Computed oil flow at the surface of the cylinder (finite length cylinder)

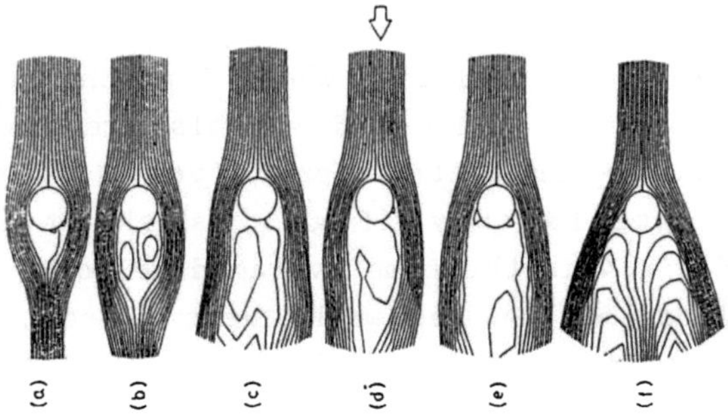

Fig.7 Time-averaged streamline in various cross sections (finite length cylinder)

structure from our calculated results. We can find secondary flow which goes downstream along the cylinder. In order to see this fact more clearly, we show the velocity fields at the plane of $\theta = 0^{\circ}$ and 180° in Fig.5. Fig.6 is the oil flow around the surface of the cylinder showing the separation point of the flow. It suggests the behavior of the flow in the wake region. We can see strong axial flow at upper region which gradually becomes weak at lower region. This fact is consistent to the width (which becomes wider at lower region) of the wake region, is shown in Fig.7.

Conclusion

Although we show only the part of the results, we can conclude as follows from our three dimensional computations around a yawed circular cylinder at Reynolds number 2000:

1) Our computational results show good agreement with the experiments.
2) Two dimensionality is well established for perpendicular cylinder.
3) For yawed circular cylinder of finite length, strong three dimensionality are observed not only near the end-plate but also at the middle of the cylinder. The existence of the end-plate greatly affect the location of the separation point and therefore the wake of the cylinder.
4) Weak three dimensionality can be seen even in the flow around the yawed cylinder of infinite length.

References

1. Harlow, F.H. and Welch. J.E., Phys. Fluid, 8-12 (1965), 2182
2. Kawamura, T. and Kuwahara, K., AIAA paper, 84-0340 (1984).

Vorticity Shedding From a Lentil at Large Incidence in Uniform Flow

G. Riccardi and R. Piva
Dip. di Meccanica e Aeronautica
Universitá di Roma "La Sapienza"
via Eudossiana 18 - 00185 Roma, ITALY

Abstract

We study the $2D$ flow about a lentil shaped body at high incidence in uniform stream at large Reynolds numbers. The separation points are fixed on the body edges and the vorticity released through a Kutta-like condition is spatially discretized by point vortices. Their Lagrangian *inviscid* dynamics is described by using both a *conformal mapping* and a *boundary elements* method.

The effects of different Kutta conditions and of several choice of the generation times are investigated. Different amalgamation procedures in the far wake are used to show that global quantities (*Strouhal number*, shed circulation) are rather sensible to these treatment of the vortex street.

1 Introduction

The flow about a sharp edged body in uniform stream has been investigated both numerically and experimentally, being the simplest separated flow having the separation points in fixed positions. As in previous numerical simulations on this subject, the Reynolds number has been assumed so high that the vorticity dynamics can be considered inviscid and the effects of the viscosity are accounted for implicitly by modelling the flux of vorticity from the boundary layers into the flow field via a Kutta-like condition. In the present work we analyze the flow about a lentil at large incidence and, in order to guarantee that our results do not depend on the specific numerical technique adopted, we compare two different descriptions of the vorticity field: the *conformal mapping* and the *boundary elements* techniques.

The Kutta condition is usually applied in a conformal mapping context by imposing a *finite* velocity value on the body edge. Two ways are possible: given the position of the nascent vortex its strength can be obtained (fixed position metod in [3]); alternatively, given the circulation of the nascent vortex by evaluating the vorticity fluxes shed from the body edge, its position can be defined (variable position method in [5]). In the present work a third form of the Kutta condition is employed in comparison with the previous ones, namely a form more consistent with a boundary integral formulation of the flow field. Moreover, in order to reduce the quadratic growth of the calculation time, some procedure of *amalgamation* of the far wake are frequently used: a certain number of vortices or an entire vortex cluster in a region far from the body is replaced by a point vortex of equal strength in its center of vorticity. The present work shows that this procedure can change substantially the Strouhal number up to a reasonable agreement with the experimental value.

The previous numerical investigation of Kiya and Arie [3] shows that the Strouhal number of the numerical simulation is not in agreement with the old experiment conducted by Fage and Johansen [2], whereas in the results of Sarpkaya the Strouhal number is rather similar to the experimental one but the wake has an unsatisfactory behavior, as underlined in [3]: e. g. the dipole reorganization of the far wake is not observed.

A more recent experimental investigation, for the normal flat plate [1], finds lower values for the Strouhal number, hence more consistent with the computational findings. We expect an analogous result also for the inclined flat plate. Two main features of the model might explain in large part the disagreement between the numerical and the experimental results, namely the way in which the vorticity is generated and its treatment in the far wake region. We analyze both aspects in the present paper. Two furter physical aspects not accounted for by the present semplified model are the action of viscosity and the possible rising of $3D$ secondary motions. They certainly have same influence on the global quantities considered for comparison with the experiments, but different models should be considered for their analysis. In particular a model including the diffusion of the vorticity is under investigation.

The sharp edged lentil is defined by a *Karmann-Trefftz transformation* with a parameter p which maps the unitary circle in the transformed plane into a *biconvex symmetrical lentil* of uniform curvature ($p \in (1,2)$) and into a *flat plate* ($p = 2$) at incidence α with respect to the uniform stream in the physical plane.

As previously mentioned, two different numerical techniques are used to study the vorticity dynamics: the *conformal mapping* and the *boundary elements* methods. In the first approach the potential flow around the lentil is mapped in a transformed plane where the presence of the body (circular cylinder) is considered by introducing the image vortices into the body, the velocity of the vortices are calculated by excluding the self-induced contribution *in the physical plane*. The production of the vorticity is performed by imposing vanishing velocity on the image points of the body edges in the transformed ζ-plane. In the second approach we use a velocity field representation in terms of its boundary values. An integral equation is obtained by taking the limit to the body surface and, after solving this equation, the velocity is known everywhere. The vorticity production is performed by integrating the trajectory of a *neutral particle*, leaving the edge at time t, along a suitable time Δt. Then the particle becomes a vortex and its strength is related to the total vorticity shed from the edge between t and $t + \Delta t$. Further details about the numerical techniques are given in a more extended paper on this subject [4].

In section 2 we consider the influence of the generation mechanism on the Strouhal numbers and some numerical results *without* any treatment of the wake are analyzed. In section 3 we focus our attention on the effects of the amalgamation procedure and we consider for the same flow conditions different values of the massing parameters.

2 Numerical Results and Comparison with Experiments

Our results are described in terms of the *Strouhal numbers* of the wake separation related to the lentil length ($2p$) given by $St = 2pf$ where f is the detachment frequency. The effects of the geometry (thickness, curvature) are described by performing the calculations with different parameters p. At a fixed incidence of $\alpha = 60°$ they can summarized as it follows

Different lentil geometry with incidence of $\alpha = 60°$

p	length	max. thickness	T	f	St_l	St_f
1.5	3	1.243	20.30	0.049	0.148	0.128
1.6	3.2	1.040	21.97	0.046	0.146	0.126
1.7	3.4	0.816	24.20	0.041	0.141	0.122
1.8	3.6	0.570	27.92	0.036	0.129	0.112
1.9	3.8	0.299	30.27	0.033	0.126	0.109
2	4	0	33.19	0.030	0.121	0.104

The decrease of the Strouhal number for larger values of p is due essentially to the increasing length of the body, hence to a larger distance between the two shedding edges. In fact for the same p, if we consider different incidences, we obtain a Strouhal number varying with a sinusoidal law, i. e. approximately the same Strouhal number, if referred to the frontal section height.

The most important result is for the flat plate ($p = 2$) in disagreement with the experimental results of Fage and Johansen [2] giving $St \simeq 0.172$.

We analyze now the effects on the Strouhal number of different vorticity production mechanisms. From the analysis of the numerical simulations we may reach a quite significant conclusion: two *different* conditions as the fixed position method and the generation mechanism usually employed in the boundary elements formulationsgive the same results, both in terms of the total shed circulation and the Strouhal number. In other words the flow is rather insensitive to the way in which the vorticity is generated, while it appears strongly dependent on the details of the vorticity dynamics, also in regions far from the body. Therefore the vorticity generation mechanism doesn't seem to be responsible for the disagreement with the experimental results.

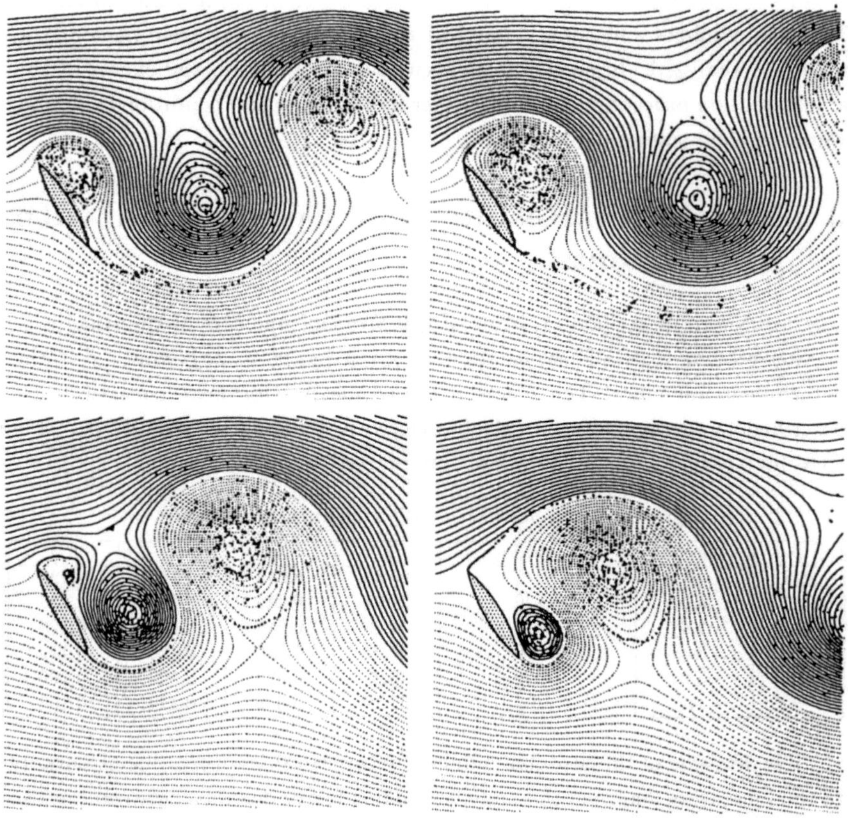

Stream function field for $p = 1.8$ *and* $\alpha = 60°$ *(the time runs clockwise)*

Instead the choice of the ratio λ between the generation time step and the numerical integration step is rather critical: for a flat plate at an incidence of 60° we find for $\lambda = 2$ the same value $St \simeq 0.139$ reported by Kiya and Arie [3] while $\lambda = 1$ gives $St \simeq 0.121$. In most of our simulations we assume $\lambda = 1$

to retain the consistency between the discretizzation of the dynamics and the generation of the vorticity, whereas in the calculation of sarpkaya [5] the value $\lambda = 5$ has been choosen.

Let us consider now another interesting conclusion provided by the numerical simulations: the substitution of a whole cluster far from the body with *one* point vortex leads to unexpected variations of the Strouhal number. The *massing* procedure cam be summarized as follows: fixed a limit distance d_l between two vortices (successively released) as a given multiple of the mean distance d_m (that is $d_l = \beta d_m$ with $\beta > 1$ fixed) we mark all the vortex pairs formed by vortices that are at a distance greater than d_l. We suppose that to each of these pairs corresponds a wake detachment, then a cluster is the set of all vortices generated between of two consecutive pairs of this kind. When the center of vorticity of the cluster has a distance from the origin greater than a fixed one L we substitute the cluster with a single vortex, placed in its center of vorticity, having the same circulation of the whole cluster. The application of this procedure (with $\lambda = 2$ in order to compare with [3]) gives for the flat plate ($p = 2$) at an incidence $\alpha = 60°$ a Strouhal number varying from the value of 0.139 for $\beta = 0$ (i. e. without massing) up to a value of 0.169 for $\beta = 4$. Hence by an *adequate* calibration of the massing procedure it is possible to obtain Strouhal numbers up to values matching with the experimental ones.

This behavior of the numerical flow suggests that the far wake dynamics plays an important role in the periodic wake detachment. If this behavior is supported by a physical meaning, like for example the merging of equal signed vortices, the effects of diffusion should be included in the model for a better description of the real motion in the far wake.

3 Concluding Remarks

In this work we have analyzed the application of the *Vortex Methods* for the study of the two-dimensional flow past a sharp edged body at large incidence. Whichever approximated model is adopted for the flow field and for the vorticity dynamics, is very important to retain the consistency of the time step between the generation and the dynamics of the vorticity. On the one hand the numerical results show that the modellization of the vorticity shedding from the body edges with a suitable "*Kutta condition*" has a weak influence on the resulting flow field. On the other hand any *amalgamation* procedure of the far wake, i. e. the *massing* a cluster in one point vortex, frequently used as a numerical tool, may leads to large variations of the Strouhal number. Finally it appears necessary, for a deeper evaluation of these methods, to investigate the effect of viscosity in terms of *secondary separations* and *amalgamation* of the vorticity clusters: these phenomena, that are completely neglected by the present model, might have a large effect on the Strouhal number. Also the presence of relevant $3D$ motions, shown by the experimental vortex street, could play a significant role in the periodic wake separation.

References

[1] Chua, K.; Lisoski, D.; Leonard, A.; Roshko, A. *A numerical and experimental investigation of separated flow past an oscillating flat plate*, Proceedings of the International Symposium on Nonsteady Fluid Dynamics, **92** (1990) 455 – 464

[2] Fage, A.; Johansen, F.C. *On the flow of air behind an inclined flat plate of infinite span*, Proc Roy. Soc., A **116** (1927)

[3] Kiya, M.; Arie, M. *A contribution to an inviscid vortex-shedding model for an inclined flat plate in uniform flow*, J. Fluid Mech., **82** part 2 (1977) 223 – 240

[4] Riccardi, G.; Iafrati, A.; Piva, R. *Vorticity dynamics in the wake of a lentil at large incidence*, submitted to Meccanica (1992)

[5] Sarpkaya, T. *An inviscid model of two-dimensional vortex shedding for transient and asymptotically steady separated flow over an inclined flat plate*, J. Fluid Mech., **68** part 1 (1975) 109 – 128

The Influence of a Downstream Body on Wake Formation Behind a Bluff Body

M.K. BULL and J.M. PICKLES

Department of Mechanical Engineering
University of Adelaide
Adelaide S.A. 5000, Australia

Summary

The stability of the shear-layers separating from the trailing edge of a thick plate is shown to be strongly affected by the presence of a downstream plate. Under some conditions of plate spacing and Reynolds number vortex formation can be inhibited. The plate spacing at which this occurs is found to be strongly dependent on Reynolds number.

Introduction

Following earlier work on the general character of vortex flows over various arrays of flat plates [1,2], an investigation has been made of vortex-wake formation behind a thick plate with an elliptic leading-edge profile, under the influence of a rectangular-section downstream plate of the same thickness, as shown in figure 1. The vortex-generating plate and the downstream plate had ratios of chord c to thickness t of $c_1/t = c_2/t = 6.1$. The range of Reynolds number based on plate thickness and free-stream velocity, $Re_t = U_1 t/\nu$, in the experiments was $500 < Re_t < 8 \times 10^4$. The boundary layers separating from the vortex generator were in all cases laminar. Streamwise velocity fluctuations near the outer edge of the separated shear-layer were measured by means of a hot-wire anemometer. Tests at the lower end of the Reynolds number range were made in a low-speed wind tunnel, and those at the higher end in a free-jet rig. The tests show that two quite different regimes of flow over the plate combination can occur, depending on the Reynolds number and the ratio $G = g/t$ of the gap-width g between the plates to the thickness. These are identified as the vortex-street regime and the trapped-vortex regime.

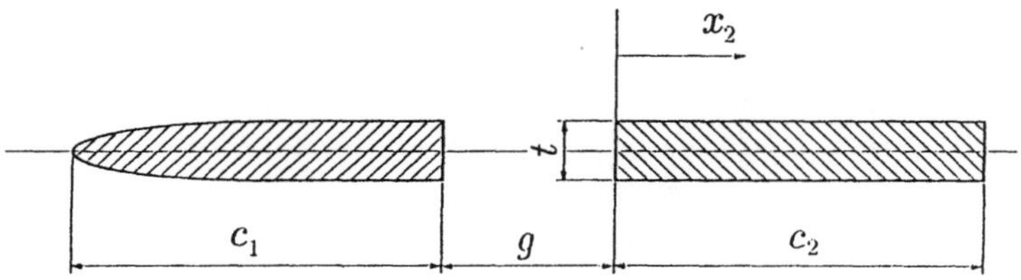

Figure 1 Plate Geometry.

Vortex-Street Regime

For the leading plate in isolation, the wake has the form of a well-defined vortex street. Introduction of the second plate at a very large distance downstream does not change the character of the wake, and, as the gap between the plates is reduced, the same general character persists until G falls to a value in the region of 3. Flow visualisation shows that in this regime vortices are shed alternately from the top and bottom corners of the square trailing edge of the upstream plate. Vortices formed in the separated shear-layer from the top of the upstream plate impinge on the leading edge of the downstream plate and then pass over its upper surface, and similarly vortices from the bottom of the upstream plate pass over the lower surface of the downstream plate. In this regime the spectrum of velocity fluctuations in the separated shear-layer exhibits a single large dominant peak at the vortex-shedding frequency.

Trapped-Vortex Regime

As G is reduced, the Strouhal number of vortex-shedding falls slowly and progressively. The fall becomes more rapid as G approaches a value of about 3. (The form of variation for $Re_t > 13\times10^3$ is shown later in figure 5.) For G in the range of 2 to 3, vortex formation is suppressed and the nature of the near-wake velocity-fluctuation spectrum changes dramatically: the single dominant peak gives way to a large number of peaks at frequencies which are harmonics of a fundamental frequency f_1. This behaviour is similar to that observed by Unal and Rockwell [3] in their experiments with a wedge-shaped body in the wake of a circular cylinder. The spectral peak with the largest amplitude is in general at a higher frequency than that of the fundamental. Flow visualisation shows that in this state the flow is characterised by two large counter-rotating standing vortices in the gap between the two plates, and that the shear- layer which separates from the upstream plate reattaches to the downstream plate without vortex formation occurring.

Transition between Flow Regimes

From tests at various values of G over the range of Reynolds number achievable in the wind-tunnel, the form of the boundary between these two regimes in the $Re_t - G$ plane has been established (figure 2). The location of the boundary depends on whether the boundary is approached from the vortex-shedding regime, shown as a full line in figure 2, or from the trapped-vortex regime, shown by a broken line in figure 2. At fixed G, transition from the vortex-street to the trapped-vortex regime with decreasing Reynolds number is sharply defined. For the reverse transition, the trapped-vortex regime persists to a higher Reynolds number, but the flow then appears to be in a metastable state and sensitive to small perturbations. It appears that,for combinations of G and Re_t lying in the region below the boundary, the separated shear-layer is subject to the instabilities which lead to vortex formation in an unimpeded wake, but without vortex formation taking place. The apparent discontinuity in the boundary between $G = 2.8$ and 2.9 is at present unexplained.

The transition between the two flow regimes is also reflected in the variation of surface-pressure

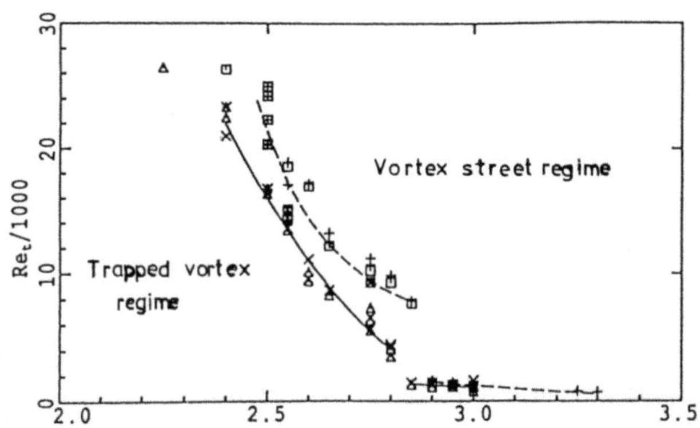

Figure 2 Boundaries between shedding and no-shedding vortex regimes. - - - Re_t increasing: ⊓ Trapped-vortex regime, + Vortex-street regime; — Re_t decreasing: △ Trapped-vortex regime, × Vortex-street regime.

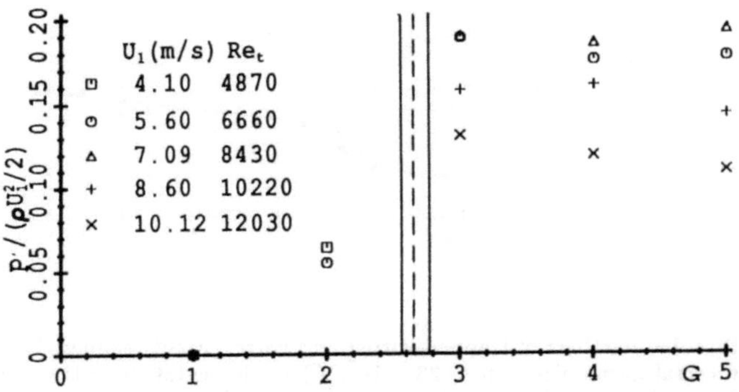

Figure 3 Dependence of rms pressure fluctuation on downstream plate on gap width. $x_2/c_2 = 0.083$. Vertical lines indicate position of transition between flow regimes at $Re_t = 12.0 \times 10^3$ (left-hand pair of lines) and $Re_t = 4.9 \times 10^3$ (right-hand line), from figure 2.

fluctuations on the downstream plate with the spacing between the plates. Figure 3 (from [4]) shows the rms pressure fluctuation p' (in terms of the free-stream dynamic pressure q) at $x_2/c_2 = 0.083$. There is a rapid rise in p'/q as the gap approaches the value at which vortex formation takes place, followed by a slow fall when this value is exceeded.

Instability and Vortex-Shedding Frequencies

The variation of the frequencies of various spectral peaks with flow velocity in the trapped-vortex regime is shown in figure 4 for $G = 2.5$. At this particular gap-width, vortex-shedding at the higher Reynolds numbers takes place at essentially the same Strouhal number ($St = ft/U_1$) as that corresponding to the fundamental instability frequency f_1 at lower Reynolds numbers. But, for smaller gap-widths, as can be seen from figure 5, there is a significant difference between the Strouhal number of the low-Reynolds-number instability and that of the high-Reynolds-number vortex-shedding.

Further, in the trapped-vortex regime, the peak in the velocity-fluctuation spectrum at the fundamental instability frequency is in general quite small. A much larger spectral peak occurs at a somewhat higher frequency f_x, which is not related to the sequence of harmonics of the fundamental instability frequency in any obvious way. The dependence of f_x on flow velocity for

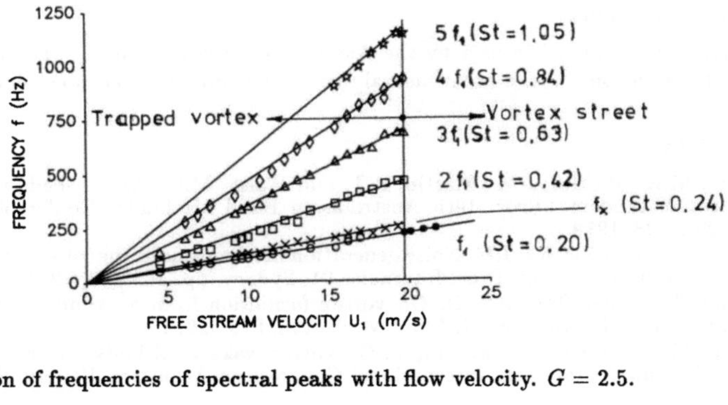

Figure 4 Variation of frequencies of spectral peaks with flow velocity. $G = 2.5$.

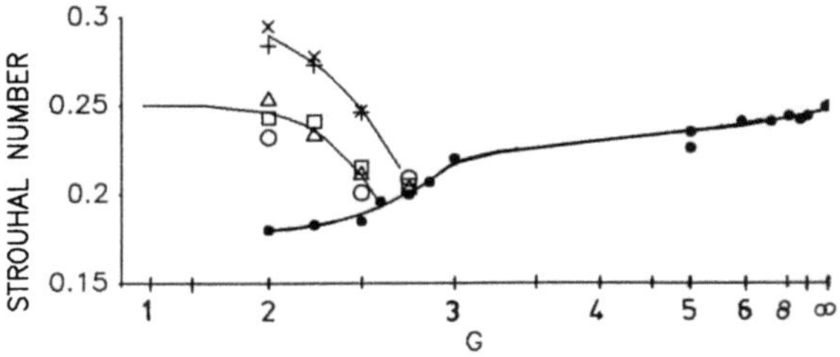

Figure 5 Strouhal numbers of trapped-vortex and vortex-street regimes. No vortex shedding, spectral harmonics, $Re_t < 28 \times 10^3$: ○ fundamental shear-layer instability; ⊓ first harmonic (÷2); △ second harmonic (÷3); × f_x; + first f_x harmonic (÷2). Vortex shedding, no spectral harmonics, $13 \times 10^3 < Re_t < 80 \times 10^3$ •.

$G = 2.5$ is shown in figure 4. At smaller gap-widths, harmonics of this frequency also occur. Figure 5 also shows Strouhal numbers corresponding to f_x and its harmonics, as a function of G. The origin of the additional strong low-frequency peak is not at present understood.

Conclusions

The conclusions which can be drawn from the work which has been done to date are as follows:

1. The stability of the shear-layers separating from the blunt trailing edge of a thick plate which has a faired leading edge can be significantly modified by the introduction of a downstream plate of the same thickness.
2. If the gap between the two plates is large (typically G greater than about 3 to 4), vortex formation still occurs with the same character as that from the isolated plate, but the Strouhal number of shedding is reduced.
3. If the gap between the two plates is small (typically G less than about 3) and the Reynolds number sufficiently low, vortex formation is suppressed.
4. The gap size at which transition takes place from a wake which persists as a disturbed separated free shear-layer to a wake in which the shear-layer instability leads to vortex formation is strongly dependent on Reynolds number.
5. Spectral evidence indicates that, for a given gap width, the Strouhal number of the fundamental instability in the separated shear- layer when vortex formation is suppressed is greater than the Strouhal number of vortex shedding which occurs at higher Reynolds number.

Acknowledgements

This research is funded in part by the Australian Research Council. The assistance of Mr N. Petticrew with some of the experimental work is also gratefully acknowledged.

References

1. Bull, M.K., Pickles, J.M., Martin, B.T. and Welsh, M.C. Vortex shedding from rectangular plates in tandem. Proc. 10th Australasian Fluid Mechanics Conference, Melbourne, pp. 11.25-11.28, 1989.
2. Bull,M.K. and Pickles, J.M. Noise generation mechanisms in the interaction between a vortex wake and a bluff body. Proc. Internoise-91, Sydney, pp. 403-406, 1991.
3. Unal, M.F. and Rockwell, D. On vortex formation from a cylinder. Part 2. Control by splitter-plate interference. J. Fluid Mech. 190, 513-529, 1988.
4. Bull, M.K.,Pickles, J.M. and Li, Y. On vortex-wake/bluff-body interactions and noise from circular saws. Proc. 14th International Congress on Acoustics, Beijing, Sept. 3-10,1992.

Numerical Simulation of a Steady Flow Behind a Body of Revolution at Moderate and High Reynolds Numbers, Using a Spectral Method

Yoshihiro MOCHIMARU

Dept. of Mechanical Engineering, Tokyo Institute of Technology, Tokyo 152, Japan

Summary

Steady-state axisymmetric laminar flow fields past an ellipsoid of revolution are analyzed numerically. Using a Fourier spectral method through substantially doubly exponential transformation in space, the equation of vorticity transport can be integrated to give a steady-state solution through a time marching way even for relatively high Reynolds numbers. Not only drag coefficients and streamlines but also isobars are presented.

Introduction and Analysis of a Flow Field

Numerical analyses for intermediate Reynolds number cases were treated elsewhere, e.g. by Pitter et al.[1]. Here, it is assumed that the velocity vector sufficiently far from a body of revolution [an oblate spheroid, a sphere, or a prolate spheroid], i.e. the undisturbed velocity vector, is uniform and parallel to the symmetrical axis, and that the fluid is incompressible, Newtonian, and of constant viscosity. The velocity field treated is limited to an axially symmetric one with respect to the symmetical axis and is assumed to be laminar. Hereafter, length, velocity, pressure, and time are assumed to be made dimensionless with respect to a (1/2 of the symmetric axis), U_∞ (free stream velocity), ρU_∞^2, and a/U_∞ respectively. Using a cylindrical polar coordinate system (the z -axis being coincident with the symmetric axis of the body, the r -axis being perpendicular to the z -axis, θ being the azimuth angle),the vorticity vector can be given through the only one non-zero θ -component, ζ. In addition, to match large variation of velocity near the surface, introduced is a boundary fitted coordinate (α, β) [$\alpha = \alpha_0$ corresponds to the body surface] specified by the transformation of the type $z + ir = f(\alpha + i\beta)$ in the $z - r$ plane, where f is a complex function, either one of $(1/\cosh\alpha_0)\cdot\cosh$, $\exp(-\alpha_0)\cdot\exp$, or $(1/\sinh\alpha_0)\cdot\sinh$, according to the prolate speroid ($\alpha_0 = \tanh^{-1}(b/a)$), the sphere($\alpha_0 = 0$), or the oblate spheroid($\alpha_0 = \tanh^{-1}(a/b)$) respectively, and b is 1/2 of the axis normal to the z -axis. Thus the equation of vorticity transport is expressed with the aid of an axially symmetric stream function ψ as

$$\left\{\left(\frac{\partial r}{\partial\alpha}\right)^2 + \left(\frac{\partial r}{\partial\beta}\right)^2\right\}\frac{\partial\zeta}{\partial t} + \frac{\partial(\zeta,\psi/r)}{\partial(\alpha,\beta)} + \frac{\partial}{\partial\alpha}\left(\frac{1}{r}\frac{\partial r}{\partial\beta}\zeta\frac{\psi}{r}\right) - \frac{\partial}{\partial\beta}\left(\frac{1}{r}\frac{\partial r}{\partial\alpha}\zeta\frac{\psi}{r}\right)$$

$$= \frac{1}{Re}\left\{\left(\frac{\partial^2}{\partial\alpha^2} + \frac{\partial^2}{\partial\beta^2}\right)\zeta + \frac{\partial}{\partial\alpha}\left(\frac{1}{r}\frac{\partial r}{\partial\alpha}\zeta\right) + \frac{\partial}{\partial\beta}\left(\frac{1}{r}\frac{\partial r}{\partial\beta}\zeta\right)\right\} , \tag{1}$$

supplemented with

$$-\left\{\left(\frac{\partial r}{\partial\alpha}\right)^2 + \left(\frac{\partial r}{\partial\beta}\right)^2\right\}\zeta = \left(\frac{\partial^2}{\partial\alpha^2} + \frac{\partial^2}{\partial\beta^2}\right)\left(\frac{\psi}{r}\right) + \frac{\partial}{\partial\alpha}\left(\frac{1}{r}\frac{\partial r}{\partial\alpha}\frac{\psi}{r}\right) + \frac{\partial}{\partial\beta}\left(\frac{1}{r}\frac{\partial r}{\partial\beta}\frac{\psi}{r}\right) . \tag{2}$$

In Eq.(1) $Re \equiv U_\infty a/\nu$ (a Reynolds number.) Asymptotic solutions for ψ and ζ in the wake under a steady-state [$\beta = 0$ corresponds to the wake side] are given by

$$\psi(\alpha_\infty,\beta) = \frac{1}{2}r^2 - \frac{C_D}{2\pi}\left\{\frac{1+\cos\beta}{2} - \exp\left(-Re\, R\sin^2\frac{\beta}{2}\right)\right\} , \tag{3}$$

$$\zeta(\alpha_\infty,\beta) = -\frac{Re^2\, C_D}{8\pi R}\sin\beta\,\exp\left(-Re\, R\sin^2\frac{\beta}{2}\right) , \tag{4}$$

$$C_D = \frac{\pi}{Re}\left[\int_0^\pi r^3\frac{\partial}{\partial\alpha}\left(\frac{\zeta}{r}\right)\mathrm{d}\beta\right]_{\alpha=\alpha_0} , \tag{5}$$

where C_D is a drag coefficient based on $\rho U_\infty^2 a^2$, $R = r/\sin\beta$,and $\exp\alpha_\infty \gg 1$. As boundary conditions, no slip flow on the surface is used:

$$\frac{\partial\psi}{\partial\alpha}(\alpha_0,\beta) = 0 \quad , \quad \psi(\alpha_0,\beta) = 0 \ . \tag{6}$$

The former is approximated by

$$\frac{2}{h^2}\left(\frac{\psi}{r}\right)_{\alpha=\alpha_0+h} + \left\{\left(\frac{\partial r}{\partial\alpha}\right)^2 + \left(\frac{\partial r}{\partial\beta}\right)^2\right\}_{\alpha=\alpha_0} \times (\zeta)_{\alpha=\alpha_0+h} = 0 , \tag{7}$$

where $\alpha_0 + h$ is the coordinate at the adjacent grid to the surface.

Numerical Solution Procedure

A Fourier spectral finite difference scheme is applied to Eqs. (1) and (2) (including boundary conditions) to decompose them into Fourier components; in this case ζ and a combined variable $\psi^*(\equiv \psi/r)$ can be expressed as

$$\psi^* = \sum_{n=1}^{\infty}\psi_n(\alpha,t)\ \sin n\beta \ , \qquad \zeta = \sum_{n=1}^{\infty}\zeta_n(\alpha,t)\ \sin n\beta \ , \tag{8}$$

and use is made of the addition formulae of trigonometric functions etc. Then, the set of decomposed equations can be discretized in time and space by a finite difference approximation, using the following non-uniform grid spacing in α. [The series in Eq.(8) can be truncated up to a certain order.] That is, the coordinate α_n for the n -th grid point is defined as

$$\alpha_n = \alpha_0 + h\left\{\frac{\sinh\ (n-1)\gamma}{\sinh\gamma} + 1\right\}, \tag{9}$$

where γ is a suitable positive constant. This results in substantially doubly-exponential transformation in the physical space. By assuming, as an initial field, a developing boundary-layer flow field superimposed on a potential flow, such discretized equations can be integrated implicitly with respect to time to get a steady-state solution .

Numerical Results

Three examples of the relation between the drag coefficients C_{DN} [beased on the dynamic pressure and the projection area ; $= (2/\pi)(a/b)^2 C_D$] and the Reynolds number Re_D are shown in Figs.1, 2, and 3 for b/a =1 (a sphere), 5/9 (a prolate spheroid), and 1/6 (a prolate spheroid) respectively, where $Re_D(\equiv 2(b/a)\times Re)$ is based on U_∞ and $2b$. Pressure p itself can be obtained from the following total differential equation:

$$\mathrm{d}\left[p + \frac{1}{r^2}\left\{\left(\frac{\partial\psi}{\partial r}\right)^2 + \left(\frac{\partial\psi}{\partial z}\right)^2\right\}\right] = -\frac{\zeta}{r}\left(\frac{\partial\psi}{\partial z}\mathrm{d}z + \frac{\partial\psi}{\partial r}\mathrm{d}r\right) + \frac{1}{Re}\left\{\frac{\partial\zeta}{\partial z}\mathrm{d}r - \left(\frac{\partial\zeta}{\partial r} + \frac{\zeta}{r}\right)\mathrm{d}z\right\} . \tag{10}$$

Pressure distribution along the surface (meridian) of a sphere at a supercritical state is shown in Fig.4. In case of $b/a = 5$, streamlines and pressure contours at $Re_D = 300$ are shown in Figs.5 and 6 respectively. Characteristics of wake length for $b/a = 5$ and 50 are shown in Fig.7. From Figs.1 and 3, it is inferred that flow features change at $Re_D \approx (400 \sim 800)$, above which nature of a bundary layer becomes remarkable. In case of $b/a = 5$, it is very hard to get steady-state solutions for larger Reynolds numbers than shown in Fig.7.

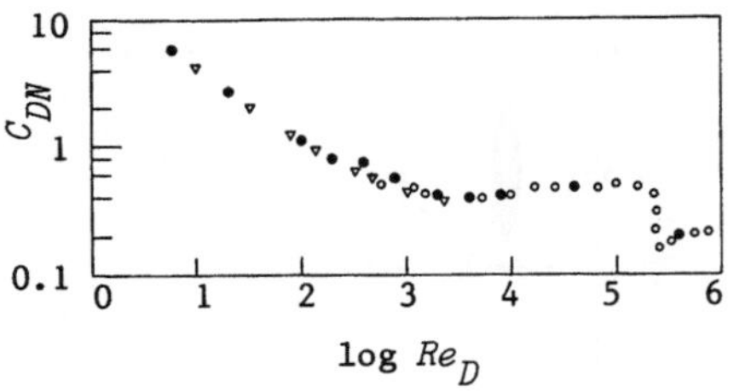

Fig.1. Drag coefficients for a sphere ($b/a = 1$). ●: present numerical data, ○: experimental data by Wieselsberger[2], ▽: experimental data by Liebster[3].

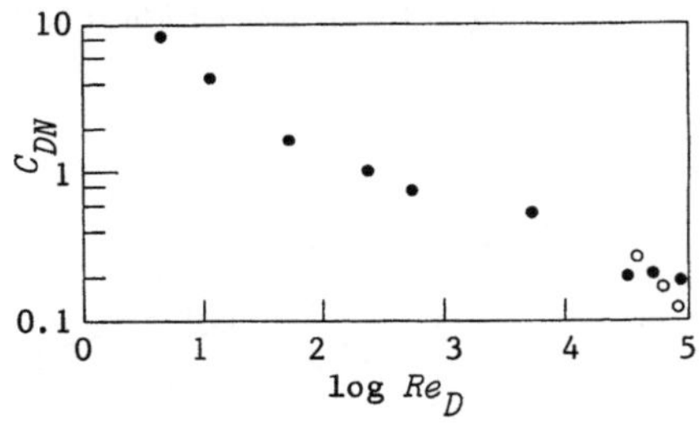

Fig.2. Drag coefficients for a prolate spheroid of $b/a = 5/9$. ●: present numerical data, ○: experimental data by Zham et al.[4].

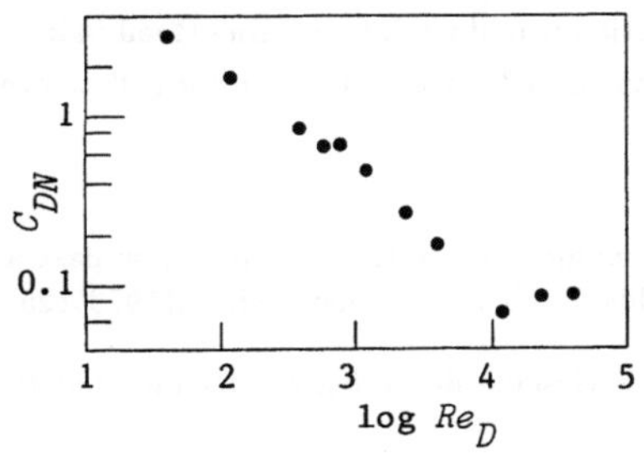

Fig.3. Drag coefficients for a prolate spheroid of $b/a = 1/6$. ●: present numerical data.

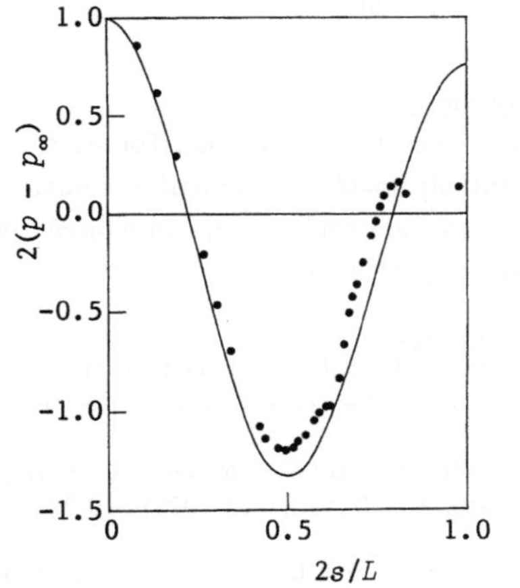

Fig.4. Pressure distribution along the surface of a sphere ($b/a = 1$). —: present numerical data ($Re_D = 400,000$), ●: experimental data by Fage[5] ($Re_D = 424,500$). s: arc length measured from the forward stagnation point, L : perimeter, p_∞: free stream pressure.

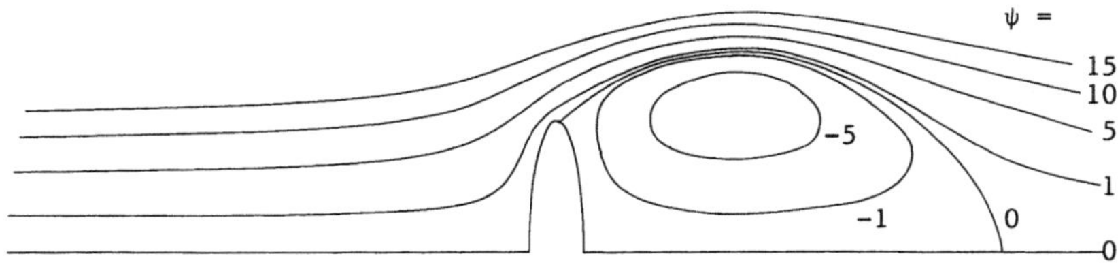

Fig.5 Streamlines for $b/a = 5, Re_D = 300$. Fluid flows from left to right.

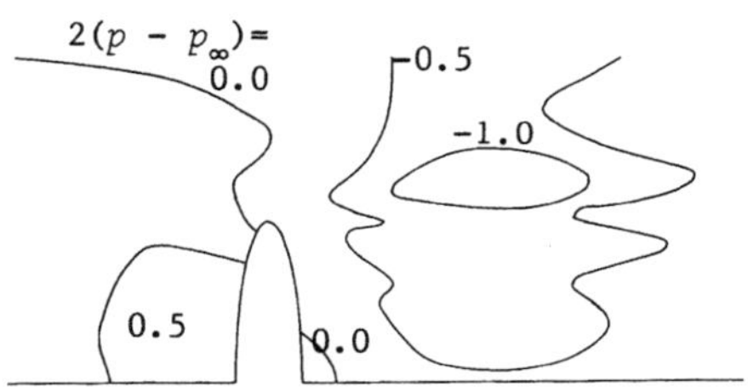

Fig.6. Pressure contours for b/a=5, $Re_D = 300$.

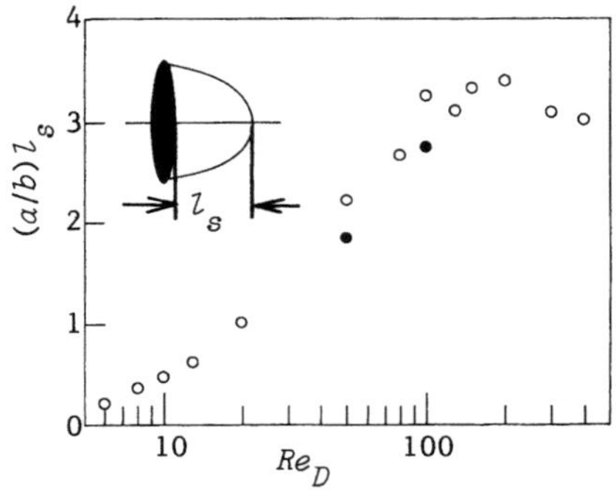

Fig.7. Wake length for an oblate spheroid. ○:$b/a = 5$, ●:$b/a = 50$.

Conclusion

It is found that the current Fourier spectral finite difference method in conjunction with substantially doubly-exponential coordinate transformation would work for getting steady-state flow and pressure fields around an ellipsoid of revolution placed parallel to a uniform flow even at high Reynolds numbers.

References

1. Pitter,R.L..;Pruppacher,H.R.;Hamielec,A.E.: A numerical study of viscous flow past a thin oblate spheroid at low and intermediate Reynolds numbers. J.Atmos.Sci. 30(1973)125-134.

2. Wieselsberger,C.: Weitere Festellungen über die Gesetze des Flüssigkeits- und Lufttwiderstandes. Phys.Z.. 23(1922) 219-224.

3. Liebster, H.:Über den Wiederstand von Kugeln.. Ann.Physik 82(1927)541-562.

4. Zahm,A.F.;Smith,R.H.;Louden,F.A.:Spheroids.Aerodynamische Versuchsanstalt Göttingen. Ergebnisse II(1932).

5. Fage,A.: Experiments on a sphere at critical Reynolds numbers. Aero. Res. Council Gt. Brit.,R.&M.. 1766(1936).

Sound Genearation Mechanism Around a Square Prism and an Airfoil Arranged in Tandem in Transonic FLows

T. NAKAGAWA

Kanazawa Insitute of Technology, Kanazawa, 921 Japan

Summary

A study has been done to elucidate the sound generation mechanism around a square prism and airfoil. Several acoustic waves are accidentally recorded on interferograms. It is found that when a vortex strikes at the shear layer an acoustic wave is generated, and is propagating to the upstream direction. Among numeral noise sources only acoustic waves due to the vortex-shear layer interaction have been identified on interferograms. Hance, this new noise source must be comparable with the well-known noise sources such as blade-vortex interaction or high-speed impulse in an aerodynamic and hydraulic machinery.

Introduction

In any aerodynamic or hydraulic machinery containing blades such as helicopter, aeroplane, windmill, fan, pump, power generator, and the screw, the loudest sound may be generated upon the collision of a vortex and blade, one of the primary noise sources of preset-day life [1]. A significant part of the incident vortex energy is converted into sound, and energy of the secondary vortices on the airfoil. However, several acoustic waves, which seem to have new origin, have been accidentally founded on the interferograms during the analyses of the previous records taken during an extensive study on vortex shedding from a square prism [2]. This prompts the present study.

An experimental study has been done to elucidate the sound generation mechanism around a square prism(side length D=20 mm) and an airfoil(NACA 0018, chord length 20 mm) arranged in tandem, at free stream Mach numbers between 0.1526 and 0.9081. The spacing ratio of the central distance L between the square prism and airfoil to the side length D of the square prism is varied from 1.125 to 5.5.

Experiment

Fig. 1 shows a schematic diagram of the experiment. The experiment was conducted at the Max-Planck-Institut für Strömungsforschung, Göttingen, using the transonic wind tunnel. The test section is 620 mm long, 330 mm high and 100 mm wide. The air is sucked from the laboratory and then flows into the vacuum tank of 132 m^3 through the test section. The velocity in the test section is controlled by adjusting the openings of the diffusor.

A two-dimensional square prism for the generation of the periodic vortices and an airfoil are installed between the two transparent glass windows. The spacing ratio L/D is one of the parameters in this study. The glass windows have the diameter of 230 mm and are interferometric quality. A Mach-Zehnder interferometer(Zeiss) is used for the high-speed flow visualization. This optical method is based on the principle of interference and provides the density field directly. When air flow is steady and isentropic, so that the viscous effects are neglected, the density ratio $\tilde{\rho}$ of the local density to stagnation density is related to the pressure ratio $\tilde{p}$ of the local pressure to stagnation pressure by $\tilde{p}=\tilde{\rho}^{\gamma}$, where γ is the ratio of specific heats, and is 1.401 for dry air at a temperature of 15°C and atmospheric pressure. When the test section is in vacuum, the interferogram shows a total of 96 fringes in this optical arrangement, where each fringe is shown as either a white or black line. The density ratio $\tilde{\rho}$ on each of the interferogram is given by $\tilde{\rho}=1-0.01058N$, where N is the fringe order from 0 either at a white or black line including the stagnation point, to 96 at a vacuum point in the flow. It is therefore possible to obtain the pressure ratio at any point in the flow, by simply counting the number of fringes of order N on the interferogram and then substituting the foregoing relations. The light source is a flash bulb(Fruengel). High-speed recordings of the interferogram are made with a rotating drum camera; the film frame rate is about 7 kHz.

Discussion

In the upstream travel the propagation velocity of acoustic waves, to be observed by a stationary observer, may be roughly estimated by the expression c-u, where c is the sound speed and u the local flow velocity. Hence, the propagation velocity of acoustic waves may be extreamly slow if the following condition is satisfied, viz., $u \lesssim c$. This must be the reason why acoustic waves are recorded on the interferograms.

Fig. 2 illustrates the high-speed flow visualization around a square prism and an airfoil arranged in tandem at Mach number M=0.6339, t denotes the time sequence starting from t=1, and f the fundamental frequency of the vortex shedding from the square prism. At t=1, the upper main vortex interacts with the upper separating shear layer. As the result, two acoustic waves are generated over the vortex formation region, and are propagating upstream. It may be of significance that among the many known noise sources in the experiment only the acoustic waves generated by the vortex-shear layer interaction are recorded on the interferograms. Hence, the vortex-shear layer interaction must be a comparable noise source with those due to a blade-vortex interaction and/or high-speed impulse.

References

1. Nakagawa, T.:Conversion of vortex energy into acoustic energy. Naturwissenschaften, 74 (1987) 338-339.
2. Nakagawa, T.:Vortex shedding behind a square cylinder in transonic flows. J. Fluid Mech. 178 (1987) 303-323.

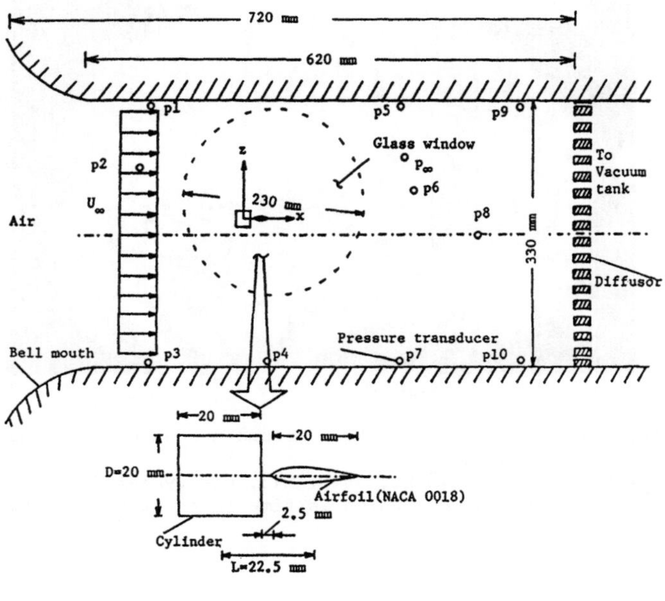

Fig.1 Schematic diagram of the experiment

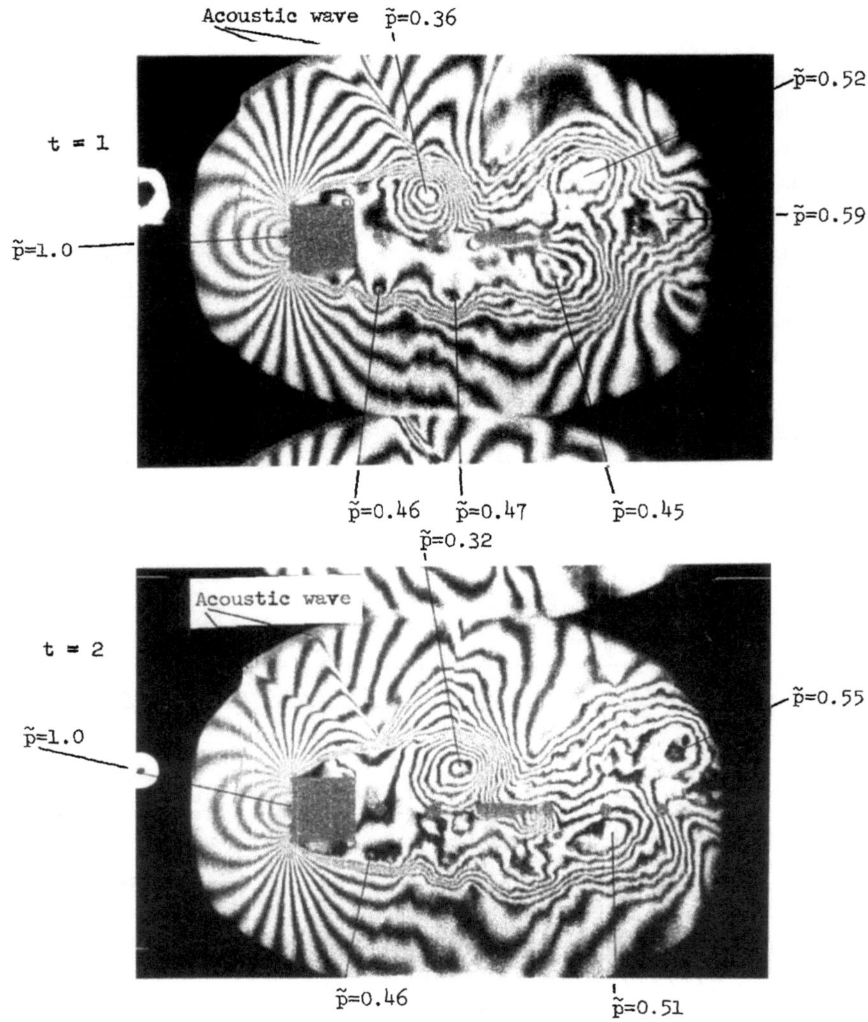

Fig.2 High-speed flow visualization around a square prism and airfoil arranged in tandem.
U_∞=216.6 m/s, f=1348 Hz, St=0.1245, Re=2.923×10^5, M=0.6339, L/D=2.875, Δt=143 μs, which is time interval between two frames.

A Unified Approach to Galloping of Rectangular Cylinders with or Without a Stationary Splitter Plate

Y. Nakamura and K. Hirata
Research Institute for Applied Mechanics
Kyushu University
Kasuga, 816 Japan

1. Introduction

Galloping is a single-degree-of-freedom flutter of a bluff body in translation in a cross-flow direction. Despite the great success of the quasi-steady theory [1], two important problems related to high-speed galloping remain unsolved. The first one concerns the effects of decreasing reduced velocity on galloping. Obviously, the quasi-steady galloping theory is only valid in the high-speed range for which $\overline{U} >> \overline{U}r$. Here, $\overline{U}$ and $\overline{U}r$ are defined respectively as $\overline{U}=U/fh$ and $\overline{U}r=U/f_Vh$, where U is the flow velocity, h is the length scale of a bluff body, f is the body frequency, and f_V is the frequency of vortex shedding for the body at rest under the same flow conditions. The effects of two flow modules, i.e., wake undulation and vortex resonance, are both neglected in the quasi-steady galloping theory. Wake undulation becomes increasingly more influential on galloping as $\overline{U}$ is lowered, and furthermore, as $\overline{U}r$, the vortex-resonance velocity is approached, strong interaction between galloping and vortex excitation can take place.

The second problem is the aerodynamic excitation mechanism of galloping. On rectangular cylinders, soft high-speed galloping, i.e., high-speed galloping starting spontaneously from rest can occur for $0.75 < d/h < 3.0$, approximately [1], where d is the section depth (streamise dimension) and h is the section height (cross-flow dimension). This assumes a static transverse force coefficient over the range of d/h that is positive (downward) at a positive (clockwise) incidence. A question is raised as to how this is aerodynamically possible. If pressure inside the near wake is everywhere uniform, no such aerodynamic force is created so that the question must be closely related to the complicated secondary flow that occurs in the near wake.

2. Galloping and vortex excitation of rectangular cylinders with or without a stationary splitter plate

Figure 1 shows the stability boundaries for galloping and vortex excitation of rectangular cylinders without a splitter plate, plotted as functions of d/h and $\overline{U}$, together with curves for the critical depth and the vortex resonance velocity [2]. It shows that high-speed galloping can occur for long rectangular cylinders just beyond the critical depth. Here, the critical depth for rectangular cylinders, either at rest or oscillating, is defined as a particular cylinder for which the time mean base suction takes a maximum as d/h is increased from zero (see Nakamura and Hirata [3] for details). The close relation between the onset of high-speed galloping and the critical depth was first noticed by Parkinson [1] and later more systematically studied by Nakamura and Tomonari [4] under various flow conditions.

A further interesting result in Figure 1 is that galloping can also occur at low $\overline{U}$ well below $\overline{U}r$ on short rectangular cylinders for which high-speed galloping can not occur, or in other words, for short rectangular cylinders approximately below the critical depth.

Figure 2 shows the stability boundary for galloping of rectangular cylinders with a splitter plate [2]. As can be seen, galloping can occur over a wide range of d/h, covering from d/h=0 up to d/h=6.0, approximately. Because of the absence of regular vortex shedding, the results are much simpler than those for cylinders without a splitter plate. Another point of interest is a close similarity in the low-speed galloping of rectangular cylinders with or without a splitter plate. The two stability boundaries in the low-speed range, shown in Figures 1 and 2, nearly overlap on each other.

3. Unified approach to high- and low-speed galloping of rectangular cylinders with or without a stationary splitter plate

Consider a rectangular cylinder, shown in Figure 3, which is moving downward with a velocity y(t) in the flow of velocity U; the effective angle of incidence is given by $\alpha = \arctan \dot{y}(t)/U$. The shear layer on the lower side moves closer to the body and hence is more highly curved while the corresponding layer on the upper side moves further away from the body and is less curved. The pressure on the lower side is then smaller than that on the upper side, thus leading to a net downward pressure force across the side faces. This gives the basic aerodynamic excitation mechanism of galloping.

As Owen [5] remarked, however, the explanation for galloping just described is not quite persuasive. No such pressure force is observed experimentally for short inclined rectangular cylinders at rest. The upper and lower flows can communicate with each other through the wake to adjust such a pressure difference; the flow in the near wake is normally slow so that it can not sustain any appreciable pressure difference. Accordingly, no (time mean) pressure force can be produced across the side faces for short rectangular cylinders at rest.

For the basic mechanism to work effectively, the communication between the side flows must be interrupted with any measures, either artificial or natural. We mention three such possible examples among others. These are the flow around rectangular cylinders with a long splitter plate, the low-speed oscillatory flow around rectangular cylinders and the high-speed oscillatory flow around long rectangular cylinders with d/h beyond the critical. They are sufficient to explain the onset of galloping shown in Figures 1 and 2. The first example is easiest to understand; a splitter plate can make the upper and lower flow almost separate with each other. The second example is supported by our recent finding [6] that the effect of the wake undulation on the flow around an oscillating cylinder vanishes as $U \rightarrow 0$. Under such a condition, the flows around an oscillating cylinder with or without a splitter plate are almost identical, which presumes independence of the upper and lower flows. The third example is most sophisticated of all. The onset of the shear-layer/edge direct interaction for long cylinders beyond the critical is an embryo of flow reattachment, which is characterized by a reattachment-type pressure distribution on the side [3]. Accordingly, a cylinder moving downward at high $\overline{U}$ can experinence a downward pressure force which is contributed from a pressure difference in the front half of the pressure distribution.

References

[1] Parkinson, G. V. 1974 IUTAM Symp. Flow Induced Structural Vibrations (ed. E. Naudascher), Karlsruhe.

[2] Nakamura, Y., Hirata, K. and Urabe, T. 1991 Proc. 5th Int. Conf. Flow Induced Vibrations, Nashville.

[3] Nakamura, Y. and Hirata, K. 1989 J. Fluid M. 208.

[4] Nakamura, Y. and Tomonari, Y. 1977 J. Sound and Vibration 52.

[5]Owen, P. R. 1973 Aero. Q. 77.

[6] Nakamura, Y., Hirata, K. and Urabe, T. 1992 J. Fluids and Structures 5.

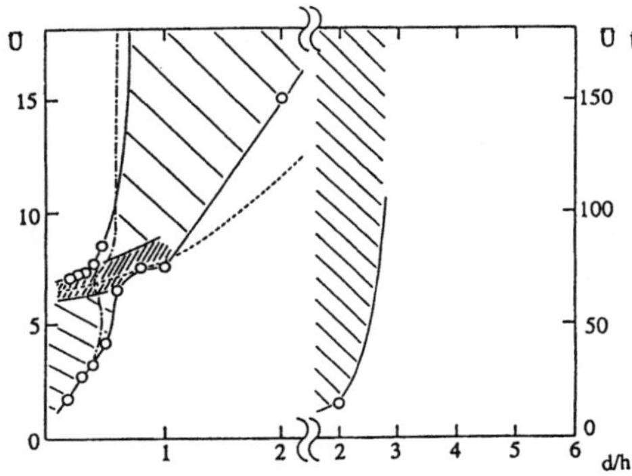

Fig. 1 Stability boundary of galloping and vortex excitation for rectangular cylinders without a splitter plate.
-------, Vortex-resonance velocity;
—-—-—-—, Critical depth.

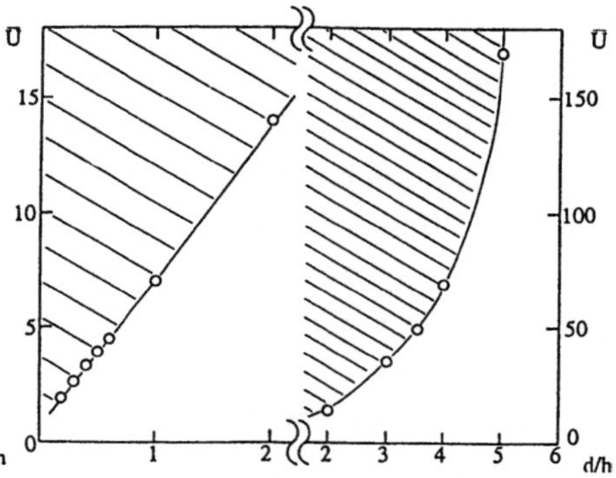

Fig. 2 Stability boundary of galloping for rectangular cylinders with a splitter plate.

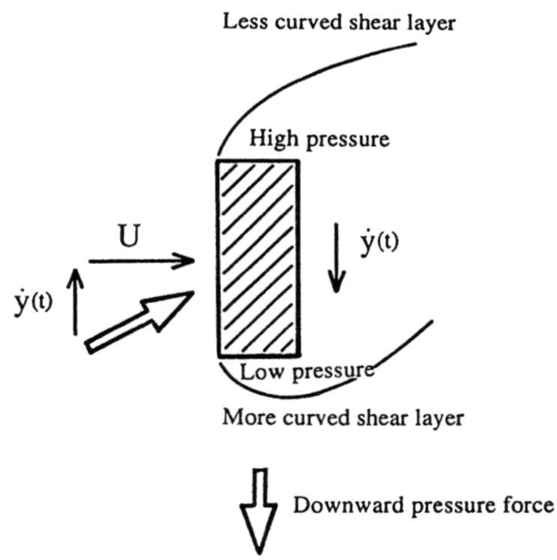

Fig. 3 Basic mechanism of galloping.

Observations of the Flow Over Prismatic Obstacles in a Fully Developed Channel Flow

R. Martinuzzi, C. Tropea, J. Volkert

Lehrstuhl für Strömungsmechanik
Universität Erlangen-Nürnberg
Cauerstr. 4, 8520 Erlangen, Germany

The flow over prismatic obstacles in fully developed channel flow, as shown schematically in Figure 1, is representative of a large number of practical flow situations in industrial processes and of complex turbulent flows in general. Previous investigations of this flow provide descriptions of the time averaged velocity field as a function of Reynolds number and obstacle dimensions, in particular the width-to-height ratio [1,2]. In the present paper, emphasis is placed on the time dependent behaviour of the recirculation zones and their interaction. Observations have been made for the Reynolds number range 5000 - 90000 using local smoke injection and a laser light sheet illumination. More detailed velocity measurements have been performed at Re = 80000 using a two-component laser Doppler anemometer. The follow results are presented for the cube obstacle.

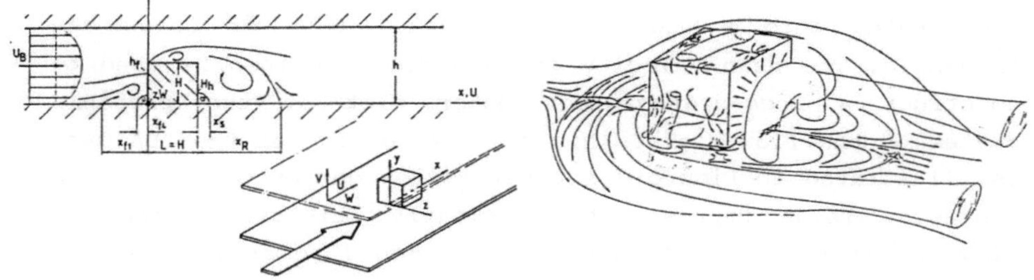

Fig.1. Obstacle geometry and derived mean flow pattern around the cube obstacle

Both video sequences of the flow visualization and the probability density distributions (PDF) of the U and V components of velocity, indicate that the flow in front of and on top of the obstacle is preferentially in one of two states or modes. This behaviour has been observed previously for other geometries [3] and is characterised by a bi-modal velocity distribution, as illustrated in Figure 2. The power spectrum of the velocity fluctuations, also shown in Figure 2, indicates however, that these velocity fluctuations do not occur at a preferred frequency.

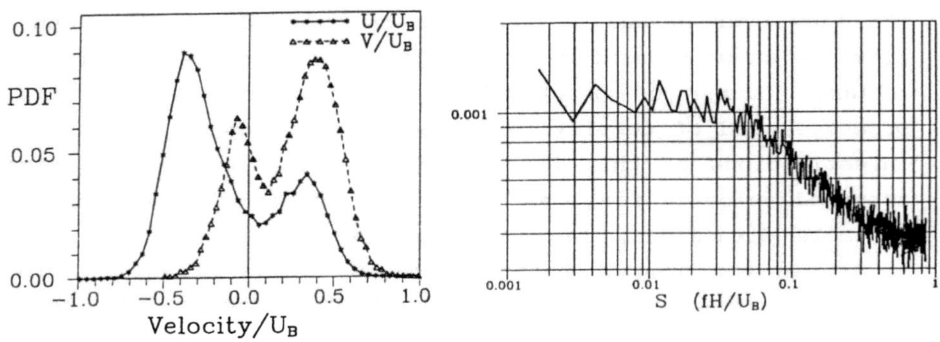

Fig.2. a) Typical PDF of velocity fluctuations on top of obstacle b) Power spectral density of velocity fluctuations at same location (U_B - mean channel velocity, H - cube height)

By using an amplitude threshold chosen at the valley of the bi-modal distributions, the instantaneous velocity can be segregrated into a mode 1 or 2, i.e. above or below the threshold. The mean velocity of each mode can then be used to construct a *mode-segregrated* vector plot of two probable states of the flow field. This post-processing of the LDA data is illustrated in Figure 3 for the region on top of and behind obstacle. These representations clearly illustrate that the flow can instantaneously either be attached to or free of the upper cube surface.

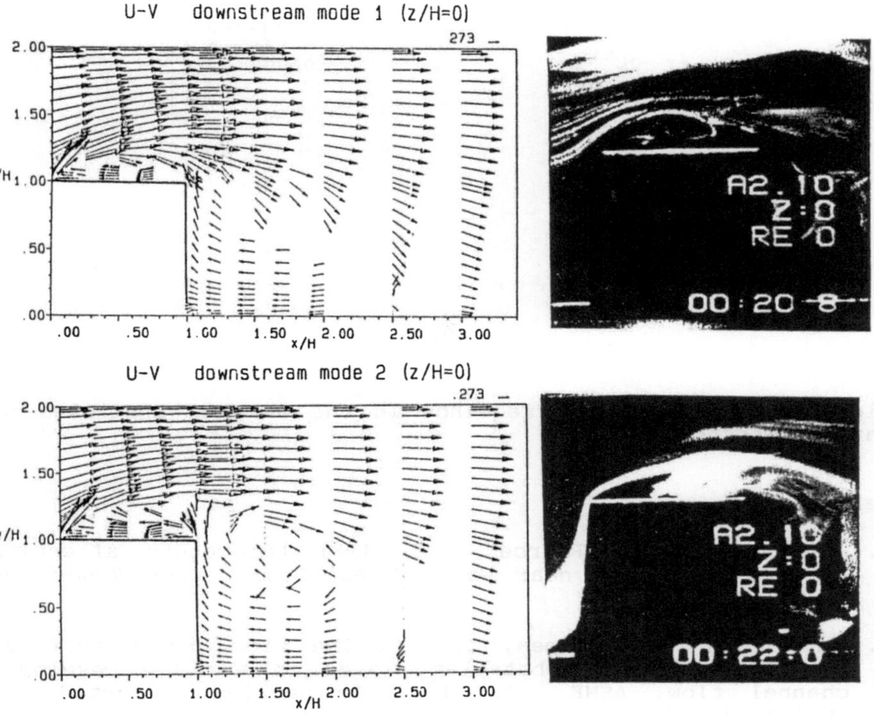

Fig.3. Mode segregated flow field with typical video visualizations

Independent of the flow fluctuations arising from a mode switching as described above, vorticies are also shed from the side of and behind the obstacle. A Strouhal number can be derived from the power spectal density of the velocity fluctuations in these regions. A summary of these measurements is contained in Figure 4, in which two distinct Strouhal number bands are observed: The upper trace corresponds to vortex shedding from the obstacle side; the lower trace is the vortex shedding from the rear of the obstacle. This figure also indicates the influence of obstacle width (crou-stream width). It is clear from these observations, that the vortex shedding in the obstacle wake is not suppressed by the instabilities observed in front of or on top of the obstacle.

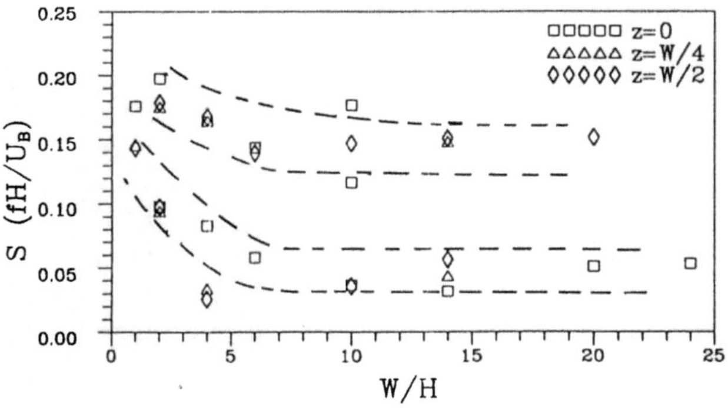

Fig.4. Vortex shedding frequency in the wake of obstacles of various widths

References

1. Fackrell, J.E.; Pearce, J.E. 1981 Parameters affecting dispersion in the near wake of buildings, CEGB Report No RD/M/1179/N81

2. Martinuzzi, R.; Tropea, C. 1992 The flow around surface-mounted, prismatic obstacles placed in a fully developed channel flow, ASME J. Fluid Engineering (accepted for publication 1993)

3. Devenport, W.J.; Simpson, R.L. 1990 Time-dependent and time-averaged turbulence structure near the nose of a wing-body junction, J. Fluid Mech. 210 23-55

Wake Flow Measurements in an Oscillating Delta Wing at High Incidence

H. STAPOUNTZIS, G. XENOPOULOS, S. RAPTOPOULOS
University of Thessaloniki, Thessaloniki 540 06, Greece.

Summary

Hot wire measurements and flow visualization on a Delta wing oscillating in pitch at high mean incidences were undertaken. The effects of oscillation on the velocity spectral characteristics and the Strouhal frequency were investigated.

Introduction

Unlike a two-dimensional airfoil, a Delta wing separates at a fairly small angle of incidence α, about 5°. For moderate α the flow round the wing is dominated by two symmetric stationary spiral leading edge vortices, which induce velocities on the suction side, in addition to the ones caused by potential flow. Thus the lift on the wing consists of the "potential flow" and the "vortex" lifts and varies non-linearly with α. What follows when the angle of incidence (attack) increases, depends partly on the Aspect Ratio, AR = 2b/c, where b is the span of the wing (at the trailing edge) and c the root chord. For AR = O(1), the vortices break down at some distance x_B from the apex, x_B decreasing as α increases. Ayoub and McLachlan [1] found that there were two modes of breakdown, the spiraling type and the burst type which in fact alternated periodically on either side of the wing at dimensionless frequencies $f_B c/U_0 = 0.1$, 0.13, 0.17, 0.15 at α = 40°, 45°, 50°, 55° respectively (U_0 is the free stream speed and f the frequency in Hz). With further increasing α (i.e. $\alpha \geq 60°$) the suction surface of the wing has globally separated, the leading edge vortices cannot be held in place and thus start to shed.

The precise manner of shedding is still unclear. Far wake measurements, see Rediniotis et al. [2] and Stapountzis et al. [3], by means of two hot wires placed on either side of a wing center plane resulted in same velocity spectral peaks at the shedding frequency f_v, in phase (ϕ = 0°) for transverse distances close to the apex and out of phase (ϕ = 180°) otherwise. The latter indicates alternate periodic shedding at a Strouhal frequency $S = f_v b/U_0$, while the former a periodically pulsating bubble type flow behind the wing. Those types of shedding at high angles α, (up to 90°) have no equivalent to the usual shedding behind a two dimensional bluff body. It was also found that S decreased with α, unlike the previous regime of vortex breakdown.

Vortex flow asymmetry, breakdown and "natural" vortex shedding can cause unsteady loading on the wing as also do oncoming gusts and/or sudden wing motion. The last is a requirement of high maneuverability aircraft, often flying at very high incidences. Aircraft performance can either benefit or be humbered by the special features in the unsteady flow. Numerous wind tunnel tests on Delta wings oscillating in pitch e.g. Brandon and Shah [4], Gad-el-Hak and Ho [5] show that hysteresis effects are present at reduced frequencies, $k = \pi f_0 c/U_0$, as low as 0.02 (f_0 = wing oscillation frequency) and are enhanced as the mean incidence, the amplitude of oscillation, α_0, and k increase. In particular, a lag in the vortex burst point location, of the order of up to 30 convective time scales (= c/U_{axial}), may result in persistence of unsteady aerodynamic effects for sufficient duration to enhance wing performance at high α, or provide time for breakdown control e.g. by means of tangential periodic blowing at the leading edge, Gad-el-Hak and Blackwelder [6].

So far, work on unsteady Delta wings is limited to the study of loading and flow field round the (not globally separated) wing and its very near wake (about half chord length). In the present work we examine the effects of sinusoidal pitching motions of a Delta wing on its wake characteristics at moderate and high angles of incidence. It is known that transverse, in line (with respect to the mean flow) or pitching oscillations of two dimensional bluff bodies can greatly influence the surface pressure distributions and the vortex shedding in the wake, especially when f_0 approaches f_v, see Bearman and Obasaju [7], Tokumary and Dimotakis [8]. Of course "natural" vortex shedding of Delta wings with AR ≈ 1 (like the one considered in this paper) is a much more complicated phenomenon than in a two dimensional bluff body or even a three dimensional slightly tapered cylinder (Gaster and Ponsford [9]). The interaction of vorticity, springing from the inclined leading edge, does not take place in a more or less orderly manner, as in a 2-D bluff body, and as a consequence vortex shedding is weak and irregular.

Experimental approach

The Delta wing used had a chord c = 0.124 m and span b = 0.06 m (AR ≈ 0.97) and it was tested at two chord Reynolds numbers $Re = U_0c/v = 5x10^4$ and $2.9x10^5$. The wind tunnel had a square cross section 0.3 x 0.3 m, length of working area 2 m and background turbulence level of about 0.2%. Pitching oscillations took place around the mid chord of the wing, which was also the origin of the coordinate system (x: streamwise, y: spanwise, z: vertical), Fig. 1. In this work, wake velocity measurements were undertaken using two identical hot wires placed on either side of the vertical symmetry plane (through the origin) and a crossed hot wire. The Cross Power Spectral Density (CPSD) function of the two hot wires was computed in order to locate the vortex shedding spectral peaks from the maxima (at f_v) in the individual Power Spectral Densities (PSD). That the maxima in PSD represented genuine vortex shedding and not any mechanical resonance was further confirmed by the systematically close to 1.0 values of the coherence, γ(f), the systematic proximity of the phase ϕ to 180° (alternate shedding) or 0° (simultaneous shedding, see [2] and [3]) and the continuous variation of f_v with U_0.

Results of velocity measurements and discussion

The effect of wing oscillation on the Strouhal number S is shown in Fig. 2. The reduced frequency k ranged between 0.030 and 0.160 at $Re = 2.9x10^5$ and 0.197 to 0.928 at $Re = 5x10^4$. A common feature for all α, Re and z is that, initially, at low k, the Strouhal number falls below its steady state, S_0, ($f_0 = 0$) value. As k increases, S recovers to its steady state value and for the lowest α (40°, 50°) it tends to overshoot it. The highest oscillation amplitudes about the mean, $\alpha_0 = \pm 16°$, bring about the greatest deviations of S from S_0.

A typical effect of wing oscillation on the vortex shedding spectra, coherence and phase is demonstrated by Fig. 3a,b,c. When the wing is held steadily ($f_0 = 0$) at α = 60°a broad, but well defined, velocity spectral peak is present at f_v about 90 Hz. The coherence is close to 1.0 around f_v and the phase fluctuates about zero. This agrees with our earlier findings ([2], [3]) of "simultaneous" vortex shedding from a Delta wing at a dominant frequency f_v, leading to a constant Strouhal number of about 0.15 scaled with the maximum wing span b. (This was realized for other hot wire positions as well). When the wing oscillates at $f_0 = 4.2$ Hz, (k = 0.047), Fig. 3b, the (simultaneous) vortex shedding peak decreases by almost an order of magnitude, f_v shifts to lower values (75 Hz), γ decreases, and ϕ stays close to zero. A new vortex shedding peak however emerges in the range of 103 Hz (S ≈ 0.18) corresponding to "alternate" vortex shedding as implied by the high values of γ and ϕ. This peak appears stronger at other transverse locations, not shown here. The vortex shedding peak is comparable in size to the velocity peak at f_0. The relatively slow amplitude excursion of the wing seems to produce therefore two different effects: One dynamic effect related to the lowering of the vortex shedding frequency (U_0 was maintained constant during the oscillation) and disintegration of the shedding coherence and a "quasi steady" effect leading to alternate vortex shedding. The latter was observed, [2], [3], on a steady wing placed at incidences $\alpha > 75°$. In the present case, the wing incidence occasionally exceeds this limit ($\alpha = 60°$, $\alpha_0 = \pm 16°$). At a higher oscillation frequency, $f_0 = 11$ Hz, $k = 0.122$ (Fig. 3c) both the aforementioned vortex shedding peaks became broader and are further suppressed relative to the oscillation spectral peaks, (1st and 2nd harmonics). In fact the value of the coherence at frequencies f_0, $2f_0$ is now larger than at f_v. The above results show, that external periodic forcing at time scales considearably larger than the intrinsic time scales (shedding = $1/f_v$, convection = c/U_0) can change the flow characteristics of the Delta wing. Significant hysteresis effects, reported by other workers with regard to vortex breakdown and loading, were also found in this study.

An example of flow visualization by means of a smoke wire is shown in Fig. 4. In Fig. 4a the wing is held steadily at $\alpha = 56°$, while at 4b, it is captured at its upstroke position $\alpha(t) \approx 53°$ during an oscillation about $\alpha = 43°$ ($f_0 = 4$ Hz, $k = 1.25$). Although the wing is at the same incidence in both cases, the flow picture is entirely different.

The interaction of the excitation and vortex shedding velocity peaks in the lower Reynolds number examined (5×10^4) is shown in Fig. 5. Again, the spectral power at the oscillation frequency grows as f_0 increases while vortex shedding power (originally centered about $f_v \approx 23$ Hz) is diminishing. When the highest oscillation frequency is reached, $f_0 = 14$ Hz, the vortex shedding peak is completely eliminated.

Some further results of the present investigation which are not presented here are:

The Strouhal number remained effectively constant in the streamwise direction ($4.5b < x < 10b$) with and without oscillation.

The mean velocity components (streamwise, transverse in the z direction) were not significantly altered by oscillation at $x = 4.5b$ but the RMS transverse velocity component for $f_0 = 8$ Hz increases by about 25% at the level of the trailing edge.

Conclusions

Oscillating a Delta wing in pitch at high mean incidences always reduces the vortex shedding Strouhal number, S, at low oscillation frequencies f_0. The subsequent influence on S at higher f_0 depends on the amplitude of oscillation and the mean incidence α. High α and α_0 tend to reduce S further with increasing f_0, while at low α and α_0, S can overshoot its steady state value. The velocity power in the wake and its coherence at the vortex shedding frequency, f_0, are suppressed by oscillation and hysteresis effects are observed even when f_0 and f_v are one order of magnitude different or the reduced frequency k is as low as 0.03.

References

1. Ayoub A.; McLachlan B.G. Slender Delta Wings at High Angles of Attack. AIAA-87-1230 paper, 1987.

2. Rediniotis O.; Stapountzis H.; Telionis D.P. Vortex Shedding over Delta Wings. AIAA Journ. vol. 28 (1990) 944-946.

3. Stapountzis H.; Iliadis G.; Agelakis J.; Mansour G. Vortex shedding over delta wings and plates at high angles of incidence. Z. Flugwiss. Weltraumforch. 16 (1992), in press.

4. Brandon J.M.; Shah G.H. Effect of large amplitude pitching motions on the unsteady aerodynamic characteristics of flat plate wings. AIAA 88-4331-CP paper, 1988.

5. Gad el Hak M.; Ho C.M. The pitching delta wing. AIAA Journ. vol. 23 (1985) 1660-1665.

6. Gad el Hak M.; Blackwelder R.F. Control of discrete vortices from a delta wing. AIAA Journ. vol. 25 (1987) 1042-1049.

7. Bearman P.W.; Obasaju E.D. An experimental study of pressure fluctuations om fixed and oscillating square section cylinders. Journ. Fl. Mech. vol. 119 (1982) 297-321.

8. Tokumaru P.T.; Dimotakis P.E. Rotary oscillation control of a cylinder wake. Journ Fl. Mech. (1991) 77-90.

9. Gaster M.; Ponsford P.J. The flows over tapered flat plates normal to the stream. Aeron. Journ. May 1984, 206-212.

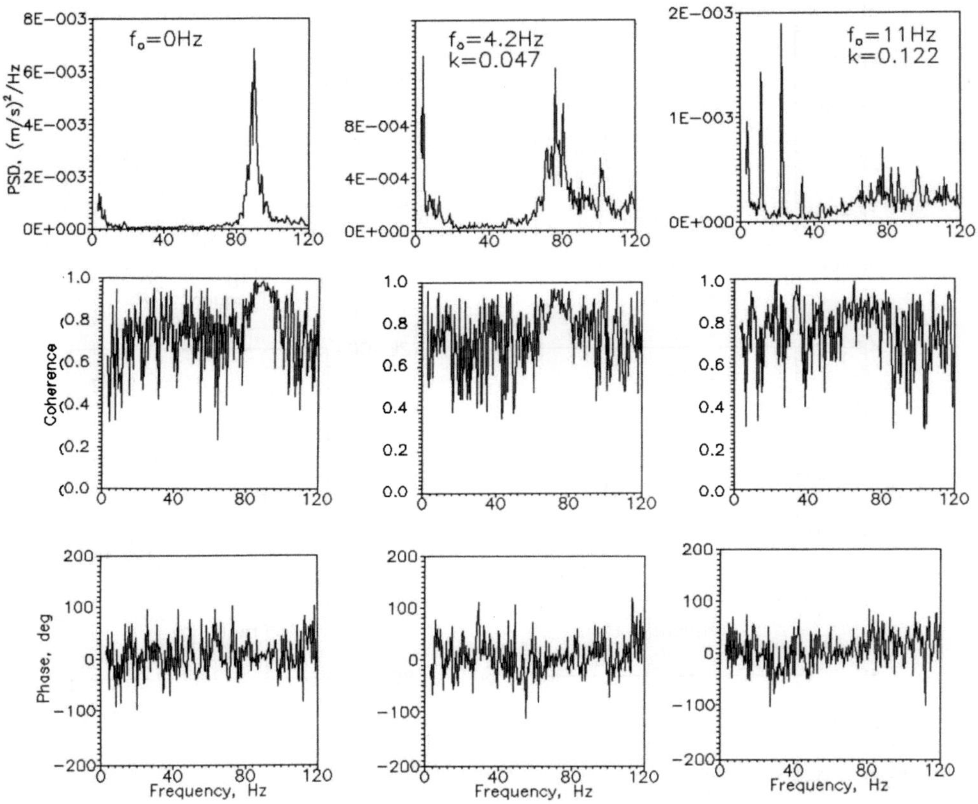

Fig.3 Spectral density, coherence and phase, as affected by wing oscillation. $x=4.5b$, $y=+1.6b$, $z=0$, $a=60°$, $a_0=+16°$

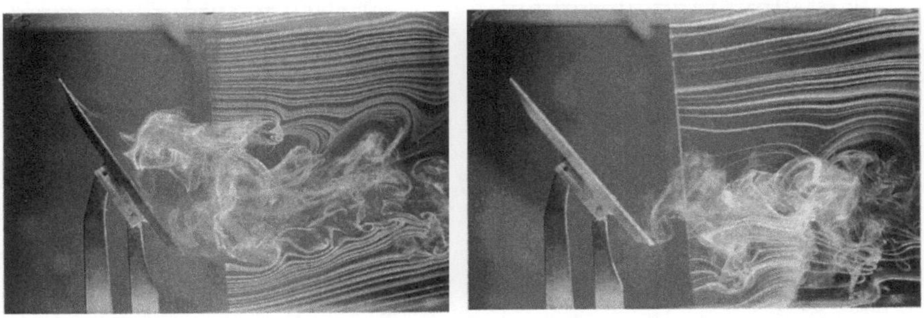

Fig.4 Smoke flow visualization
Left: $a=56°$, no oscillation
Right: $a=43°$, $f_0=4Hz$, $a_0=+10°$

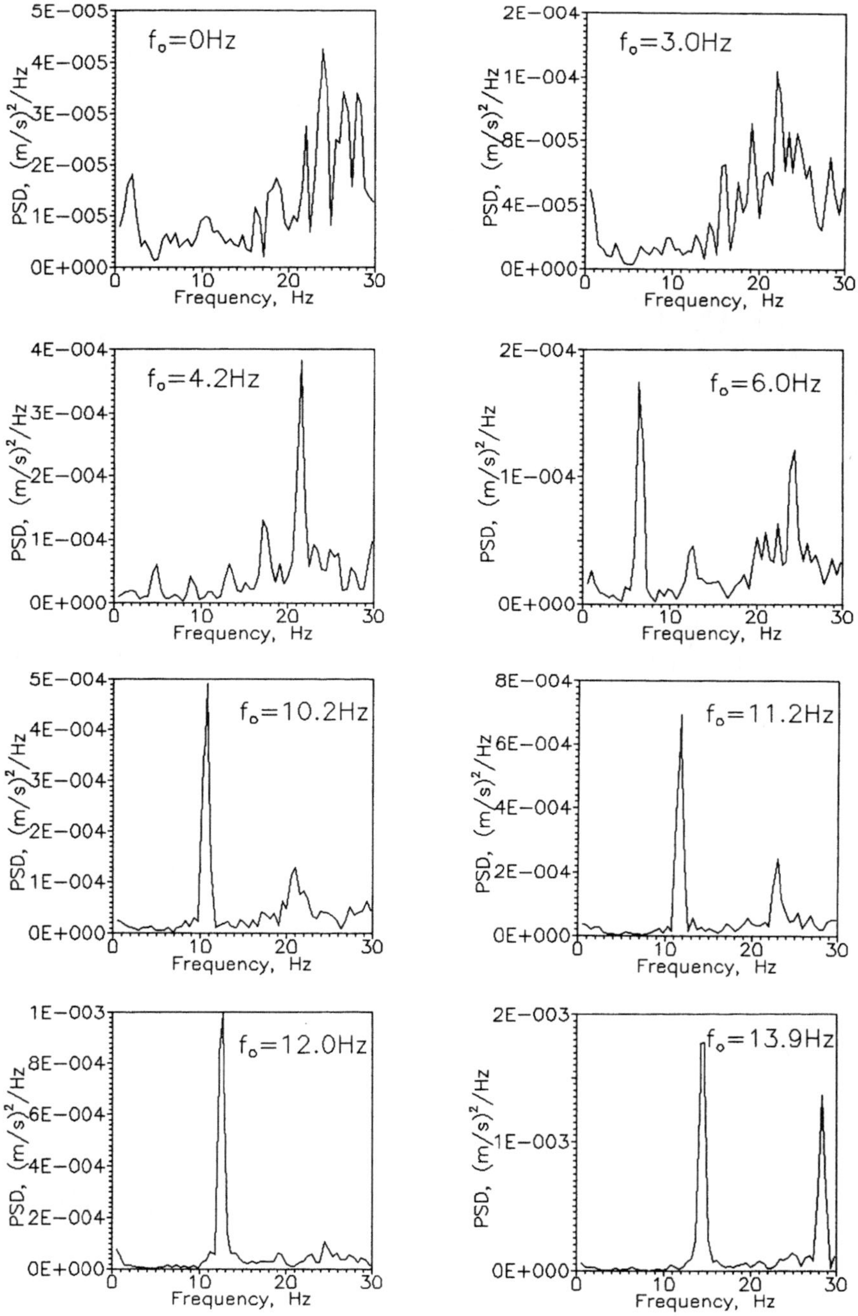

Fig.5 Velocity spectra at various oscillation frequencies
$a=60°$, $a_o=+10°$, $Re=5x10^4$, $x=4.5b$, $y=1.6b$, $z=-c/4$

Formation and Behavior of the Vortices Associated with a Transverse Jet

Y. Sugiyama
Department of Aeronautical Engineering
Nagoya University, Nagoya, Japan

Summary

Generating mechanism and growth and decay processes of the vortex pair induced by a jet issuing normal to a cross flow have been experimentally investigated in a subsonic range. The laser-sheet visualization has shown that : (1) The jet behaves like a solid cylinder for a distance from the jet exit. (2) A low energy region due to the jet diffusion spreads only laterally in jet cross-sections. The growth and decay processes are also shown using velocity vectors in the jet cross-sections. A simple infinite linear vortex model of the vortex pair has successfully explained characteristic features of the velocity field induced by the vortex pair.

Introduction

The introduction of a jet transversely into a moving stream is a basic configuration which finds applications in engineering fields. However, there is an absence of data regarding to the strength and location of the vortex bound in the jet, together with the detailed structure of the vortex system near the jet orifice. The absence is also like in growth and decay processes of the vortex pair and in a behavior of shed vortices behind the vortex centerline. There are many papers on a local strength of the vortex pair. However, these do not discuss a mechanism of the vortex pair separation and associated increase in jet cross-section.

The present paper aims mainly at understanding a mechanism forming the vortex pair. It is intended to find dependence of a roll-up process into the vortex pair on the deformation of a jet cross-section, and also to know the growth and decay processes of the vortex pair by measuring its induced velocity field.

Apparatus and Procedures

Fig.1 shows an arrangement of a working section with a rectangular cross-section, a jet supply pipe and one of probes used. There is a jet orifice (diam. $d = 8$ or 20 mm) on the channel floor, through which the jet ($W_j = 29$ or 37 m/s) issues normal to the cross flow. The jet and cross flows consist of air at a room temperature. Reynolds number based on the main-flow velocity and the jet diameter is about 10^4.

Velocity ratio m is defined as the ratio of (jet exit velocity W_j / main-flow velocity U_∞). Velocities in Figs.5 and 6 are measured with a hot-wire probe with a sensing element inclined at 45° and made of a tungsten wire of 5 μm diam., together with a hot-wire response equation. However, velocities in Figs.2 and 9 are measured by orienting an I-type hot-wire probe normal to a local flow direction previously determined with a 3 or 5 hole pitot-tube. An oil-film flow visualization is applied to the channel floor and to the jet symmetry plane replaced by an iron sheet 0.6 mm thick. The jet cross-section is visualized by mixing a small amount of water with air, together with the He-Ne laser sheet. Water is mixed with air in a first plenum chamber.

Formation of vortex pair

Fig.2 shows a comparison between the y-component U_y of velocities measured around the jet column and U_{ypot} calculated around a solid cylinder in a potential flow. The profiles of U_y indicate that the jet acts like a solid cylinder in $z/d <$ (1 to 2). This is supposed us a formation of a horseshoe vortex around the jet column, but Fig.3 indicates no sign of the horseshoe vortex. This means that the horseshoe vortex is entrained with fluid of the cross flow into the jet. In the just lee side of the jet orifice we find two steps in passages of the flow entrained into the jet. The one is that in the y-direction into the wake region of the jet column (Fig.3). The other is that in the z-direction in the wake region of the jet column (Fig.4). The latter occurs a flow along the jet symmetry plane (Figs.5(c)-(g)), and velocity vectors with a large pitch angle (Fig.6). However, we cannot show still a reason why the flow rolls up into the vortex pair.

Figs.7 and 8 show an application of the flow visualization by the laser sheet to the cross-section. Fig.7 shows that in the upstream part within the jet orifice ($-0.5 < x/d < 0$) the jet column acts like a solid cylinder in $z/d < 1$ where the jet spread is small. In $z/d > 1$ the jet spread increases steeply laterally. In the downstream part within the jet orifice ($0 < x/d < 0.5$) spread of the jet column in $z/d > 1$ is steeper than in the upstream part within the jet orifice and the jet cross-section varies from the circle to the kidney shape. With formation of the kidney shaped cross-section the jet cross-section extends like an arc from the lateral ends of the kidney shape toward the channel floor (Fig.7). Fig.8 shows jet cross-sections in the plane parallel to the floor. Spread of the jet cross-section is laterally large, but small streamwise. At $z/d = 0.15$ we find a sign of the roll-up of the cross flow at the lateral end of the jet cross-section. Figs.7 and 8 suggest that i) the cross flow takes off a low energy fluid occupying the diffusion region of the jet, ii) the cross flow entraining the low energy fluid separates from the jet region like a flow around an edge and rolls up into the vortex pair. Thus, the formation of the vortex pair is not due to the entrainment by the jet, but is attributed to the separation and the succeeded roll-up of the cross flow. It is also known that the vortex centerline begins with a point where the jet column ceases from acting like a solid cylinder.

Behavior of vortex pair

A characteristic of the y-component U_y of the velocity induced by the vortex pair is discussed downstream of the jet orifice. We replace the deflected vortex centerline by an infinite linear vortex centerline which tangents to an intersect of the deflected vortex centerlines and a cross plane, on which the y-component U_y of the velocity induced by the vortex pair is calculated. The circulation along the vortex centerline is given semi-analytically [1] and the vortex centerline empirically [2]. We introduce mirror images to satisfy the boundary condition at the floor. Fig.9 shows that notwithstanding the rough vortex model the induced velocity calculated by the Biot-Savart law agrees qualitatively with the experimental induced velocity, especially near the floor ($z/d = 0$) and in the region, where U_y decreases asymptotically toward the infinity.

References

[1] Le Grives, E.: Mixing process induced by the vorticity associated with the penetration of a jet into a cross flow. Trans. ASME, J. of Engineering for Power, 100-3 (1978) 465-475.

[2] Fearn, R.; Weston, R.P.: Vorticity associated with a jet in a cross flow. AIAA J., 12-12 (1974) 1666-1671.

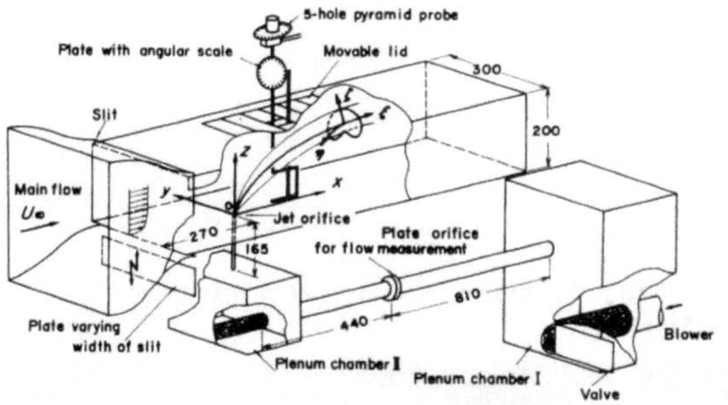

Fig.1. Sketch of working section

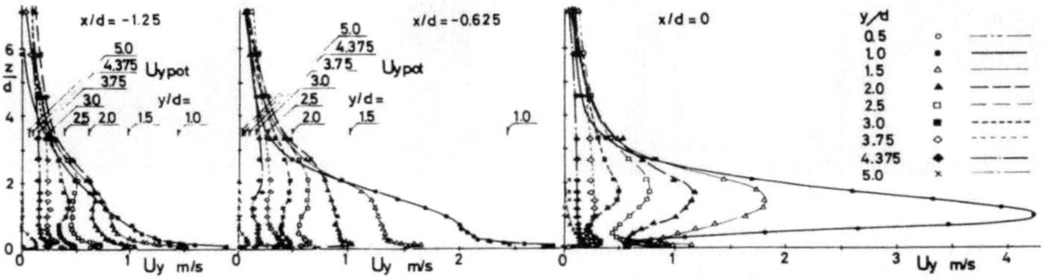

Fig.2. Profiles of lateral velocity component U_y around exit of unconfined jet (d=8 mm, W_j = 29 m/s, U_∞ = 15 m/s, m = 2), and U_{ypot} calculated around two-dimensional solid cylinder in potential flow

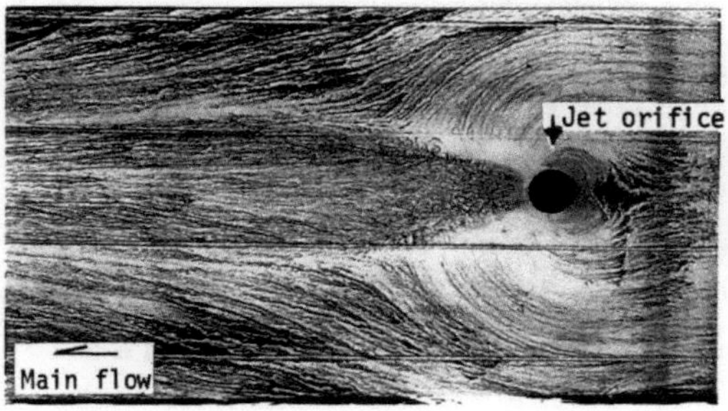

Fig.3 Oil-flow pattern on channel floor affected by confined jet (d = 20 mm, m = 4.11)

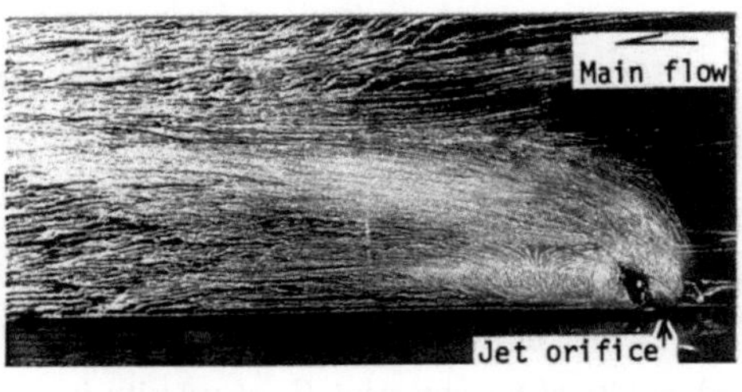

Fig.4. Oil-flow pattern in plane of symmetry of unconfined jet (d = 8 mm, m = 2)

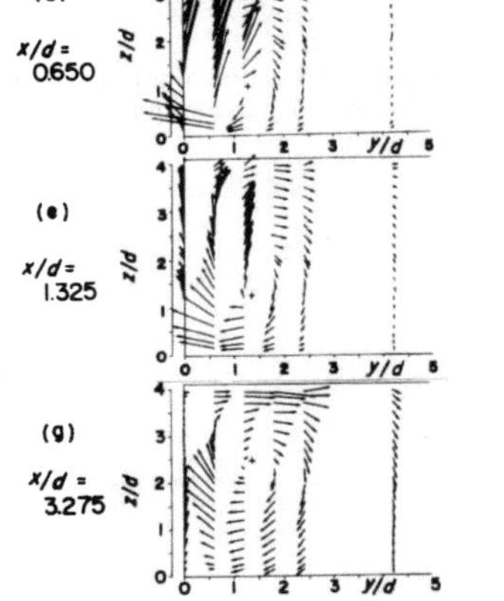

Fig.5. Velocity vectors projected on cross planes at m = 4.11 (+ : centers of vortex flow, d = 20 mm, confined jet)

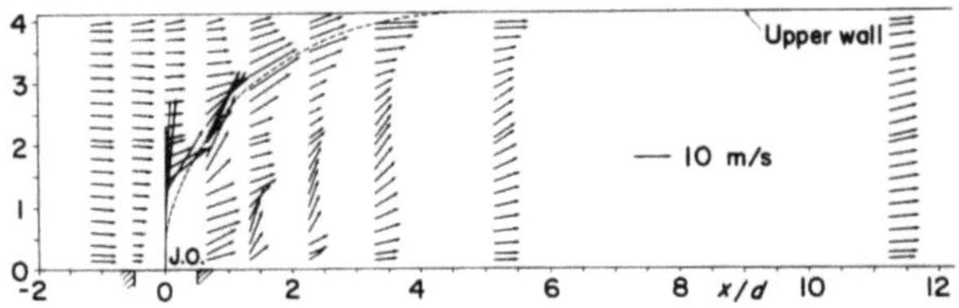

Fig.6. Velocity vectors in plane of symmetry of confined jet at $d = 20$ mm and $m = 4.11$ (- - : Jet centerline, J.O. : Jet orifice)

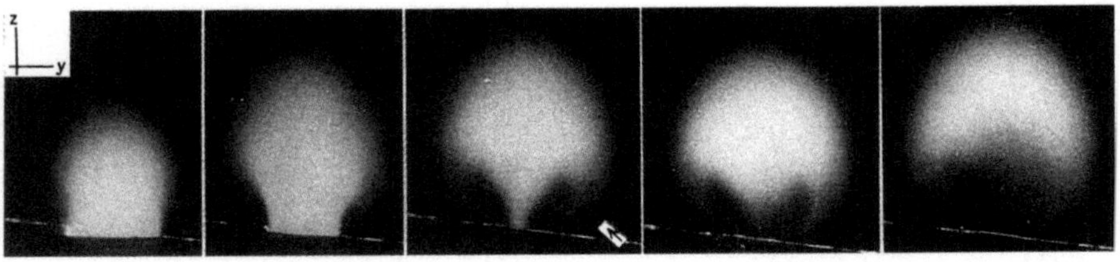

$x/d = 0$ $x/d = 0.325$ $x/d = 0.5$ $x/d = 0.525$ $x/d = 0.65$

Fig.7. Variations of jet cross-section parallel to yz-plane with streamwise distance x ($d = 20$ mm, $m = 2$, unconfined jet)

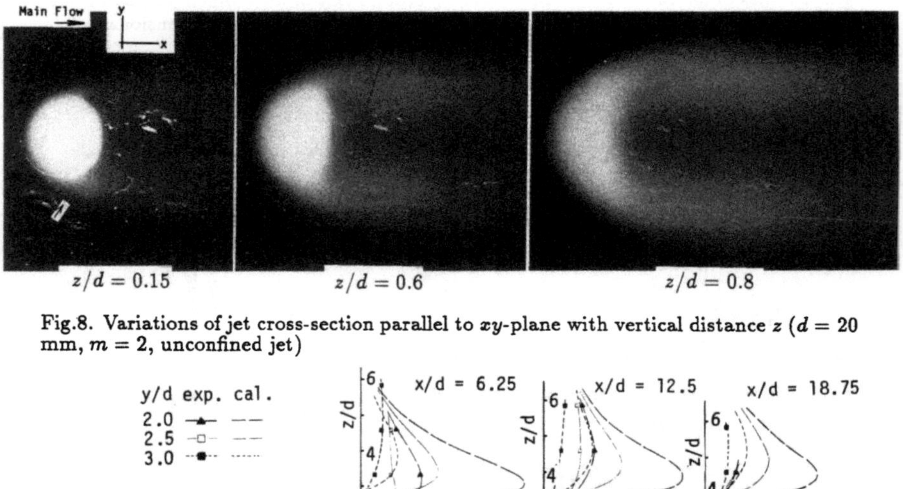

$z/d = 0.15$ $z/d = 0.6$ $z/d = 0.8$

Fig.8. Variations of jet cross-section parallel to xy-plane with vertical distance z ($d = 20$ mm, $m = 2$, unconfined jet)

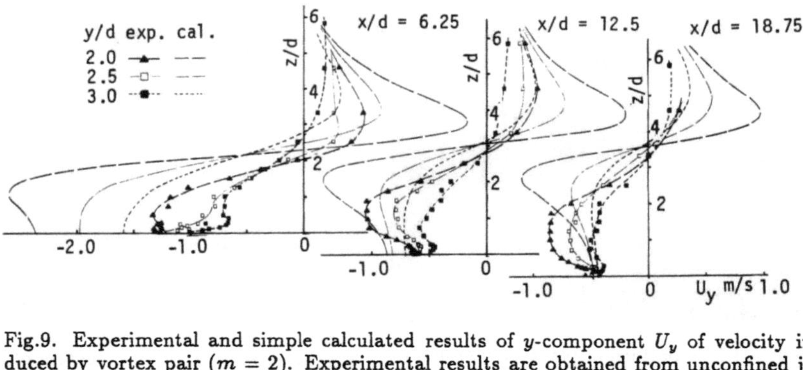

Fig.9. Experimental and simple calculated results of y-component U_y of velocity induced by vortex pair ($m = 2$). Experimental results are obtained from unconfined jet in Fig.2

Kinematical Investigation of The 2-D-Flow over an Oscillating Airfoil Under Stall

Heinrich Vollmers and Wolfgang Geißler
German Aerospace Research Establishment (DLR)

The data to be analysed have been created by a numerical simulation of two-dimensional flows over oscillating airfoils NACA 23012 and a NACA 0012 airfoil with buffet. The numerical method solves the unsteady two-dimensional Navier-Stokes equations on a dynamically adapted C-grid with 157*59 points in space using finite differences and a conventional Baldwin/Lomax eddy viscosity turbulence model.

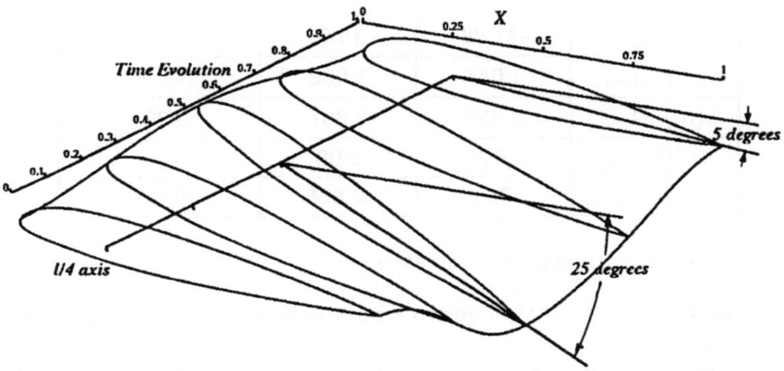

Figure 1. Sketch of configuration with sinusoidal pitching motion in space-time domain.

As sketched in Figure 1 the airfoil oscillates harmonically between 5° and 25° incidence having the spatially fixed point of rotation at a quarter of the chord. One period has been resolved by 50,000 steps to ensure accuracy in time. The free stream Mach numbers Ma of the oscillating ('deep dynamic stall') cases are 0.283 and 0.12, for the buffet case Ma = 0.75. Further details of the treated cases are listed in the table.

As conventionally done a flow case is presented as consecutive drawings of instantaneous functions of the flow field, e.g. Mach-number, curl of velocity, pressure, etc. The dynamic behaviour of streak lines and time lines is shown in a video-movie, too. Examples of pictures from the movie for case 1 are reproduced in Figure 2.

Because of the fleeting impression of the frames in a movie other possibilities for presenting the information have been investigated for their suitability. A further intention is the extraction and concentration of information for purposes of comparison and to support imagination of the mutual dependencies of effects.

A survey of the time evolution of the flow of case 1 is achieved by presentations using the time as a coordinate. In Figure 3 the evolution of pressure on the surface, of regions of separation, traces of critical (singular) points of the velocity field with vortex classification are displayed. Thus we can correlate directly the position of the pressure valley on the surface to the position of the stall vortex and fur-

	case 1	case 2	case 3 (buffet)
ω^*	0.301	0.301	(Str=0.044)
Ma_∞	0.283	0.12	0.75
Ma_{max}	1.18	0.485	1.478
$c_{L_{max}}$	2.3	2.53	0.693
$c_{L_{min}}$	0.435	0.328	0.10
$c_{D_{max}}$	1.05	1.01	0.07
$c_{D_{min}}$	-0.007	-0.001	0.014
$c_{M_{max}}$	0.134	0.061	0.132
$c_{M_{min}}$	-0.555	-0.572	0.033
$c_{p_{max}}$	1.167	1.203	1.13
$c_{p_{min}}$	-11.08	-14.59	-1.50
$(\rho/\rho_\infty)_{max}$	1.047	1.009	1.295
$(\rho/\rho_\infty)_{min}$	0.437	0.867	0.385
$(u/U_\infty)_{max}$	2.998	2.98	1.72
$(u/U_\infty)_{min}$	-2.10	-2.066	-0.826

Table. Properties of treated cases

thermore read the velocity of stall vortex from the figure: It is about 25% of the free stream velocity.

The trace of the stall vortex originates from a line of vanishing skin friction. At first the vortex moves to the leading edge, then it creates a straight trace on its way to the trailing edge. There its trajectory is strongly disturbed by the counter rotating vortex. As a reaction the stall vortex is considerably weakened and nearly decaying.

The lines of vanishing skin friction border regions of reverse flow. They are the locations of separation and reattachment.

Figure 4 shows the a mapping of time lines into a space time domain.

for references see:

H.Vollmers, The Recovering of Flow Features from Large Numerical Data Bases. in VKI-Lecture-Series on 'Computer Graphics and Flow Visualisation in Computational Fluid Dynamics' at von Karman Institute, Rhode-Saint-Genese, Belgium, Sept. 16-20, 1991

W.Geißler, Numerical Investigation of Dynamic Stall on Oscillating Airfoils. Proceedings of International Symposium on Aeroelasticity, June 3-6, 1991, Aachen, Germany

W.Geißler, H.Vollmers, H.Große, Dynamic Stall. video movie, Institut für wissenschaftlichen Film, Göttingen, to appear 1992

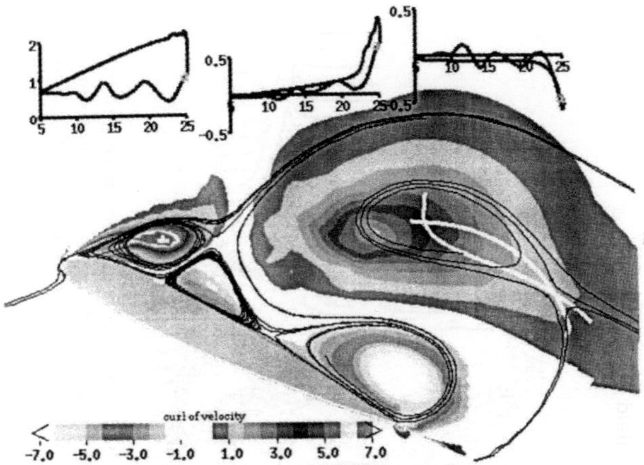

a)

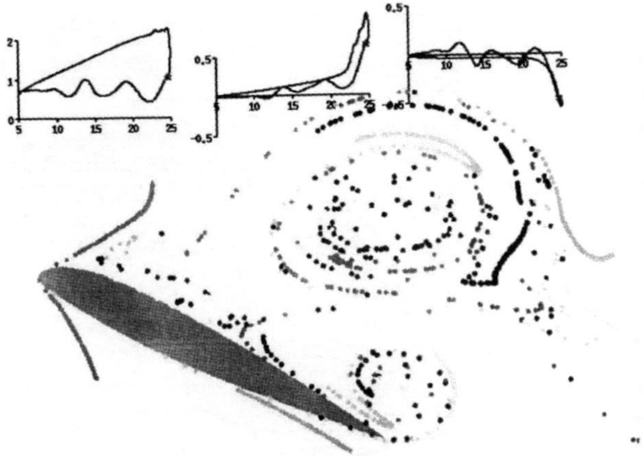

b)

The last but three line assembly is not displayed.

Figure 2.

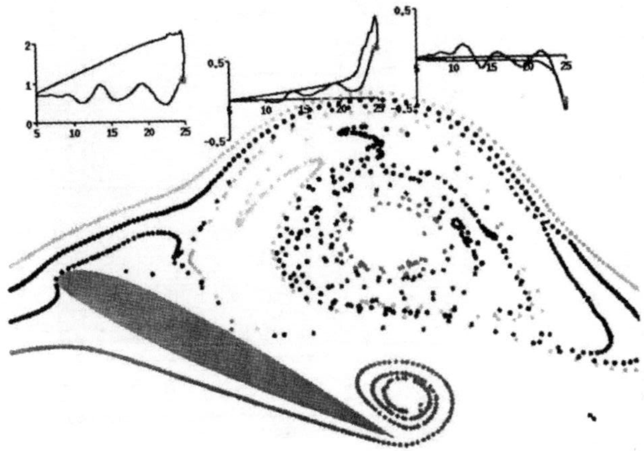

c)

Three figures describing the flow situation at the beginning of downstroke. They are taken from movie sequences. The diagrams in the upper part of the figures show lift, drag, and moment, respectively, versus angle of incidence. The gray points indicate the current position of the airfoil.

a) The background displays the the values of curl of velocity coded in gray tones. White (thick) lines (and points) are segments of zero contours of velocity components in the neighbourhood, where the respective other component is zero. The crossings indicate critical (singular) points of the velocity field. Black (thin) lines are instantaneous stream lines that represent the direction of the velocity field. From this, the classification of the critical points is obvious.

b) Time lines are simulated by particles starting on a line segment in front of the airfoil. They have initially constant distance and are emitted at constant time intervals.

c) Streak lines consist of particles being emitted from fixed points at constant time intervals.

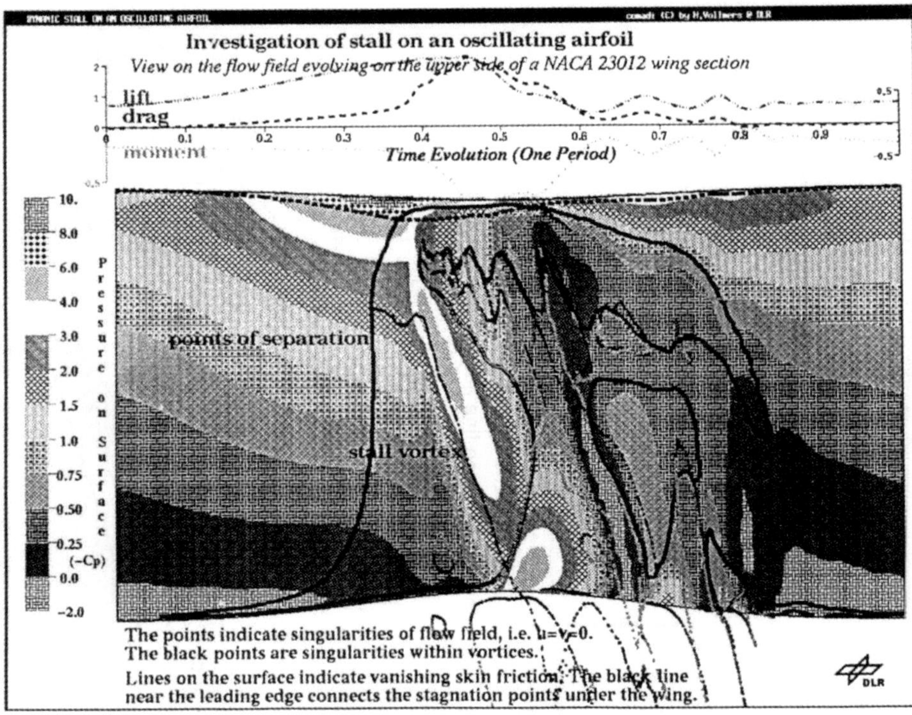

Figure 3. Survey of time evolution of the flow field of case 1.

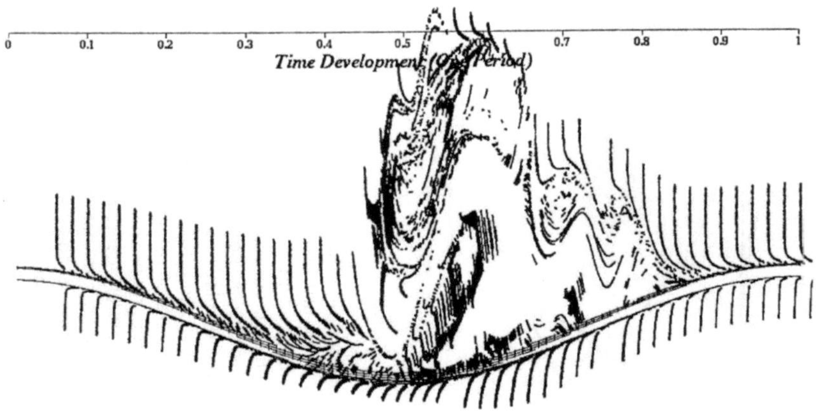

Figure 4. Cut of paths of particles being emitted from line segments in front of the airfoil at constant time intervals (cf. Figure 2 b). The line segments are mapped onto the cutting plane at x = 0.9 by the flow field. Two line assemblies are not displayed.

Wake Flow and Vortex Genearation Behind Groups of Cylinders in Tandem and Triangle Arrangement

W. KÜHN [1)], F. KLATT [2)]

[1)] Deutsche Airbus GbmH, EF 17, Hünefeldstr.1-5, 2800 Bremen 1, Germany
[2)] Strömungstechnisches Labor,KAI e.V., Rudower Ch. 6, 1199 Berlin, Germany

1. Introduction

In recent years, the wake and the vortex generation behind cylinders has been the subject of several investigations which have explored instability problems. As the main emphasis of these studies has been the laminar wake and comparatively high Reynolds numbers, there is a lack of experimental data for $Re \approx 10^3$ [1],[2]. The investigations have utilized different methods and devices in order to control the vortex shedding and the development of the wake. For example, splitter plates [3] and base bleed [1], heating of the cylinder [4] or of the wake [5] can be mentioned. By using one or two additional (control) cylinders with a diameter much smaller than the main cylinder, it is possible to suppress the vortex shedding in a low Reynolds number region [6].

2. Experimental studies

The present investigation was undertaken in order to gain the insight in the instability mechanism leading to the vortex street. As the interaction of the two shear layers of a cylinder plays an important role in vortex generation models, control cylinders have been used to affect the vortex shedding and the wake. In contrast to several other devices, such as splitter plates or grids, control cylinders do not suppress the direct interaction between both shear layers.

The measurements have been performed at Reynolds numbers $Re_{dH} \approx 1000...5000$ (dH diameter of the main cylinder). The diameter dV of the control cylinder(s) has been of the order of magnitude of dH. Therefore, in contrast to the common use of control cylinders, even the smallest cylinders have had a turbulent wake ($Re_{dV} > 300$).

Hot wire measurements were conducted in a wind tunnel having a nozzle outlet of 0.58 x 0.58 m^2 cross section. Under these conditions length-to-diameter ratios of the main cylinder l/dH>45 were possible. According to the Reynolds number under consideration, the values l/dH are sufficiently high to neglect the influence of the cylinder end-effects.

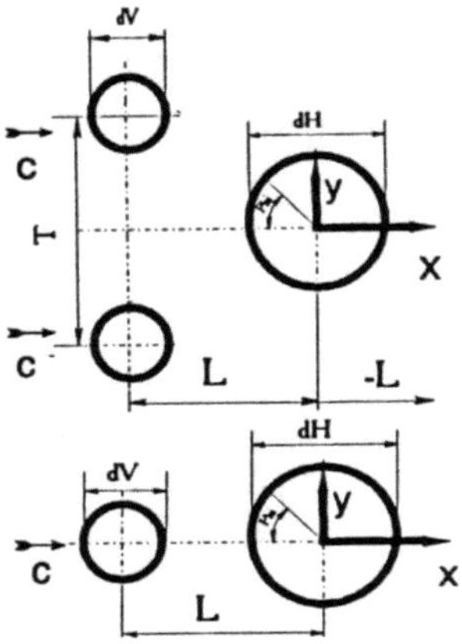

Fig.1. Sketch of tandem and triangle arrangement.

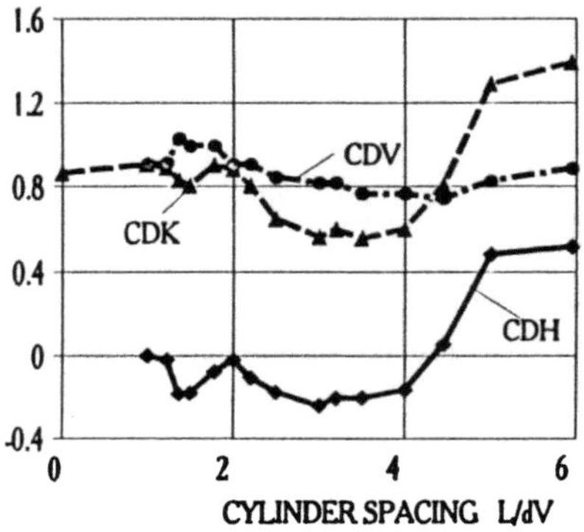

Fig.2. Form drag coefficient of a tandem arrangement. d= dV=dH. ((V) upstream, (H) downstream, (K) both cylinders).

In addition to the wind tunnel experiments, flow visualizations were performed in a water tunnel. Light sheet technique and conventional streak line visualization was used to observe the interaction between the vortices generated by the control cylinder(s) and the main cylinder.

3. Tandem arrangement

The spacing L and the diameter ratio dV/dH has been varied. The tandem arrangement shows a minimum drag (CW) and form drag (CDK) coefficient at dimensionless spacings L/dV ≈ 3, depending on the ratio dV/dH (Figs. 2 and 3). Minimum values CW(L/dV,dV/dH) ≈ 0.4CW(0,1) were obtained, where CW(0,1) represents the drag coefficient of a single cylinder with a diameter equal to dH (Fig.3). The minimum drag spacing is characterized by an impingement of the vortices generated by the control cylinder on the upstream side of the main cylinder at $\phi_H \approx \pm 60..70°$. Then a dominant twin eddy shedding can be

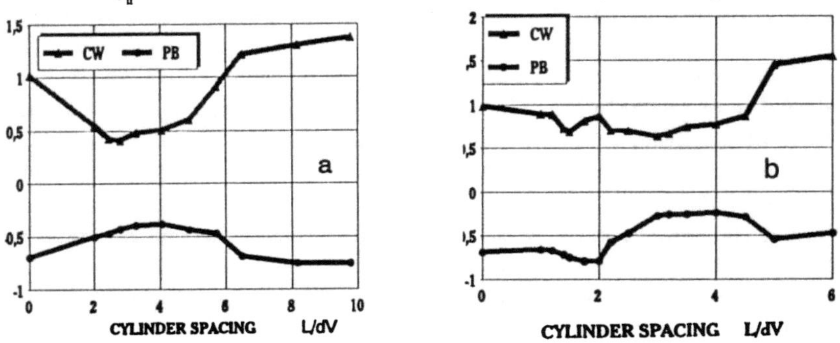

Fig. 3. Drag coefficient and dimensionless base pressure pB of the main cylinder. a: dV=dH. b: dV/dH=8/13.

detected behind the main cylinder in the wind tunnel as well as in the water tunnel experiments (Fig.4), but the vortex generation can stochastically switch over to the common alternating street. In accordance with the low drag coefficients observed for L/dV ≈ 3, a narrow wake exists during the formation of a symmetrical vortex street.

4. Triangle arrangement

Symmetrical triangle arrangements of two control cylinders (dV= 6 mm) in front of (L>0) and behind (L<0) a main cylinder (dH=14 mm) have been investigated. The spacing between the main cylinder and the control cylinders in flow direction (L) and between the two control cylinders (T) has been varied.

The most interesting results are obtained when a strong interaction takes place between the wakes of the cylinders. When the control cylinders are located in front of the main cylinder (L/dV≈1,T/dH≈1), a vortex street is generated which consists of large alternating vortices. In contrast to this result, the vortex shedding behind the main cylinder is suppressed for L/dH≈-1

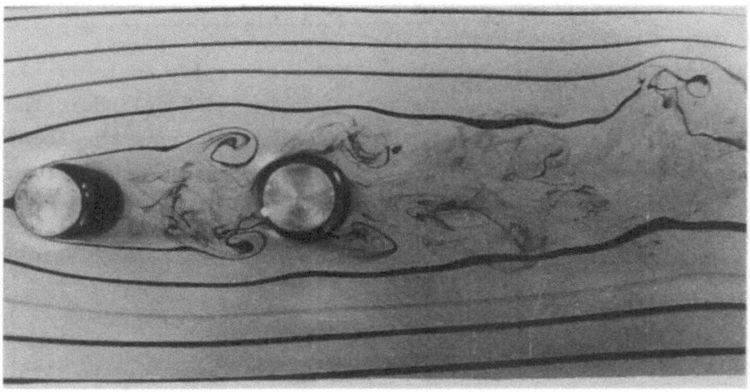

Fig.4. Vortex generation behind a tandem arrangement. dV=dH, L/dV=3.5.

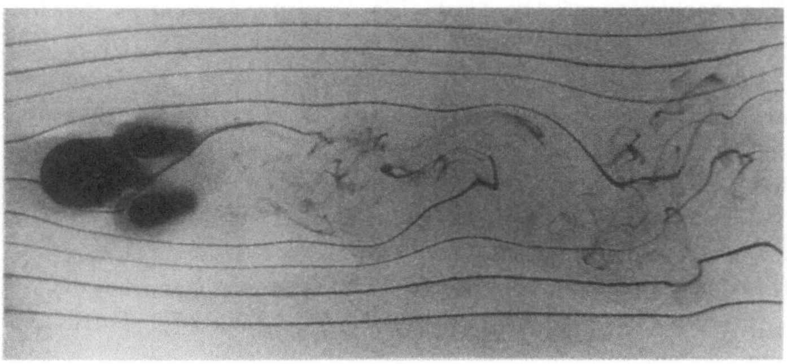

Fig.5. Vortex generation behind a triangle arrangement. L/dH=-8/7, T/dH=8/7.

and T/dH≈1. Under these circumstances a narrow wake with weak disturbances occurs. The drag of the whole arrangement is not very different from CW(0,1). Its value depends on the existing type of the wake flow, which is very sensitive to the geometry of the cylinder configuration. Therefore we have failed to manage a symmetrical wake without lift forces (Fig.5). But, in contrast to the intermittency of the vortex street behind low drag tandem arrangements, the type of wake "selected" by the flow has not changed during a running experiment.

5. Conclusions

The tandem as well as the triangle configuration show features for $Re_{dH} \approx 10^3$ which are also characteristic in case of a laminar wake. Nevertheless, the vortex generation is of different nature behind these two arrangements.
The low drag tandem arrangement is characterized by an intermittent formation of the vortex street. The vortices are shed by the control cylinder. They disturb the boundary layer of the main cylinder, but the separation point from the downstream cylinder moves only a few degrees if the spacing L/dV is varied. A comparatively high base pressure pB (Figs. 3a and 3b) is an important consequence of the symmetrical vortex shedding from the main cylinder, causing mainly the drag reduction observed.
The triangle arrangement provides the opportunity to create large scale vortex streets as well as to suppress the vortex shedding from the main cylinder. The suppression takes place if the control cylinders are located near the shear layers separated from the main cylinder. With reference to the effects in laminar wakes, the control cylinders should cause a reduction of the mean velocity gradients and, consequently, of the mean vorticity .

6. References

1. Berger, E.; Wille, R.: Periodic flow phenomena. Ann. Rev. Fluid Mech. 4(1972) 313-340.
2. Chen, S.-S.: Flow induced vibration of circular cylinder structures. Hemisphere Publ. Corp., Washington, New York, London 1987.
3. Gerrard, J.H.: The mechanics of the formation region of vortices behind bluff bodies. J. Fluid Mech. 25(1966) 401-413.
4. Lecordier, J.C.; Hamma, L.; Paranthoen, P.: The control of vortex shedding behind heated circular cylinders at low Reynolds numbers. Exp. Fluids 10(1991) 224- 229.
5. Mori, Y.; Hijikata, K.; Nobuhara, T.: A fundamental study of symmetric vortex generation behind a cylinder by wake heating or by splitter plate or mesh. Int. J. Heat Mass Transfer 29(1986) 1193-1201.
6. Strykowski, P.J.; Sreenivasan, K.R.: On the formation and suppression of vortex shedding at low Reynolds numbers. J. Fluid Mech. 218(1990) 71-107.

Base Bleed Efficiency Criterion

V.L. ZHDANOV
Byelorussian Academy of Science, Minsk, Byelorussia
H. ECKELMANN
Max-Planck-Institut für Strömungsforschung, Göttingen, Germany
B. STASICKI
DLR Institut für Experimentelle Strömungsmechanik, Göttingen, Germany

Summary

The base bleed effect of a two-dimensional body with different slot widths at the trailing edge is examined. It is shown that the value of the maximum base pressure depends on the slot width but in all cases corresponds to the same value of the momentum flux density. This parameter is determined only by the body drag. Turbulent characteristics of the wake behind the body are measured at the optimum value of the momentum flux density. Visualization of the flow pattern immediately behind the trailing edge of the body was made in the wake without and with bleeding.

Introduction

Blowing a certain amount of fluid from the rear part of the body into the flow field leads to a change of the base pressure. For a relatively small bleed rate the base pressure is increased [1-3], while for a high bleed rate it is decreased [4]. So there is an optimum value of the bleed rate which corresponds to the maximum base pressure. Even for the same body but with a different width of the slot at the trailing edge through which the bleeding takes place, the value of the base pressure depends on the slot width at the same bleed rate [1-3,5]. I.e. the maximum of the base pressure is a function of the bleed rate as well as of the width of the slot. It has been shown [4] that for models with narrow slots the maximum base pressure occured at about the same value of the momentum flux density of the jet. By applying the known experimental findings to the models with wide slots [1-3] we did not obtain the same result. We guess that this is connected with some differences of the bleed effect for thin and thick jets. Hence systematic investigation of models with narrow and wide slots was necessary.

It is known that the flow pattern behind a body is determined mainly by the development of the base pressure distribution. There are only a few investigtions available where base pressure measurements and flow visualization [1,3] or turbulence characteristics of the wake were simultaneously recorded [4,5]. Therefore it can be expected that the measurements of the mean and fluctuation velocities in the bleeding regime corresponding to the maximum base pressure and a flow visualization of the near region of the wake will improve the understanding of the interaction of the bleeding jet with the shear layers behind the model.

Experimental Setup

The experiments were carried out in an open return wind tunnel. The mean velocity of the potential flow in the test section was 14 m/s. The test section has a cross section of 280×1400 mm^2. The two-dimensional model, which is described in detail in [5], has a semicircular leading edge and two parallel side walls. It is closed by a blunt trailing edge. The cord is 200 mm, the hight $H = 40$ mm, and the length 280 mm. The slot width h in the rear surface of the body can be varied. Models with eight different slot widths were investigated: $h/H = 0.02$ (mod.1), 0.05 (mod.2), 0.07 (mod.3), 0.1 (mod.4), 0.18 (mod.5), 0.26 (mod.6), 0.4 (mod.7), and 0.75 (mod.8).

The flow rate of the bleeding jet was measured by a gas meter with an accuracy of 0.5%. The shedding frequency, mean velocity, longitudinal and transverse velocity fluctuation profiles were determined by both single wire and X- wire probes connected to "DANTEC" hot-wire anemometers. Fog was used for the visualization of the wake. The fog generator (Schaidt Foggy 2000) was placed outside of the wind tunnel. Fog was injected from a tube in to the settling chamber of the wind tunnel. The tube was mounted so that the fog impinged on the leading edge of the body symmetrically. The model and wake were illuminated by a xenon flash lamp placed on the upper side of the tunnel. The light pulse duration was about 50 μs. The pulse energy was varied between one and three Joules. The flow pattern was observed by means of a CCD video camera (Sony XC-77CE) connected to a frame grabber card (ARCO FG 2000) installed in a PC 486. The frames were displayed onto a video monitor and printed out on a laser printer. The flash lamp control circuit (Drello Scop 1018) was driven by a Trigger Conditioner 889 [6] to obtain a proper synchronization of the system.

Results and Discussion

The present investigations show that for models 1 to 5 the maximum base pressure is achieved at a momentum flux density of $C_q(V_j/U_o) = 0.04 \pm 0.001$. For models 6 to 8, the optimum value of $C_q(V_j/U_o)$ decreases with increasing slot width (Fig.1). For these models this trend also correlates well with the growth of the base pressure for zero bleeding ($C_q = 0$). The base pressure of the model 8 is about 16% higher than for models 1-5. This means that the bodies with narrow and wide slots have different drag coefficients. So, there is no optimum value of the momentum flux density for all investigated models and this is the reason why there is no single value of this parameter for the models with wide slots [1-3]. The momentum flux density is also more sensitive to pressure distribution at the base of the model in comparison with the bleed rate.

Shedding frequency measurements for models 6 to 8 show that the relation between the Strouhal number and the bleed rate C_q is identical to the one in the ref. [1]. The Strouhal number of the models 1 to 4 increases with increasing bleed rate up to a certain rate above which it becomes impossible to observe. At zero bleed the shedding frequency is highest for the models 6 to 8 which are characterized by the lowest body drag. Using the momentum flux density as the independent variable allows a collapse all experimental results into two frequency dependency curves (Fig.2). This means that there must be a difference between the interaction of narrow and of wide jets with shear layers.

The turbulence characteristics of the wakes of models 1 to 4 have already been investigated [4]. The characteristics of the wake of models 6 to 8 are presented here. The measurements are made for six different streamwise positions ($x/H = 3, 7, 10, 15, 20, 30$) at zero base bleed and at the optimum value of the momentum flux density. Without bleed the turbulent structure of the wake has the same qualitative features as the models 1 to 4. Quantitative differences are associated with a larger mean velocity defect and a higher level of the transverse fluctuations. The base bleed modifies the flow field structure so that the power-law dependencies peculiar to a pure wake are now valid for all investigated parameters: $\Delta U = u_1 = u_2 \simeq (x)^{-0.5}$, $\delta_U = \delta_1 = \delta_2 \simeq (x)^{0.5}$ (Figs.3 to 5). It is worth noting that behind models 1 to 5 the decay of the transverse velocity fluctuations follow the power law $u_2 \simeq (x)^{-1}$. Taking into account that the difference between longitudinal and transverse fluctuations decreases when the width of the bleeding jet is increased we can draw the conclusion that the thick jet reduces the flow anisotropy.

Flow visualization of the near wake region shows that the narrow jet does not prevent an interaction of the shear layers at the optimum value of the momentum flux density (Figs. 6a and 6b). The shear layers generate vortices just behind the trailing edge ($x/H \simeq 1$) and the process is stable. The thicker the jet, the further the first vortex is created downstream. When the bleed jet becomes wider (Fig. 6c), vortices form only when the shear layers have merged ($x/H \simeq 2$). The shear layers behind the models 7 and 8 merge at a distance $x/H \simeq 4$ to 6. Vortices are visible at a distance $x/H \simeq 6$ (Fig. 6d) and they become random at a larger distance (Fig.6e).

References

[1] Bearman, P.W. 1966 Investigation into the effect of base bleed on the flow behind a two-dimensional model with a blunt trailing edge. AGARD-CP-4, 485-507.

[2] Wood, C.J. 1967 Visualization of an incompressible wake with base bleed. J. Fluid Mech. **29**, 259-272.

[3] Wood, C.J. 1964 The effect of base bleed on a periodic wake. J. Roy. Aero. Soc. **68**, 477-482.

[4] Zhdanov, V.L. & Eckelmann, H. 1991 An experimental investigation of base bleed effect on the wake turbulent structure behind two-dimensional blunt model. MPI für Strömungsforschung Göttingen, Rep. 9/1991

[5] Zhdanov, V.L., Eckelmann, H. 1990 The effect of jet bleed on base pressure distribution, shedding frequency and mean velocity profiles in the wake behind a two-dimensional blunt model. MPI für Strömungsforschung Göttingen, Rep. 9/1990.

[6] Bretthauer, B., Meier, G.E.A., Stasicki, B. 1991 An electronic Cranz-Schardin camera. Rev. Sci. Instrum. **62**, 364-369.

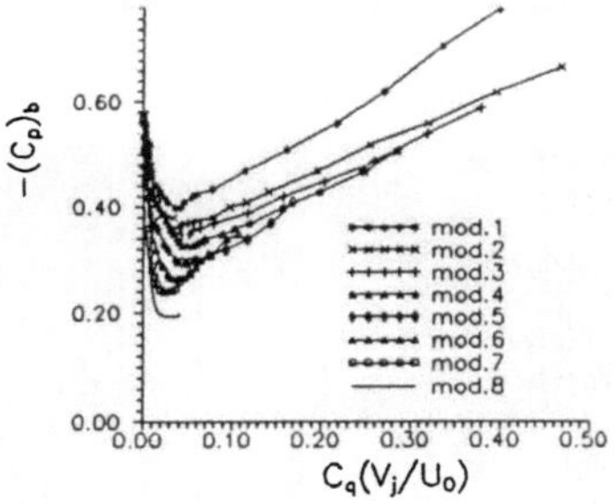

Fig. 1 **Base pressure versus momentum flux density of the jet**

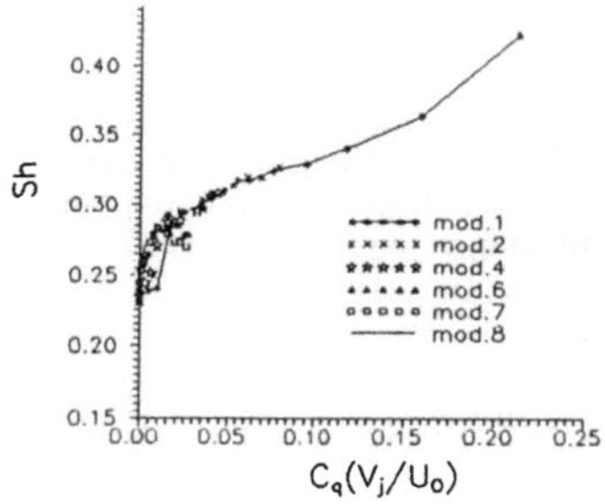

Fig. 2 **Strouhal number versus momentum flux density of the jet**

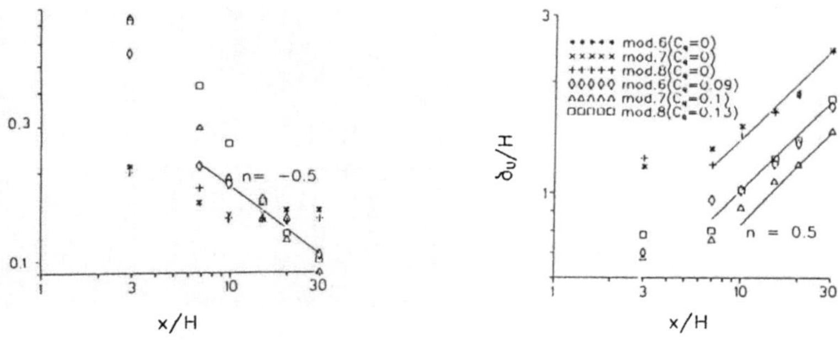

Fig. 3 **Decay of the mean velocity in the wake center (left) and spreading of the velocity profile**

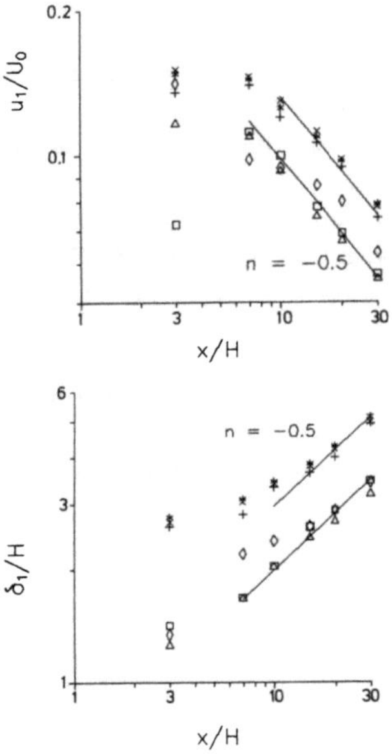

Fig. 4. Decay of the longitudinal velocity fluctuations in the wake center (top) and spreading of the fluctuation profile

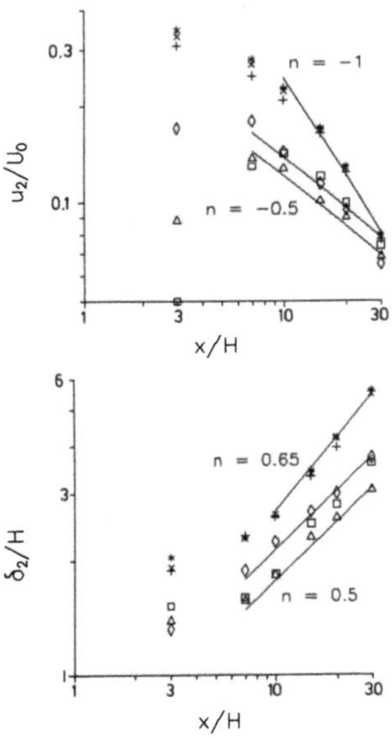

Fig. 5. Same as Fig.4 for the transversal velocity fluctuations

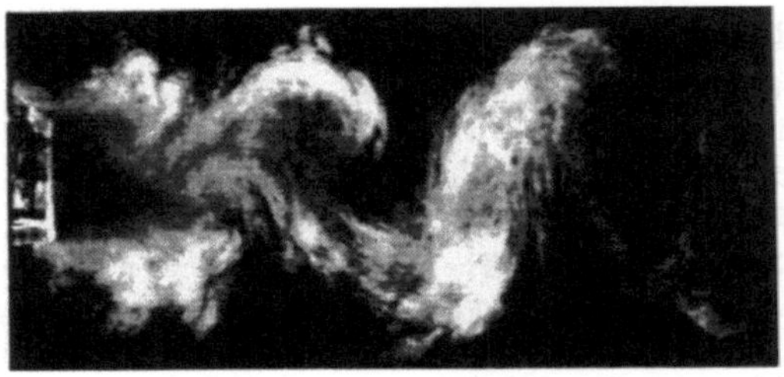

a) mod. 1 h/H=0.02

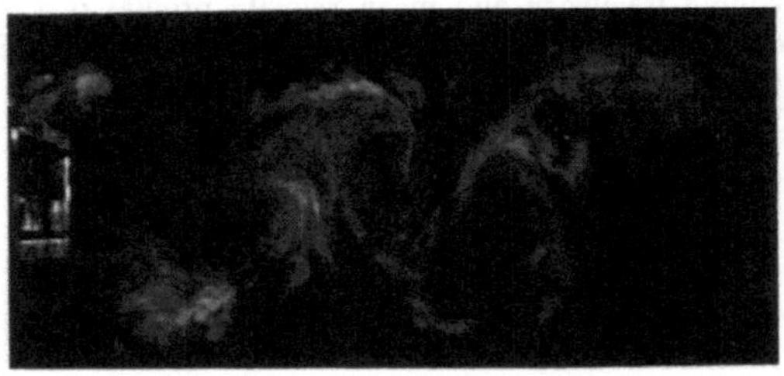

b) mod. 4 h/H=0.1

c) mod. 6 h/H=0.26

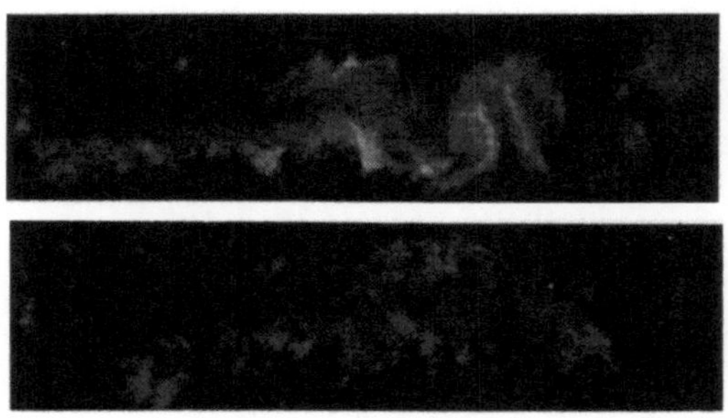

d) *(top)* and e) mod. 8 h/H=0.75

Fig. 6. Wakes behind the models at the optimum value of the momentum flux density of the jet

Wake Control

PETER A. MONKEWITZ

Department of Mechanical, Aerospace & Nuclear Engineering
University of California, Los Angeles, CA 90024-1597, USA

Summary

In this overview of wake control or, more precisely of the control of vortex shedding from bluff objects, an attempt is made at illuminating and interpreting different observations from the viewpoint of hydrodynamic stability theory. Given the limited space, the presentation does not address other equally valid viewpoints and no attempt has been made at completeness either. After a brief introduction and a summary of the concepts of absolute and global instability, a classification of control measures into three categories is proposed. They are: the steady modification of boundary conditions, unsteady open-loop forcing and feedback control. For each of the categories, selected examples are presented: in the first group, base-bleed and the effect of a disc mounted behind an axisymmetric "torpedo" are discussed. In the second and third group, transverse cylinder oscillations are examined in more detail both open-loop and as part of a proportional feedback loop.

1. Introduction - A Classification of Control Measures

When talking about control, one has to first state its purpose. In the case of wake control this has in most cases been the suppression of vortex shedding, such as Kármán shedding behind a circular cylinder, which in turn reduces or eliminates unsteady loads on the bluff body and in many cases also reduces the steady form drag. However, suppression of vortex shedding is not the only control objective worthy of consideration. One can very well imagine cases in which an enhancement of vortex shedding activity is desirable to enhance, for instance, some mixing process.

For the following discussion it will be useful to distinguish between three different categories of control measures which call for different interpretations. The first, and by far the most common method of modifying the wake dynamics consists of a suitable STEADY OR QUASI-STEADY MODIFICATION OF BOUNDARY CONDITIONS. A typical example is the insertion of a splitter plate behind a cylinder which will be discussed in section 3, along with more recent results. In cases where the control objective varies with time, this first method is often unsuitable. The simplest approach with built-in flexibility consists in UNSTEADY OPEN-LOOP FORCING of the flow. This requires one or more actuators as well as a complete "look-up map" relating actuator parameters (frequency, amplitude, wave form, etc.) to relevant flow quantities. This approach is considered in section 4. In section 5, finally, CLOSED-LOOP (FEEDBACK) CONTROL is discussed, where the "map" for the open-loop forcing is replaced by a control law that generates forcing inputs in function of the state of the wake which must be monitored in real time.

Before moving on to various examples for the three different kinds of control, some concepts of stability theory essential to the discussion are briefly summarized.

2. A Brief Summary of Concepts from Instability Theory

The concepts of absolute/convective instability and of global instability will be used to present a unifying view of the effect of different control measures. Absolute and convective instability are defined according to the long-time behavior of the impulse response of a PARALLEL mean profile of velocity, density, etc., i.e. a profile which does not change with streamwise distance [1,2]. In short, if in a given frame of reference the linear impulse response is unbounded as $t\to\infty$, i.e. "swamps" the entire parallel flow, one speaks of absolute instability. If, on the other hand, the impulse response eventually decays to zero everywhere, i.e. if all growing disturbances are "washed out" of the flow domain, the instability is called convective. In wake flows, which are nonparallel because of pressure gradients and viscous forces, this concept can therefore only be applied to LOCAL velocity profiles, i.e. to an artificial parallel flow with a velocity profile equal to the streamwise velocity profile at some fixed station of the nonparallel wake flow. In order for the instability properties derived from such a LOCALLY-PARALLEL approach to be relevant to the actual nonparallel flow, one has to require that the mean flow is weakly nonparallel. This is conveniently expressed by the smallness of the ratio $\epsilon=\lambda_{typ}/\mathcal{L} \ll 1$ of a typical instability wave length and the mean-flow evolution length $\mathcal{L}=[d(\ln w/dx)]^{-1}_{typ}$ with $w(x)$ being the local width of the wake.

It has been shown recently (see [2-4] & the paper by LeDizès et al. in this volume) that the temporal stability of a weakly nonparallel flow can be analyzed by asymptotic methods. Under specific assumptions regarding the spatial evolution of local stability properties, the two-dimensional, linear global modes have been shown to be driven by a "wavemaker" at the streamwise location x^t, where the local wave number k^t and the local frequency $\omega^t=\omega(k^t)$ are determined by $\partial\omega/\partial k=0$, i.e. by the condition of vanishing group velocity, and $\partial\omega/\partial x=0$. In a neighborhood of x^t, a weakly nonlinear stability analysis of global modes has been carried out under rather restrictive assumptions for two-dimensional, i.e. z-independent wake flows. Nonetheless it appears that the theory remains useful as a conceptual model well outside its range of validity.

The essence of the analysis is a two-stage scenario for the evolution of the NONPARALLEL impulse response near x^t, i.e. $|\xi|\leq O(1)$, where $\xi=\epsilon^{1/2}(x-x^t)$: In a first stage with characteristic time scale $\tau=\epsilon t$ the initial impulse evolves into the most unstable LINEAR global mode, described by $B(\tau,\xi)$. This is followed by exponential time-growth of the linear global mode AS A WHOLE, i.e. with the same growth rate at every point in space, before saturation occurs on the slower time scale $T=\epsilon^{3/2}t$. The assumption of this slow saturation time scale is equivalent to requiring that the Reynolds number is close to the critical Re_{cr} where the first global mode becomes unstable, i.e. $(Re/Re_{cr}-1)=O(\epsilon^{3/2})$. Hence, as usual in weakly nonlinear theory, the nonlinear processes can be described by a modulation $\mathbf{A}(T)$ of the linear mode B. Allowing in addition for slow spanwise amplitude variations on the scale $Z=\epsilon^{3/4}z$, any disturbance quantity Q may therefore be written as

$$Q = \mathbf{A}(T,Z)\, B(\tau,\xi)\, C(y;x^t)\, \exp[ik^t x - i\omega^t t] \tag{1}$$

The cross-stream normalized mode shape $C(y;x^t)$ is obtained from the local Rayleigh equation at x^t. The streamwise linear mode envelope $B(\tau,\xi)$ is governed by the linearized, variable-coefficient Ginzburg-Landau (G-L) equation [2,3]

$$\partial_\tau B + P_0\partial_\xi^2 B + P_1(\xi)\partial_\xi B + P_2(\xi)B = 0 \;; \quad B(|\xi|\to\infty) = 0 \,. \tag{2}$$

Here, the P_i are i-th degree polynomials in ξ. Fully established global modes correspond to solutions of the form $B_n(\tau,\xi)=\hat{B}_n(\xi)\exp(-i\Omega_n\tau)$. $\hat{B}_n(\xi)$ is the normalized global mode shape, which is a simple Gaussian for the fundamental $n=0$ [2,3], and the associated linear global frequencies are given by $\omega_{Gn}\sim\omega^t+\epsilon\Omega_n$. As soon as the most unstable global mode, say $n=0$, is established, its evolution is completely described by the SINGLE CHARACTERISTIC AMPLITUDE $\mathbf{A}$ which, in turn, is governed by a spanwise G-L equation with constant complex coefficients [4]:

$$\partial_T\mathbf{A} - \bar{\mu}\partial_Z^2\mathbf{A} + \bar{\ell}|\mathbf{A}|^2\mathbf{A} = 0 \,. \tag{3}$$

For $\partial_Z=0$ and $n=0$ (3) reduces to the well-known Stuart-Landau (S-L) equation for the amplitude $A(t) = \epsilon^{3/4}\mathbf{A}(T)\exp(-i\omega_{G0}t)$ in which the entire time dependence of Q has been included. Setting at this point $(-i\omega_{G0})=\sigma$, $\ell=\bar{\ell}\exp(-2i\omega_{G0,i}t)$ and reverting to the unscaled time $t=\epsilon^{-3/2}T$, one obtains evolution equations for the modulus and phase of $A=|A|\exp(i\varphi)$ as well as the (steady) saturation values for the amplitude $|A|$ and the frequency $\partial_t\varphi$ [5-7]:

$$\partial_t|A| = \sigma_r|A| - \ell_r|A|^3 \quad \Rightarrow \quad |A|^2_{sat} = \sigma_r/\ell_r \ ; \tag{4a}$$

$$\partial_t\varphi = \sigma_i - \ell_i|A|^2 \quad \Rightarrow \quad (\partial_t\varphi)_{sat} = \sigma_i - \sigma_r(\ell_i/\ell_r) \tag{4b}$$

Here and in the following, the subscripts "r" and "i" denote real and imaginary parts, respectively. At the level of approximation implied by (1), the coefficient σ in (4) may be considered a linear function of Reynolds number, while ℓ is a constant.

3. The Steady Modification of Boundary Conditions

When controlling a wake by modifying the boundary conditions, one simply CHANGES THE FLOW PROBLEM into one that satisfies the control objective. If the objective is to suppress vortex shedding we show that in most cases the physical modifications clearly correspond to a reduction or elimination of local absolute instability in the wake (see also [8,9]). Limiting the discussion to nominally two-dimensional and axisymmetric situations the following suppression methods have been found to be effective:

SPLITTER PLATES: they have been investigated systematically by Roshko [10] and found to both suppress vortex shedding in the near-wake and reduce the drag. This does not mean however that the entire wake becomes laminar - in this and all the following cases the far-wake remains convectively unstable and amplifies external noise. This results in oscillations that LOOK very similar to the self-excited vortex shedding driven by the near-wake, but are entirely noise-sustained and do not significantly affect the pressure distribution on the body since they develop sufficiently far downstream. For the splitter plate on the center-plane y=0 behind a 2-D body the connection to absolute instability is obvious as it forces the transverse velocity to be zero at y=0. This eliminates the sinuous or Kármán instability mode which is the only one associated with absolute instability (AI) in ordinary wakes [11].

SPIRAL PROTRUSIONS ON THE CYLINDER SURFACE (CABLES, SMOOTH CYLINDER WITH SPIRAL FENCES, ETC.): a woolen thread wound around a cylinder in a spiral or some other spiral protrusion is very effective in suppressing vortex shedding and is widely used to reduce unsteady loads on high smoke stacks for instance. One may argue that the function of the protrusion is to fix the separation lines. However, as in the case of splitter plates, it seems more illuminating to see the primary function of the protrusions as breaking the symmetries (reflection at y=0 and translation in the spanwise direction) of the mean flow and thus preventing a pure 2-D sinuous mode from developing.

SMALL CONTROL CYLINDER PARALLEL TO THE MAIN CYLINDER: the effect of small control cylinders in the near-wake shear layers has been discovered and investigated by Strykowski & Sreenivasan [12]. Again, their numerical simulations strongly suggest that the primary effect of the control cylinders is to break the mean flow symmetry.

NEAR-WAKE HEATING: it has been found that heating the cylinder or placing heating wires into the near-wake can suppress Kármán vortex shedding (see [13-15], [7]). While Lecordier et al. [15] argue that the effect of heating in air is to raise the Reynolds number, Yu & Monkewitz [16] suggested that the primary effect in an inertially dominated wake is a reduction of the density in the near-wake which leads to a reduction of absolute instability.

POROUS BLUFF BODIES, BASE-BLEED AND BASE-SUCTION: it has been known for some time that vortex shedding can be easily suppressed over a wide range of Reynolds numbers, i.e. not only close to critical, by bleeding fluid from the blunt base [17,18,7]. The same effect is achieved with porous bodies [19]. For this class of measures, the effect on local AI is very clear: as the centerline velocity in the near wake is increased, the reverse flow is eliminated and the near-wake AI disappears, i.e. the entire wake becomes convectively unstable [20,11,8]. The technique of base-bleed or suction has recently also been used for transient experiments [7]: after a supercritical base-bleed is switched off, vortex shedding builds up and saturates in close agreement with the S-L equation (4). This is shown visually on figure 1 which pertains to a rectangular flat plate of thickness D=4mm and the same proportions as in Hannemann & Oertel's [21] computations. The side view early in the transient gives a good impression of the linear global mode shape while the planview clearly shows that the transient is essentially two-dimensional, i.e. corresponds to parallel shedding [22,4]. This serves as justification for the reduction of the spanwise G-L equ. (3) to the S-L equ. (4). The problem of end-effects will therefore not be addressed in this paper and the reader is referred to [22], [4] and [7] for a thorough discussion.

AFTERBODIES: when investigating the effect of afterbodies and rear-mounted discs behind axisymmetric bluff bodies, Mair [23] (see also [24]) found that small changes in the geometry could induce dramatic changes in the drag. His results for the changes in drag coefficient of a "torpedo" with a

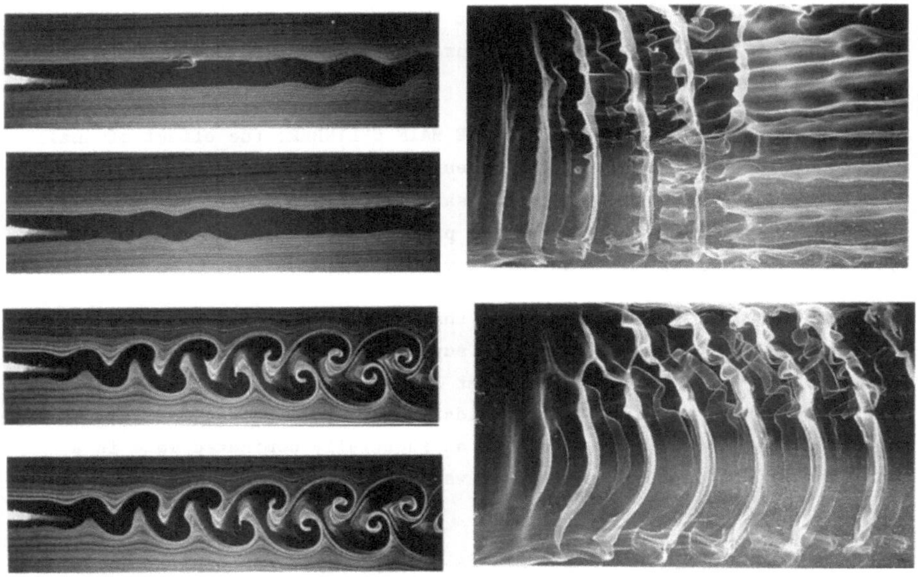

(a) (b)

Figure 1. Transient in the wake of a rectangular plate (D=4mm, chord Γ=60mm, Re=200) after switching off supercritical base bleed. a) side view with the first 3 frames 200ms apart and the last frame corresponding to saturation. Visualization by smoke wire with flow from left to right. b) plan views of the early transient and the saturated state (from [7]).

single rear-mounted disc of d=0.8D as a function of the mounting distance x from the "torpedo" base is replotted as figure 2a. It shows a dramatic "drag spike" at x/D=0.3, followed by a 35% drag reduction compared to the reference case without disc. These experiments have been repeated [25] with emphasis on the vortex shedding activity. With a pair of hot wires on opposite sides of the wake at $x_{HW}/D=2.6$ and $r_{HW}/D=0.4$ it was demonstrated that the transition between drag reduction and drag increase can be clearly associated with a Hopf bifurcation to SPIRAL vortex shedding at a Strouhal number of about 0.4 . As evidence, the power spectral density is fitted by the relation (4a), i.e. $|A|^2_{sat} = \sigma_r/\ell_r = \alpha(x_{cr}-x)/D$ with α a constant. Here, the control parameter is not the usual Reynolds number but the disc position and hence the linear growth rate σ_r is Taylor-expanded around the critical distance $x_{cr}/D=0.5$. The connection between these dramatic changes of vortex shedding activity and local stability properties is not clear at this time, but we note that the only mode to become AI in the axisymmetric wake is the first spiral mode [26] which is consistent with the observations reported here.

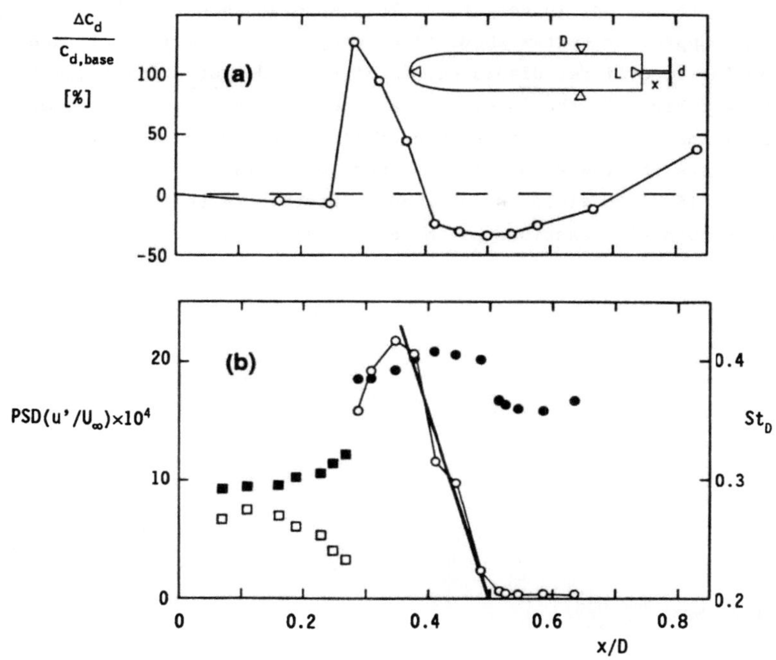

Figure 2. a) Percentage drag change versus disc position referred to drag without disc at Re=149,000 with geometry shown as insert (from [23]). b) Power spectral density (open symbols) of main spectral peak and associated Strouhal no. (solid symbols) versus x/D at Re=24,000 (from [25]).

4. Unsteady Open Loop Forcing Aimed at Modifying Vortex Shedding in the Wake

Open loop unsteady forcing in practical applications represents a significant increase in complexity vis a vis the steady modification of boundary conditions and appears only justified when the control objective varies with time or if it is energetically advantageous (in comparison to steady base-bleed for instance). From a theoretical standpoint the simple connection to global instability is lost, as the linear or nonlinear coupling between the forcing and the global modes plays an integral part in the dynamics. Examples for this class of control are transverse cylinder oscillations [27,28], rotational cylinder oscillations [29] and unsteady blowing/suction near the separation lines [30].

In the following, a few results are presented which pertain to the effect of transverse oscillations of an oblong cylinder (D=0.69mm, chord T=1.68mm) at a frequency of about 1.8 times the natural shedding frequency. The setup is identical to the one in Berger [27,28]. As discussed in [7], this type of forcing suppresses vortex shedding already at very small amplitudes, corresponding to cylinder displacements of the order of a few percent of D. Suddenly switching off the forcing at a supercritical Reynolds number therefore leads to a transient growth of Kármán shedding. This is shown on fig. 3 where the raw signal (a) has been demodulated by the techniques discussed in [6,7] to yield instantaneous growth rate (b) and frequency (c) which are accurately described by the S-L equation (4).

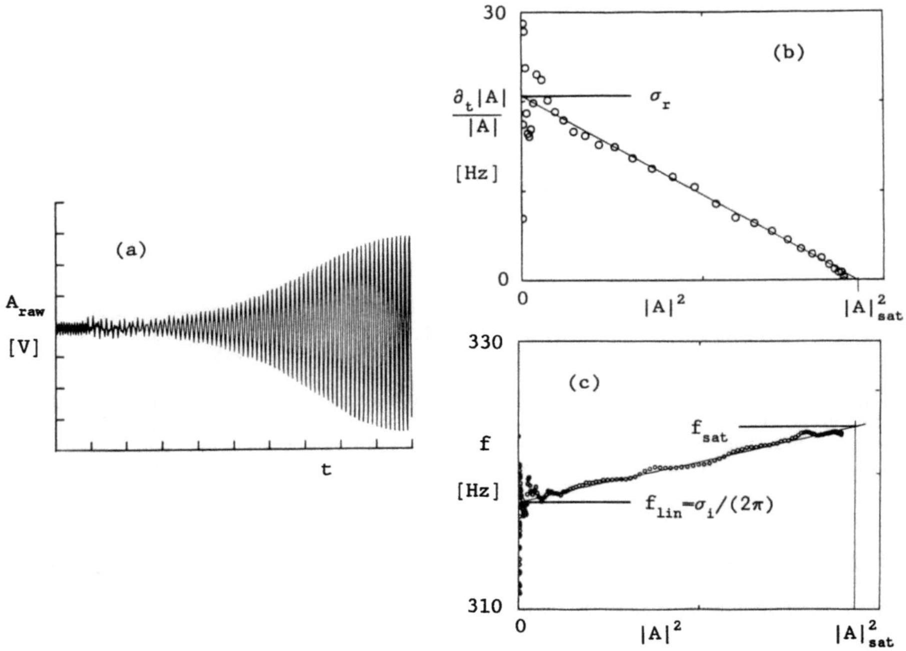

Figure 3. Supercritical (Re=84.5) transient measured by a hot wire at (x/D,y/D)=(10,1) behind the oblong cylinder (from [7]): a) Raw transient. b) Instantaneous growth rate and c) instantaneous frequency versus $|A|^2$.

A set of transients at different Reynolds numbers then allow to find [7] the coefficients of equ. (4) pertaining to the oblong cylinder as: $\sigma_r D^2/\nu = [0.116 \pm 0.002]\,(Re-Re_{cr})$; $\sigma_i D^2/\nu = [58.1 \pm 0.3] + [0.84 \pm 0.01]\,(Re-Re_{cr})$; $\ell_i/\ell_r = -[1.85 \pm 0.35]$, with $Re_{cr} = 79.2 \pm 0.2$. More importantly, by measuring these parameters at different spatial locations one can validate the "two-stage scenario" for the development of global modes. A linear global growth rate in particular, which is independent of measuring location, signals the existence of a period during which the normalized global mode shape $\hat{B}(\xi)$ does not change while the LINEAR GLOBAL MODE GROWS EXPONENTIALLY AS A WHOLE. This independence is indeed seen on figure 4 which therefore lends support to the stability-theoretical viewpoint of in this paper.

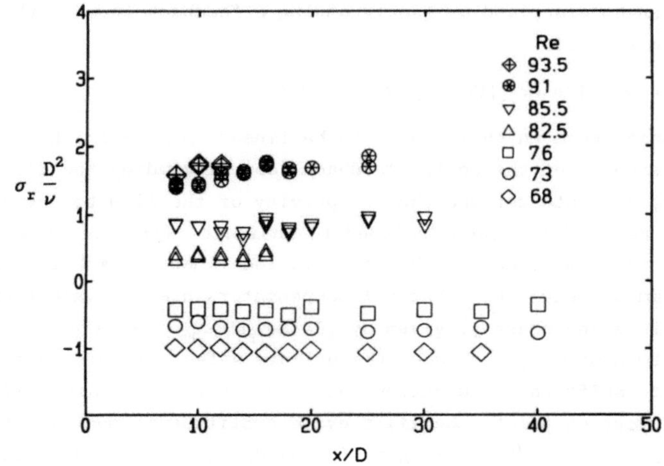

Figure 4. Linear global growth rates for the oblong cylinder versus x/D at y/D=1 and different Reynolds numbers (from [7]).

5. Closed-Loop (Feedback) Control

Renewed interest in feedback control of the wake is due to Ffowcs Williams & Zhao [31] who were able to reduce oscillations at their sensor location behind a cylinder at Reynolds numbers of several hundred. This was at odds with the theoretical analysis of this author [32], who predicted that global suppression of vortex shedding by proportional feedback is only possible close to Re_{cr}. Since then, Roussopoulos [33] has fully clarified the situation by showing that at the higher Reynolds numbers the suppression is achieved ONLY AT THE SENSOR LOCATION. In the following, feedback control near Re_{cr} is discussed in the context of Berger's experiments [27,28] which have recently been taken up again [34]. The bluff body is the same oblong cylinder as described in section 4 which is oscillated in the transverse direction according to a control law. In other words, the cylinder itself serves as actuator, while a single hot wire at (x/D,y/D)=(10,1) senses the state of the wake.

When considering theoretical models of feedback control, the issues of actuator amplitude relative to the natural limit cycle amplitude and of mode separation has to be addressed. In the simplest case the most amplified global mode with frequency ω_{G0} is well separated from higher global modes, i.e. $\omega_{G0,i} \gg \omega_{Gn,i}$, and the actuator is operated at low gain. Under these assumptions the global mode frequency $\omega_{G0,r}$ and the associated mode shape are at leading order not affected by the control. Hence this situation can be described by simply adding a feedback term to the S-L equ. (4) as shown in [35]:

$$\partial_t A = [\sigma - |Re-Re_{cr}|F(\omega_{G0,r})]A - \ell|A|^2A . \tag{5}$$

In the feedback term which is taken to be linear in A, F is the complex frequency response of the entire feedback loop, including the linear compensator, the actuator and the receptivity of the flow to the actuator. The factor $|Re-Re_{cr}|$ has been included to indicate explicitly that the gain is small of the same order as the uncontrolled linear growth rate σ_r. From (5) it is immediately clear that all compensators are equivalent which have the same gain g and phase γ, given by $|Re-Re_{cr}|F(\omega_{G0,r}) = g\exp(i\gamma)$, AT THE CRITICAL FREQUENCY $\omega_{G0,r}$. Hence, the primary effect of this type of control is to shift the bifurcation point to where $[\sigma_r - g\cos(\gamma)]=0$. Beyond the new bifurcation point, the limit cycle amplitude is modified by the factor $[1-g\cos(\gamma)/\sigma_r]^{1/2}$. As long as (5) holds, any limit cycle can be suppressed given enough control authority. As we will show, the simple model breaks down in the wake because the feedback brings higher global modes into play. This is already presaged by the asymptotic result $\omega_{Gn} \sim \omega^t + \epsilon\Omega_n$ quoted in section 2, which means that higher linear global frequencies are only separated by $O(\epsilon)$ from the fundamental n=0.

Because of the problem with higher modes, the next step in modeling the wake with feedback is to go back to the linear G-L equ. (2) which contains ALL global modes. Nonlinearity is not considered at this point as we discuss only the effect of feedback on the global instability BOUNDARIES. With proportional feedback from a sensor at ξ_S to an actuator at ξ_A equ. (2) is modified to

$$\partial_\tau B + P_0\partial_\xi^2 B + P_1(\xi)\partial_\xi B + P_2(\xi)B = g\exp(i\gamma)\, \delta(\xi-\xi_A)\, B(\tau,\xi_S) . \tag{6}$$

The main result of the investigation [32] of (6) is that above Re_{cr} a minimum compensator gain $g_{min}(\gamma_{opt})$ at the optimum phase is necessary to suppress the unstable fundamental global mode n=0. As in the simple model (5), g_{min} increases with distance from Re_{cr}. The new feature is a MAXIMUM gain g_{max} beyond which there exists no phase for which ALL global modes can be stabilized. Since g_{max} is found to DECREASE with Re, $g_{min}(Re)$ and $g_{max}(Re)$

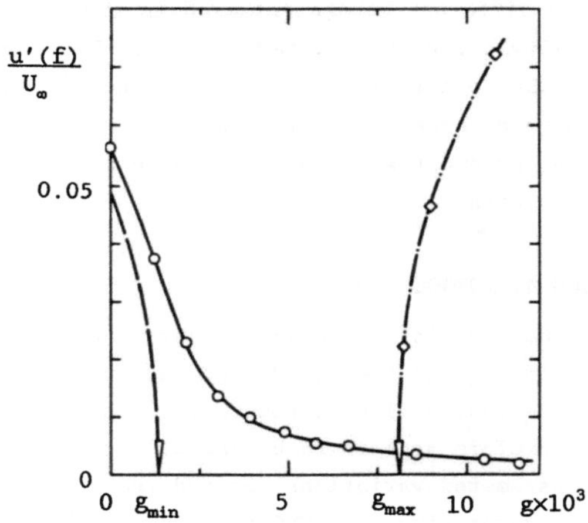

Figure 5. Dominant spectral peak of streamwise velocity at (x/D,y/D)=(10,1) versus gain g=v'(cylinder)/u'(sensor). o , Kármán mode at 320 Hz. ◇ , higher mode at 298 Hz (from [34]).

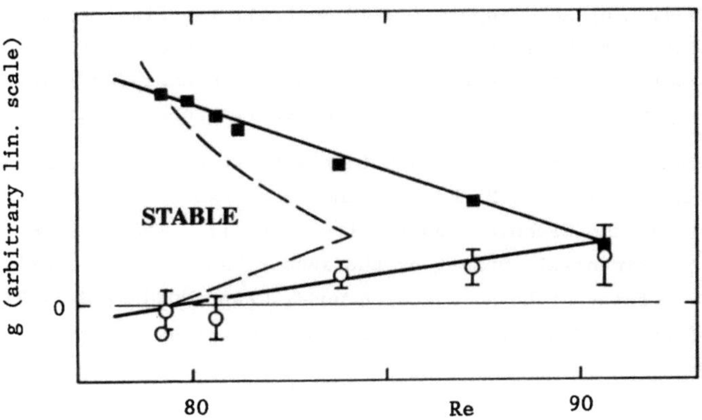

Figure 6. Stability boundaries of the oblong cylinder wake with proportional feedback for $Re>Re_{cr}=79.2$ (from [36]). o , g_{min} ; ■ , g_{max} . — — — , stability boundaries of equ. (6).

become equal at some $Re=Re_{max}$ beyond which the wake can no longer be stabilized by proportional feedback. This has been verified experimentally [34]. Figure 5 shows a typical result at constant Re and γ with stabilization of the Kármán mode above g_{min} and destabilization of a higher mode beyond g_{max}. The measured compensator gains (not including the actuator) g_{min} and g_{max} at different Re are gathered in the last figure 6, together with the stability boundary computed from (6). The computation [36,37] was performed with a priori estimates of the coefficients of (6) and shows good qualitative agreement with the measured stability boundaries. Quantitative agreement would be easy to obtain by some adjustment of the coefficients.

6. Concluding Remarks and Acknowledgments

It has been shown that the correlation between experimental suppression and enhancement of vortex shedding and the trends deduced from a simple analysis of local absolute instability is remarkable. The stability-theoretical viewpoint therefore appears to be useful for quick surveys of the effect of multiple parameter variations. Beyond that, the concept of global modes growing and saturating AS A WHOLE, which has been developed theoretically under rather restrictive conditions, is validated experimentally for situations of practical interest. The G-L model is shown to allow semi-quantitative predictions of the effect of proportional feedback control and is therefore a good candidate of a simple "plant model" for the design of more sophisticated controllers. Finally, the reader is reminded that the suppression of limit-cycle oscillations in the NEAR-wake is associated with the reduction of drag and unsteady loads, but does not imply suppression of noise-sustained oscillations in the FAR-wake.

This work is based on research collaborations with Stephane LeDizès and Chris Lee of UCLA and with Dr. Michael Schumm, Professors Eberhard Berger and Hans Fernholz of the Technical University in Berlin, who deserve most of the credit. The financial support of the author has been generously provided by the Office of Naval Research under Grant N00014-90-J-1313 and The Air Force Office of Scientific Research under Grant 89-0421.

References

1. Briggs, R.J.: Electron-stream interaction with plasmas, MIT Press 1964.
2. Huerre, P. & Monkewitz, P.A.: Local and global instabilities in spatially developing flows. Ann. Rev. Fluid Mech. **22** (1990), 473-537.
3. Monkewitz, P.A., Huerre, P. & Chomaz, J.M.: Global linear stability analysis of weakly nonparallel shear flows. To appear in J. Fluid Mech. (1993).
4. Albarede, P. & Monkewitz, P.A.: A model for the formation of oblique shedding and "chevron" patterns in cylinder wakes. Phys. Fluids **A 4** (1992), 744-756.
5. Provansal, M., Mathis, C. & Boyer, L.: Bénard-von Kármán instability: transient and forced regimes. J. Fluid Mech. **182** (1987), 1-22.
6. Raghu, S. & Monkewitz, P.A.: The bifurcation of a hot round jet to limit-cycle oscillations. Phys. Fluids **A 3** (1991), 501-503.

7. Schumm, M., Berger, E. & Monkewitz, P.A.: Self-excited oscillations in the wake of two-dimensional bluff bodies and their control. Submitted to J. Fluid Mech. (1992). See also Schumm, M.: Experimentelle Untersuchungen zum Problem der absoluten und konvektiven Instabilitaet im Nachlauf zweidimensionaler stumpfer Koerper. Ph.D. thesis no. D83 - FB 09, Technical University Berlin. Published in VDI - Fortschrittsberichte Reihe 7, no. 196, VDI - Verlag, Duesseldorf 1991.

8. Monkewitz, P.A.: The absolute and convective nature of instability in two-dimensional wakes at low Reynolds numbers. Phys. Fluids **31** (1988), 999-1006.

9. Monkewitz, P.A.: The role of absolute and convective instability in predicting the behavior of fluid systems. Eur. J. Mech. B/Fluids **9** (1990), 395-413.

10. Roshko, A.: On the drag and shedding frequency of two-dimensional bluff bodies. NACA TN 3169 (1954).

11. Monkewitz, P.A. & Nguyen, L.N.: Absolute instability in the near wake of two-dimensional bluff bodies. J. Fluids Struct. **1** (1987), 165-184.

12. Strykowski, P.J. & Sreenivasan, K.R.: On the formation and suppression of vortex shedding at low Reynolds number. J. Fluid Mech. **218** (1990), 71-107.

13. Noto, K., Ishida, H. & Matsumoto, R.: A breakdown of the Kármán vortex street due to the natural convection. In *Flow Visualization III* (W.J. Yang, Ed.), Hemisphere Publ., Washington DC 1985, 348-352.

14. Mori, Y., Hijikata, K. & Nobuhara, T.: A fundamental study of symmetrical vortex generation behind a cylinder by wake heating or by splitter plate or mesh. Int. J. Heat Mass Transfer **29** (1986), 1193-1201.

15. Lecordier, J.C., Hamma, L. & Paranthoen, P.: The control of vortex shedding behind heated cylinders at low Reynolds numbers. Exp. Fluids **10** (1991), 224-229.

16. Yu, M.H. & Monkewitz, P.A.: The effect of nonuniform density on the absolute instability of two-dimensional inertial jets and wakes. Phys. Fluids **A 2** (1990), 1175-1181.

17. Wood, C.J.: The effect of base bleed on a periodic wake. J. Roy. Aeronautical Soc. **68** (1964), 477-482.

18. Bearman, P.W.: The effect of base bleed on the flow behind a two-dimensional model with a blunt trailing edge. The Aeronautical Quarterly **18** (1967), 207-224.

19. Inoue, O.: A new approach to flow problems past a porous plate. AIAA J. **23** (1985), 1916-1921.

20. Betchov, R. & Criminale, W.O.: Spatial instability of the inviscid jet and wake. Phys. Fluids **9** (1966), 359-362.

21. Hannemann, K. & Oertel, H.: Numerical simulation of the absolutely and convectively unstable wake. J. Fluid Mech. **199** (1989), 55-88.

22. Williamson, C.H.K.: Oblique and parallel modes of vortex shedding in the wake of a circular cylinder at low Reynolds numbers. J. Fluid Mech. **206** (1989), 579-627.

23. Mair, W.A.: The effect of a rear-mounted disc on the drag of a blunt-based body of revolution. The Aeronautical Quart. **16** (1965), 350-360.

24. Little Jr., B.H. & Whipkey, R.R.: Locked vortex afterbodies. J. Aircraft **16** (1979), 296-302.

25. MONKEWITZ, P.A. & FERNHOLZ, H.: Self-excited vortex shedding behind an axisymmetric bluff body and its control. Bull. Am. Phys. Soc. **37/8** (1992), 1811.

26. MONKEWITZ, P.A.: A note on vortex shedding from axisymmetric bluff bodies. J. Fluid Mech. **192** (1988), 561-575.

27. BERGER, E.: Unterdrueckung der laminaren Wirbelstroemung und des Turbulenzeinsatzes der Kármán'schen Wirbelstrasse im Nachlauf eines schwingenden Zylinders. 1964 Jahrbuch der WGLR (H. Blenk, Ed.), 164-172, Friedr. Vieweg, Braunschweig.

28. BERGER, E.: Suppression of vortex shedding and turbulence behind oscillating cylinders. Phys. Fluids **10** (1967), S191-S193 (Suppl.).

29. TOKUMARU, P.T. & DIMOTAKIS, P.E.: Rotary oscillatory control of a cylinder wake. J. Fluid Mech. **224** (1991), 77-90.

30. WILLIAMS, D.R. & AMATO, C.W.: Unsteady pulsing of cylinder wakes. In Frontiers in Experimental Fluid Mechanics (M. Gad-El-Hak, Ed.), Lecture Notes in Engineering Vol. **46** (1989), Springer, Berlin, 337-364.

31. FFOWCS WILLIAMS, J.E. & ZHAO, B.C.: The active control of vortex shedding. J. Fluids Struct. **3** (1989), 115-122.

32. MONKEWITZ, P.A.: Feedback control of global oscillations in fluid systems. AIAA paper no. 89-0991 (1989).

33. ROUSSOPOULOS, K.: Feedback control of vortex shedding at low Reynolds numbers. J. Fluid Mech. **248** (1993), 267-296.

34. MONKEWITZ, P.A., BERGER, E. & SCHUMM, M.: The nonlinear stability of spatially inhomogeneous shear flows, including the effect of feedback. Eur. J. Mech. B/Fluids **10**, no.2 - Suppl. (1991), 295-300.

35. HENRICH, E.A., MINGORI, D.L. & MONKEWITZ, P.A.: Control of limit cycle oscillations with applications in fluid mechanics. Submitted to Int. J. Control (1992).

36. HENRICH, E.A., MONKEWITZ, P.A., SCHUMM, M. & BERGER. E.: The effect of proportional feedback control on the wake behind a cylinder. Bull. Am. Phys. Soc. **35/10** (1990), 2326.

37. LEE, C.K.: Feedback control of global oscillations in a bluff body wake. M.S. thesis, Univ. California, Los Angeles (1992).

No-Feedback Control of Kármán Vortex Streets at Low Reynolds Numbers

F. OHLE and H. ECKELMANN
Institut für Angewandte Mechanik und Strömungsphysik der Universiät,
3400 Göttingen, Bunsenstr. 10, Germany

Summary

Time signals of transient Kármán vortex streets for $50 \leq Re \leq 150$ are represented as trajectories in a two-dimensional state space. From these representation a low-dimensional system of ordinary differential equations of second order is derived, which is then used as a model for the description of the temporal dynamics of the vortex street. Here the model is used twofold: first for a resonant stimulation, and second to control the vortex street to a pre-selected dynamics without feedback, by using nonlinear driving forces. A good agreement between numerical simulation and experimental observation is obtained.

Introduction

A method to control the dynamics of an experimental system to a preselected dynamics without feedback was recently developed [1]. Yet, this method was applied theoretically to the van-der-Pol oscillator, the Lorenz, and the Roessler system, and experimentally only to a simple mechanical oscillator [2]. In all these cases good results with the nonlinear no-feedback control could be obtained (see e.g. Ref. [3]). The no-feedback control is based in general on the following idea:

Assume, an experimenal system can be described by a differential equation of the form

$$\dot{\vec{x}}(t) = \vec{f}(\vec{x}, t) + \vec{F}(t), \tag{1}$$

where $\vec{F}(t)$ defines the control force. If such an equation is unknown or cannot be solved exactly, a low-dimensional model

$$\dot{\vec{x}}_M(t) = \vec{h}(\vec{x}_M, t) \tag{2}$$

has to be determined, which describes accurately the temporal dynamics of the experiment. In addition the desired behavior of the system is specified by a preselected dynamics

$$\dot{\vec{x}}_G(t) = \vec{g}(\vec{x}_G, t). \tag{3}$$

Then the appropriate driving force

$$\vec{F}(t) = \vec{g}(\vec{x}_G, t) - \vec{f}(\vec{x}, t) \tag{4}$$

can be calculated. For $\vec{x}(t_c) = \vec{x}_G(t_c)$ (t_c := initial timestep of the control) and $\vec{h}(\vec{x}_M, t) = \vec{f}(\vec{x}, t)$ the experimental system immediately locks into the goal dynamics $\vec{x}_G(t)$ and the control is accomplished. The most crucial point of this method is, if Eq. (1) is unknown or cannot be solved, that a model for the system dynamics (Eq. 2) is needed. Such a model for example can be constructed from measured time signals [4].

The application of the above procedure to fluid mechanical system - such as a Kármán vortex street in the regular range ($50 \leq Re \leq 150$) - will be discussed in the following. Since in general Eq. (1) is the Navior-Stokes equation, an accurate low-dimensional model has to be determined. In the present case, this was accomplished by extracting an ordinary differential equation of the type

$$\ddot{U} = \frac{d^2U}{dt^2} = \sum_{\substack{i=0,\dots,3 \\ j=0,\dots 3-i}} a_{ij}\, U^i\, \dot{U}^j, \tag{5}$$

from measured time signals. Here U denotes the instantaneous velocity component in the mean flow direction and a_{ij} the coefficients of the Taylor series expansion (for more details of the modeling procedure and the experimental setup used, see Ref. [4]). It was shown by the presented authers [4], that Eq. (5) reproduces the physical behavior of the vortex street and predicts its response to external sinusoidal or square wave sound excitations. The correct physical behaviour was not obtained by the van der Pol oscillator, used in ealier works to describe the temporal vortex street dynamics [5]. With the low-dimensional model (Eq. 5), two nonlinear no-feedback control methods, both developed by Hübler and Lüscher [1], can now be applied to the vortex street. One control method investigated is based on a resonant stimulation and the other one on a no-feedback control to a pre-selected dynamics. The experimental excitation used here is realised by two loudspeakers working 180° out of phase and placed outside the flow directly above and below the cylinder to superimpose sound [4].

Nonlinear resonant stimulation

In this section, it is investigated to what extend the resonant stimulation method is suited to drive the vortex street to a larger amplitude $U(t)$ or to a complete suppression of its amplitude. The force $F(t)$ used for this type of control can either be obtained from the transient dynamics of the experimental system, which has to be reflected in time, or from the relation $F(t) = (w_{01} - a_{01})\dot{U}_G$, where w_{01} is the coefficient and $\dot{U}_G$ the component of the first time derivative of the pre-selected dynamics. If $F(t)$ and $\dot{U}(t)$ are in phase (i.e. $F(t) \sim \dot{U}(t)$) or the phase difference between $F(t)$ and $U(t)$ is 90° (because of the amplitude-frequency coupling of a nonlinear oscillator), a resonant stimulation can be accomplished. The resonant stimulation of a vortex street leads to an increase of the velocity amplitude and is much more efficient than a sinusoidal stimulation with comparable amplitude [6]. With this method, however, a suppression of the amplitude $U(t)$ cannot be achieved (Fig. 1 (left)). Here, the force $F(t)$, resulting from the reflection in time of the transient dynamics, is at the beginning nearly constant, than decreases and finally converges to zero. Due to this force, the amplitude $U(t)$ first increases (resonant stimulation to a larger $U(t)$), than decreases like $F(t)$, and after some time returns back to its natural amplitude. This return results from a too small excitation force $F(t)$, leading to the lock-off of the vortex street. These experimental observations are in a good agreement with numerical simulations (Fig. 1 (right)) [6].

No-feedback control to a preselected dynamics

With the no-feedback control method, the experimental system can be driven to a preselected or goal dynamics $g(w_{ij}, U_G, \dot{U}_G, t)$. Here U_G and $\dot{U}_G$ are the velocity component

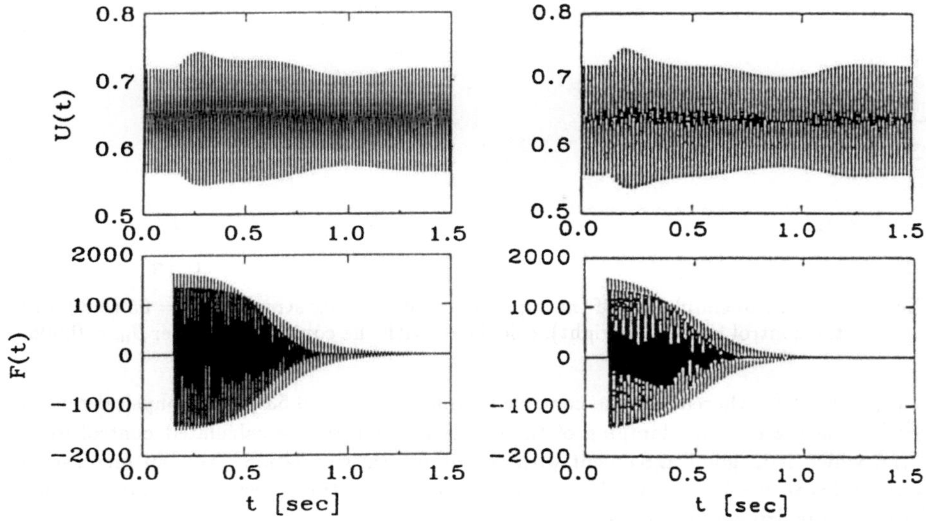

Fig. 1 Experimental (left) and simulated (right) timeseries of a resonantly stimulated vortex street for $Re = 65$ are shown on the top and the respective control force $F(t)$ on the bottom.

and its time derivative of the chosen dynamics, respectively, and w_{ij} its parameters. This dynamics can be in generell a state with a smaller or larger velocity amplitude $U(t)$, a fixpoint dynamics or a more complicated state. The necessary force $F(t)$ to achieve the pre-selected dynamics, is calculated from Eq. (4). With this type of control, a decrease of the velocity amplitude $U(t)$ of the vortex street was found, in accordance with both, experiments and numerical simulations (Fig. 2). The force $F(t)$ used here is calculated from $F(t) = (w_{02} - a_{02})\dot{U}_G^2(t)$, where $w_{02} = (1 + \beta_{02})a_{02}$ and β_{02} defines the control parameter [6]. Note, that for $Re = 65$ the mean velocity of the uncontroled vortex street

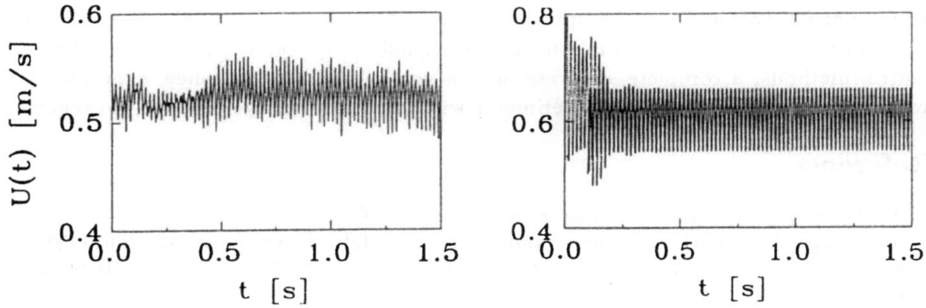

Fig. 2 Comparison of the excited vortex street at $Re = 65$ (left) taken from Roesch [7], and simulation (right) with a control parameter of $\beta_{02} = 2$. On the left the control was started at $t_c = 0.0s$, whereas on the right figure at $t_c = 0.1s$.

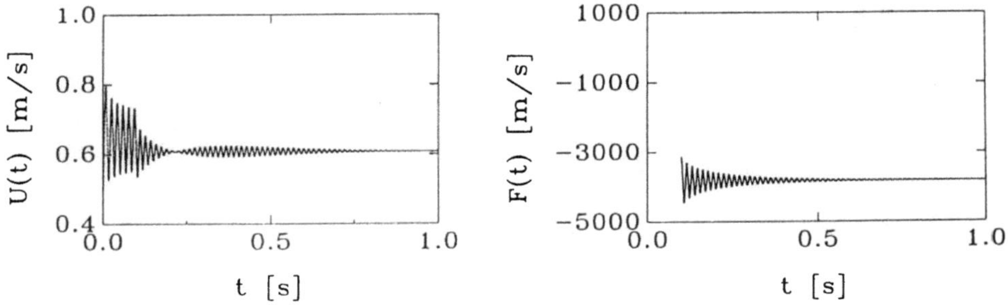

Fig. 3 Numerical simulation of the suppression of a vortex street at $Re = 65$ (left) and the control force $F(t)$ (right), calculated with the control parameter $\beta_{10} = 0.006$.

is $U_\infty \approx 0.62\,\frac{m}{s}$, whereas for the controled dynamics $U_\infty = 0.53\frac{m}{s}$. The physical reason for this decrease is the damping of the oscillation due to the calculated control force $F(t)$, which is in good agreement with the numerical simulation. Yet, an increase in amplitude can also be achieved by a pure sinusoidal excitation a decrease is only obtained by nonlinear driving forces.

Fig. 3 shows the numerical simulation of the suppression of the velocity amplitude $U(t)$ of the vortex street. Note, that in this figure the mean velocity U_∞ is also decreased. In addition, the control force $F(t)$ converges to a constant value. This convergence indicates, that by applying an external sound excitation no total suppression of $U(t)$ can be realized. To overcome this problem a method, which can apply a constant control force to the vortex street, e.g. via the pressure field right at the cylinder has to be developed. This could be realized by a circular pipe with several pressure-holes in spanwise direction on the rear side and a pinhole-microphon placed on one end of the pipe to measure the pressure-fluctuations. A control could then be possible by first applying a periodic and then a constant suction or blowing from the cylinder.

Summary and Conclusion

It was shown that the nonlinear resonant stimulation and the no-feedback control to a pre-selected dynamics can be applied to a Kármán vortex street. Since experimental results and numerical simulations are in a qualitativly agreement for the different control methods, a complete suppression of a vortex street using a new experimental arrangement can be expected. Experiments with this new arrangement are in progress.

References

1. A. Hübler and E. Lüscher, Naturwissenschaften, **76**, 67, (1989)
2. R. Georgii, W. Eberl, E. Lüscher and A. Hübler, Helv. Phys. Acta, **62**, 290, (1989)
3. E.A. Jackson, Beckman Institute, Urbana-Champaign IL, Tech. Rep. CCSR-90-8
4. F. Ohle and H. Eckelmann, Phys. Fluids A, **4**, 1707 (1992)
5. E. Roesch, F. Ohle, H. Eckelmann and A. Hübler, The Global Geometry of Turbulence, edt. J. Jiménez, Plenum Press, New York, 111 (1991)
6. F. Ohle, MPI für Strömungsforschung Göttingen, FRG, Report Nr. 17, (1992)
7. E. Roesch, Mitteil. aus MPI für Strömungsforschung Göttingen, Nr. 99, (1991)

Control of the Wake Instability at Low Reynolds Numbers by Thermal Effect

P. PARANTHOËN, J.C. LECORDIER
U.R.A.-C.N.R.S. 230 - Université de Rouen
BP 118
76134 MONT SAINT AIGNAN CEDEX (FRANCE)

I- Objective of the present work

The purpose of the present paper is to present some new experimental results concerning the possibility of control of vortex shedding at low Reynolds numbers ($40 \leq Re = UD/\nu \leq 100$) using thermal effects. Here U is the free stream velocity, ν is the kinematic viscosity of the fluid calculated at the free stream temperature and D is the diameter of the circular cylinder .

In our experiment this control can be achieved by two different ways : either by heating the main cylinder or by locating an heated secondary cylinder of diameter d in the near wake of the main cylinder. In the second method the ratio D/d between the diameters of the two cylinders was about 140, a value much higher than in the experiments described by STRYKOWSKI and SREENIVASAN (1). In these conditions the control cylinder Reynolds number was always less than 1 and the influence of the momentum defect of the heated secondary cylinder was reduced in comparison to the influence of its thermal effect.

II - Experimental set-up and measurements techniques

Experiments in air flows have been carried out with the heated cylinder located in the potential core of a laminar plane jet. The cylinder consists of a smooth stainless steel tube (D=1 mm, l/D=150) heated by Joule effect. Some experiments have been also performed with a ribbon (0.8 mm x 25 µm).
Experiments with the control cylinder were performed at the exit of a round laminar jet. The main cylinder was a tube (D=0.5 mm, l/D = 16) and the control cylinder was a fine Pt Rh 10 % wire (d=3.5 µm) mounted on a micrometer base enabling accurate two-dimensional positioning. Velocity measurements were made with TSI laser Doppler velocimeter and temperature measurements were performed with fine cold wires.
Experiments in water flows have been carried out in a water channel with a free surface. Small particles generated from silver wires located in the test section upstream the cylinder were used for visualization.

III - Experimental results

For the experiment with the single cylinder it appears that over the studied Reynolds numbers range, the control of vortex shedding can be realized owing to the heat input brought to the cylinder. As previously shown by the authors (2) in air flow, increase of the heat input in the cylinder leads to a suppression of the vortex shedding while in a liquid flowing around a heated circular cylinder (or ribbon) the effect is inverse (figure 1). This opposite behaviour found in liquids and gases suggest that this control is due to both changes of dynamic viscosity and density with temperature and is not only linked to density changes as claimed by YU and MONKEWITZ (3). Extensive velocity measurements of the near wake have put in evidence the influence of heat input on the velocity field for the case of the heated cylinder (or ribbon) PARANTHÖEN et al. (4). This influence results in slight changes of the velocity field leading to a decrease of the interaction between the two shear-layers at the end of the recirculation zone. This interaction has been characterized by the ratio between the rate of the circulation of the shear-layer and the shear-layer spacing at the end of the recirculation zone. As shown in figure

2 vortex shedding appears when this ratio reaches a threshold value. Now in presence of heating the value of this interaction term decreases with increasing heating and suppression of vortices happens when the threshold value is reached. From a practical point of view an effective Reynolds number can be calculated by taking the kinematic viscosity at the temperature of the near wake and a Strouhal number can be built using the maximum velocity and the shear layer spacing at the end of the recirculation zone.

As already found by STRYKOWSKI and SREENIVASAN (1) the use of a secondary cylinder (heated or not), located in a limited region of the near wake of the main cylinder, allows to suppress the vortex shedding phenomenon. The dimensions of this region changes with the upstream Reynolds number and with the heat input brought to the secondary cylinder. It is worth to note, as shown in figure 3 that even with a ratio D/d of about 140 the suppression of the vortex street happens without heating up to a Reynolds number of 63 ($Re-Re_c=6.7$). Now, when the Reynolds number is increased, vortex suppression can be obtained only when the cylinder is heated and is positioned close to the axis approximately 3 to 4 diameters downstream of the main cylinder. As shown in figure 4 the size of the region of suppression is reduced with decreasing heating current and can be limited to a point located on the axis. For the same heating current, the size of the region of suppression is also reduced when the Reynolds number is increased. Futhermore, as shown in figure 5, an hysteresis effect is observed on the location of the boundaries of the region of suppression and reappearance of vortices. When the control cylinder is moved to the axis, the boundary of the region of suppression of vortices is always located at a closer distance of the axis than the boundary of the region of reappearance of vortices when the control cylinder is then displaced off the axis. It is worth to note that for the same value of (Re - Rec) the power per unit length needed to suppress the vortices is about one hundred times lower in this second situation. Furthermore it appears from velocity measurements that the maximum efficiency (minimun heating) is obtained when the secondary cylinder is located at the end of the recirculation zone of the main cylinder.

(1) STRYKOWSKI P.J., SREENIVASAN K.R. (1990), J. Fluid Mech 218, 71-107.
(2) LECORDIER J.C., HAMMA L., PARANTHOËN P. (1991), Exp. in Fluids 10, 224-229.
(3) YU M.H., MONKEWITZ P.A. (1990), Phys. Fluids A (7), 1175-1181.
(4) PARANTHÖEN P. , BROWNE L.W. , LE MASSON S. , LECORDIER J.C. (1992) (in preparation)

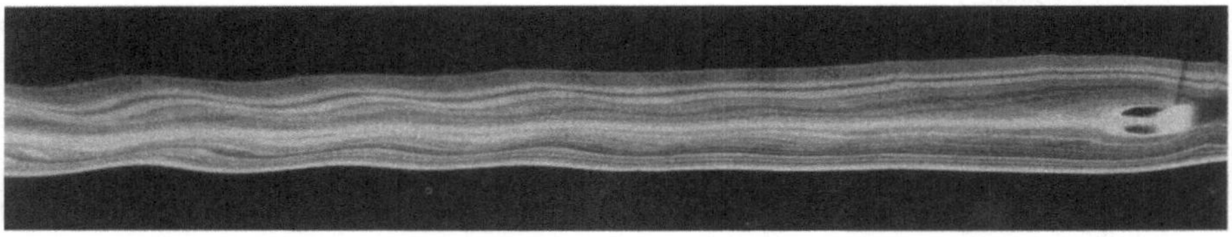

Wake of the ribbon just under the critical Reynolds number

Wake of the heated ribbon at the same velocity

Figure 1

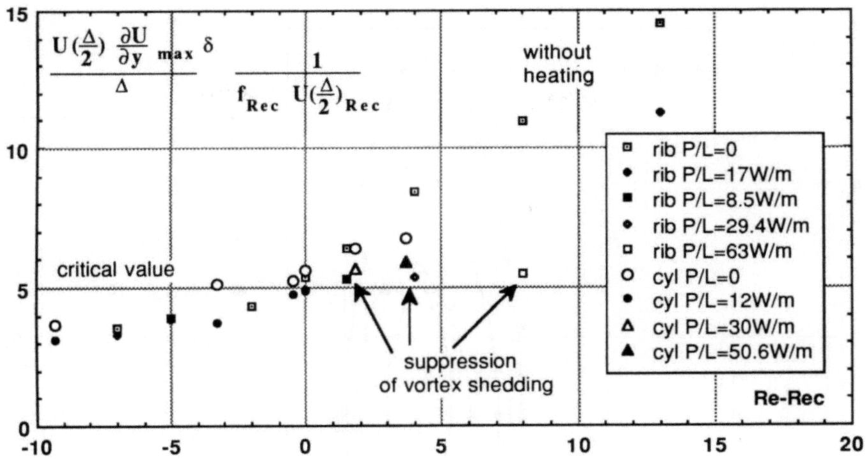

Ratio between the rate of circulation in the shear layer and the shear layer spacing at the end of the recirculation zone

Figure 2

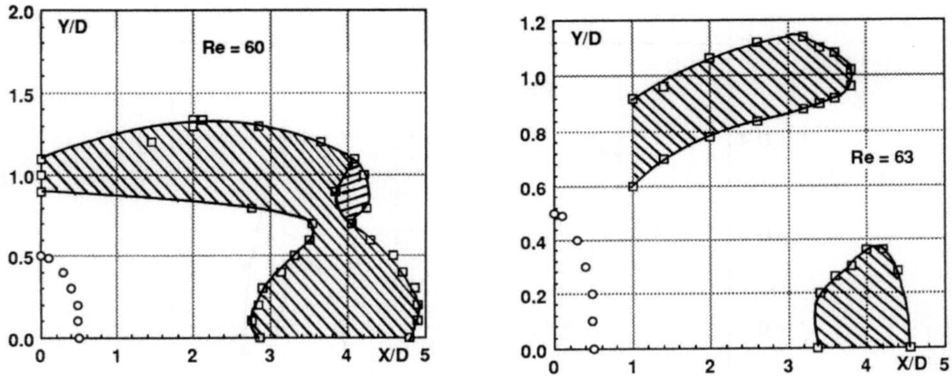

Regions of suppression of vortex shedding with the unheated control cylinder

Figure 3

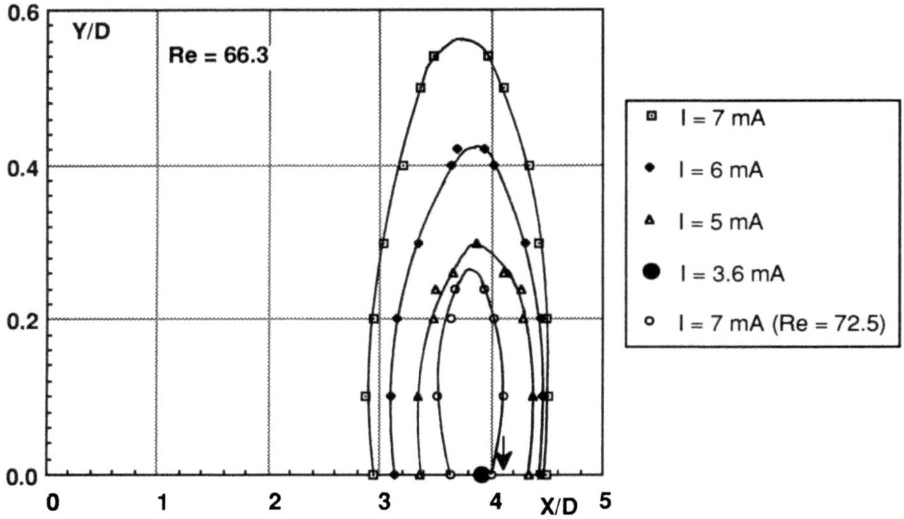

Regions of suppression of vortex shedding with the heated control cylinder.
The arrow shows the end of the recirculation zone for this Reynolds number

Figure 4

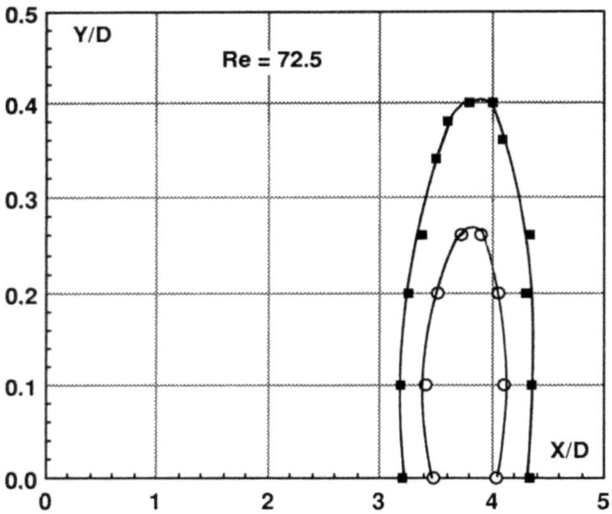

Example of hysteresis effect

Figure 5

Aspects of Bluff Body Wake Control

K. ROUSSOPOULOS
Cambridge University Engineering Department
Trumpington St, Cambridge CB2 3HU, U.K.

Abstract

This paper discusses the variation of feedback control of bluff body wakes with the feedback loop used, and the effect of the controller at points on the span away from the control sensor. In the latter case the results are compared with Ginzburg-Landau model parameters deduced by other researchers using different techniques.

Introduction

The wakes of bluff bodies exhibit periodic near wake unsteadiness at Reynolds numbers above some critical value, due to an absolute fluid mechanical instability. This unsteadiness leads to the formation of the well-known Karman vortex street in the far wake. It has been shown experimentally [1,2,3,4] and theoretically [5] that it is possible to use feedback to suppress the initial instability and so delay the onset of unsteadiness to a higher Reynolds number. This paper reports further results of the experiments described in [4], to which reference should be made for further details of the apparatus.

Experimental Method

Experiments were performed in a low turbulence open return wind tunnel. A hot wire probe (the "control sensor") was located approximately 7 diameters behind, and offset from, the axis of a circular cylinder of radius 0.775 mm mounted across the flow. The signal from this sensor, after appropriate processing, was fed to a pair of opposed loudspeakers in the tunnel wall. Experiments were performed at Reynolds numbers in the range 40–60. With this configuration it was possible to suppress the unsteadiness measured at the control sensor, as illustrated in Figure 1. A second hot wire (the "second sensor") was traversed along the span to investigate the control at other locations.

Variation of Control with Feedback Loop and Sensor Location

The control illustrated in Figure 1 was obtained with no filtering in the feedback loop. It was found that control was not possible if the feedback loop contained a filter tuned to the natural vortex shedding frequency. In addition, control was not possible if the sensor was located more than approximately 10 diameters behind the cylinder. For further discussion of these details refer to [4].

Variation of Control along Span

When the controller acts the unsteadiness is greatly reduced at the spanwise location of the control sensor. When the controller is acting successfully the actuation signal is very small (see Figure 1), and so the effect of the control at other locations along the span may be negligible. The second sensor was used to investigate the signal at other points along the span, and it was indeed found that at distant points the controller had minimal effect on the signal amplitude. It therefore is reasonable to model the controller as forcing the amplitude of shedding to zero at one spanwise location without affecting other locations.

Albarède and Monkewitz [6], using results from [7], show that a Ginzburg-Landau equation model of transverse shedding yields an explicit solution for the unsteadiness mode amplitude along a semi-infinite cylinder when the amplitude is zero at one point: this solution can conveniently be written as:

$$\frac{M(z)}{M(\infty)} = \tanh(k(\mathrm{Re} - \mathrm{Re}_c)^{1/2}(z/d)) \tag{1}$$

with (z/d) the non-dimensional distance along the span, Re_c the Reynolds number at the onset of unsteadiness, $M(z)$, the amplitude of unsteadiness at location z, zero at $z = 0$, and k a constant function of the Ginzburg-Landau equation solution coefficients.

In our experiment the second sensor was positioned a distance z from the control sensor, and the amplitude of the unsteady velocity component was recorded when the controller, inhibiting unsteadiness at the control sensor, was on and off. The value when the controller was off was used as $M(\infty)$, and the value when the controller acted as $M(z)$. The experiment was performed at several Reynolds numbers and with a large number of values of z. In Figure 2 the results are plotted as a graph of $M(z)/M(\infty)$ against $x = (\mathrm{Re} - \mathrm{Re}_c)^{1/2}(z/d)$. The dotted lines are fits to Equation 1 with k set to 0.04 and 0.067. It is seen that for $x < 20$ most of the data points fall within these limits; at larger x, the controller has greater effect than the model predicts, maybe because the controller does in fact influence the unsteadiness at these locations.

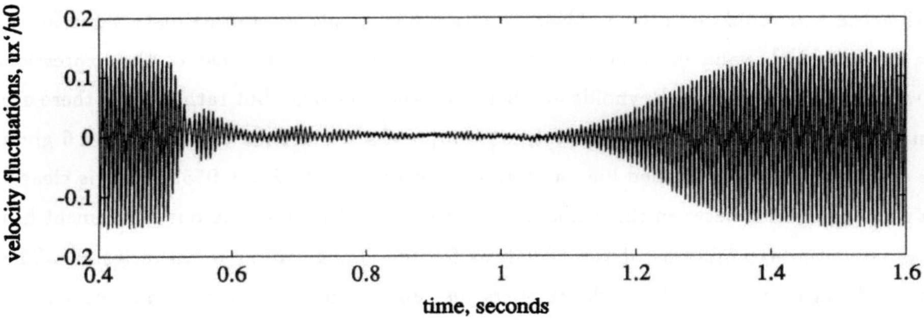

Figure 1: Illustration of vortex shedding suppression at one point in a long span at a Reynolds number of 48.9.

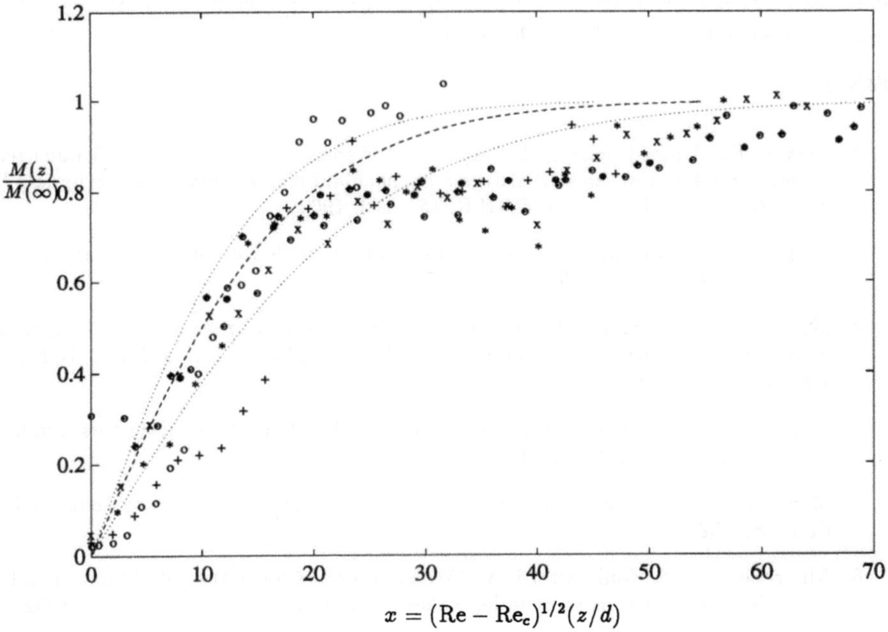

Figure 2: Effect of controller on wake unsteadiness along the cylinder span at various Reynolds numbers: 'o', 49.4, '+', 50.7, '*', 51.7, 'x', 52.7, 'ɞ', 53.8, '⊕', 54.6, '●', 59.0. $\mathrm{Re}_c = 48.4$. Data for Re=54.6, 59.0 offset on x-axis since control at z=0 not good. '...' lines: k=0.067, k=0.04: '- -' line: fit from [6]: see text.

Albarède and Monkewitz [6], in their Equation 36a, imply an approximate value for k, $k = (s/2\mu_r)^{1/2}$, valid in situations "far from critical" in their sense of the expression (which does not mean at Reynolds numbers well above critical, but rather when there are many unstable modes.) Using the values from [6] of $s = 0.2 \pm .02$ and $\mu_r = 32 \pm 6$ gives $k = 0.055 \pm .008$: the dashed line on Figure 2 corresponds to $k = 0.055$. There is clearly a good agreement between this value and our experimental data, and our experiment has validated the assumptions of their analysis for the configuration we have tested. The main deviations are at Reynolds numbers just above the onset of unsteadiness and at larger values of z/d. Why the former should be so is unclear since the data does not lie consistently above or below the theoretical curve, but may reflect the dificulty of obtaining consistent results so near the onset of unsteadiness; the latter is probably related to some residual effect of the controller on the unsteady mode.

References

1. Berger, E.: Unterdrückung der laminaren wirbelströmung und des turbulenzeinsatzes der Kármánschen wirbelstrasse im nachlauf eins schwingenden zylinders bei klein Reynolds-zahlen. Jahr. WGLR: 164–172, 1964.

2. Berger, E.: Suppression of vortex shedding and turbulence behind oscillating cylinders. Phys. Fluids **10**, S191–S193, 1967.

3. Monkewitz, P.A., Berger, E. and Schumm, M.: The non-linear stability of spatially inhomogeneous shear flows, including the effect of feedback. Eur. J. Mech. B/Fluids, **10** (2), 295–300, 1991.

4. Roussopoulos, K.: Feedback control of vortex shedding at low Reynolds numbers. Accepted by J. Fluid Mech, 1992.

5. Monkewitz, P.A.: Feedback control of global oscillations in fluid systems. AIAA Paper 89-0991.

6. Albarède, P. and Monkewitz, P.A.: A model for the formation of oblique shedding and "chevron" patterns in cylinder wakes. Phys Fluids A **4** (4), 744–756, 1992.

7. Nozaki, K. and Bekki, N.: Exact solutions of the generalised Ginzburg-Landau equation. J. Phys. Soc. Japan., **53** (5), 1581–1582, 1984.

Vortex Method Simulation of the Wake from a Rigid Cylinder at Re = 200 Subjected to Streamwise Amplitude-Modulated Flow Excitation

C. RUHI KAYKAYOGLU

Istanbul Technical University
Faculty of Aeronautics and Astronautics
Maslak, 80626, Istanbul, TURKEY

Summary

The Vortex Method incorporating the viscous extension of the vortex-in-cell method has been used to simulate the amplitude-modulated excitation of a cylinder in the streamwise direction. Amplitude-modulated in-line perturbations superimposed on the mean flow provides novel features for the near and far field Karman vortex street development including a locked-in wake structure, which is periodic at the higher multiples of the modulation frequency. The possible near wake vortex structures formed due to amplitude-modulated excitation were identified as a function of excitation frequency.

Introduction

Observation of locked-in vortex structures or destabilized vortex structures of the bluff body near wakes in a perturbed incident flow consisting of a mean flow with an oscillatory streamwise component superimposed on it is a recent area of interest. The interaction of the vortex formation and shedding from a circular cylinder under amplitude-modulated excitation in the streamwise direction will be investigated numerically by using a combined Lagrangian-Eulerian viscous vortex method of solution.

When the self-excited wake from a rigid cylinder is subjected to cross-stream sinusoidal excitation, variety of vortex patterns can occur (Ongoren and Rockwell [1], Williamson and Roshko [2]). The streamwise excitation at various modes generate a dozen of wake structures at low Reynolds numbers (Griffin and Ramberg [3], Detemple-Laake and Eckelmann [4]). Due to sinusoidal external excitations in either direction the Karman vortices are shed: at the subharmonics of the excitation frequency or at phase locked with the natural frequency of vortex formation.

Recently, Nakano and Rockwell [5] showed that cross-stream excitation frequency, amplitude and dimensionless modulation frequency influences the near wake response and locked in vortex formation can be distrubted. Nakano and Rockwell [5] demonstrated experimentally that a rich variety of wake patterns can be generated by cross-stream amplitude modulated excitations.

In this present study the author considers the streamwise amplitude modulated cylinder wake excitation and aims at carrying a numerical study to investigate the state of response of cylinder near wake as functions of various parameters. The unsteady flow topology of the near wake is the main area of interest. The incident velocity u(t) is of the following form

$$u(t)=u_\infty+u_o[1-\cos(2\pi f_m t)]\sin 2\pi f_e t \qquad (1)$$

in which f_m is the modulation frequency, f_e is the nominal excitation frequency, u_∞ is the mean incident velocity and u_o is the amplitude of the disturbance. Consideration of a range of f_m/f_e and u_o/u_∞ reveals a number of wake vortical structures.

Numerical Method

The numerical model is based on a mixed Lagrangian-Eulerian Vortex Method. The fluid is incompressible and the Reynolds number, Re, based on cylinder diameter is 200. The method solves, subject to the elliptic Vorticity-Stream function Poisson equation, the 2-D Navier Stokes Vorticity Transport equation. Viscous diffusion is carried out in a standard way by finite differences on a fixed grid. But the convection terms in the Vorticity Transport equation are simulated by convecting point vortices. A split time step approach is used to separate diffusion and convection. Further details of the method is given by Graham [6]. For the calculations shown here, the mesh had 64 x 64 mesh elements with high resolution near the cylinder surface.

Results and Discussion

Figure 1a shows the instantaneous vortex structures for the case of a natural wake shedding from a rigid cylinder at Re=200 at a dimensionless time T=100. The time trace of the induced transverse force in the form of lift coefficient is also presented (See Fig. 1b). The Strouhal number ($St=f_sD/u_\infty$) based on force coefficient variation is found to be St=.198 . A pure sinusoidal in-line wake excitation at frequencies $f_e/f_s=1.5$ and $f_e/f_s=2.0$ are depicted in Figs. 1c and 1d for a flow perturbation ratio of $u_o/u_\infty=0.5$. Clearly lock-on characteristics are evident when the excitation frequency is twice the natural shedding frequency, f_s. In contrast to the natural Karman vortex street a pair vortices is shed alternately from sides of the cylinder during one excitation cycle when f_e/f_s=1.5.

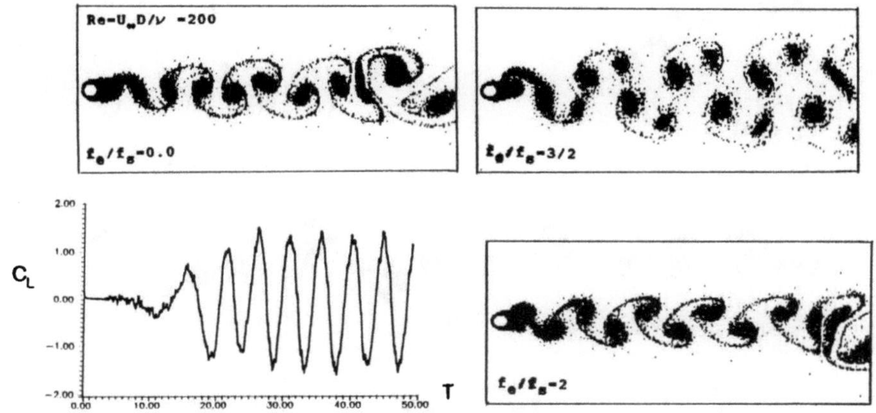

Fig. 1. The effect of pure in-line excitation frequency on the vortex wake of a cylinder at a Reynolds number of 200.

Fig. 2 shows the instantaneous vortex distributions for $2f_m$ periodic response of a cylinder wake. A fully destabilized response of wake under the excitation conditions stated in the Figure provides two separate rows each having three coherent vortices. The "S" shaped coherent structures forms with $4f_s$ period.

Concluding Remarks

The amplitude-modulated excitation of the near wake of a circular cylinder provides very rich patterns of vortex structures. The type of vortex rearrangement depends on; the amplitude modulation frequency, f_m, the nominal excitation frequency, f_e and the amplitude of flow perturbation, u_o. It is remarkable that vortex patterns are periodic with f_m or its doubled and tripled versions.

Due to space limitations only a single case was presented in this paper. The other cases shows that amplitude-excitation can be interpreted as states of new orders of destabilized vortex wakes.

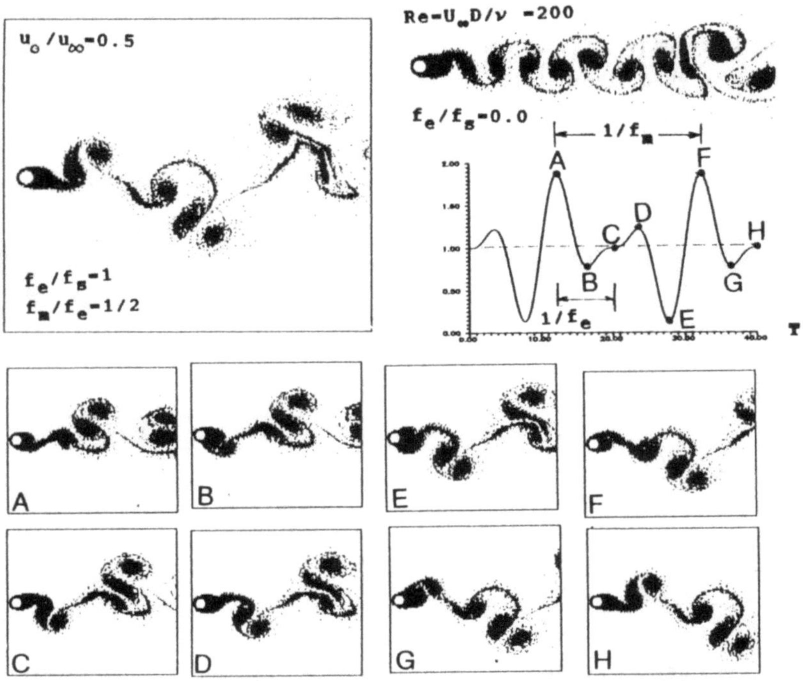

Fig. 2. $2f_m$ periodic response of a cylinder wake at an amplitude-modulated excitation.

References

1. Ongoren, A.; Rockwell, D.: Flow structure from an oscillating cylinder. Part II: Mode competition in the near-wake, Journal of Fluid Mechanics, Vol. 191, (1988) 197-223.

2. Williamson, C.; Roshko, A.: Vortex formation in the wake of an oscillating cylinder, Journal of Fluids and Structures, Vol.2, 1988 355-381.

3. Griffin, O.M. ; Ramberg, S.E.: Vortex shedding from a cylinder vibrating in line with an incident uniform flow, Journal of Fluid Mechanics, Vol.25, part 2, 1976, 257-271

4. Detemple-Laake, E.; Eckelmann, H.: Phenomenology of Karman vortex in oscillatory flow, Experiments in Fluids, Vol 7, 1989, 217-227.

5. Nakano, M.; Rockwell, D.: "The wake from a cylinder subjected to amplitude-modulated excitation", submitted for publication, 1991.

6. Graham, J.M.R., "Computation of viscous separated flow using a particle method", Num. Met. for Fluid Dyn. III, Ed.Morton W. and Baines M., Oxford Science Pub., 1988

Numerical Simulation of Control of Bluff Body Flow Using a Discrete Vortex Method Incorporating Viscous Diffusion

J. R. Meneghini and P. W. Bearman
Department of Aeronautics,
Imperial College, London SW7 2BY, UK.

Abstract
Control of bluff body flow is investigated numerically. Control is achieved by applying forced oscillation to a cylinder and cases covering a range of high amplitudes and with various frequencies have been simulated. Special attention is given to the change of phase between body displacement and lift force and the phase jump that occurs near the resonance point. The near wake structure is presented for different values of reduced velocity. One of the objectives of applying open loop control is to define possible strategies for future investigations with closed loop or feed-back control.

1 Introduction

It is known from experiments that there are a number of ways of controlling the formation and shedding of vortices behind two-dimensional bluff bodies, including splitter plates, bleed and body or incident flow oscillation. The attraction of applying numerical methods to this problem is that the fine aspects of how the flow is modified can be studied in closer detail. In the computer it is possible to investigate control strategies more readily than might be feasible in an experiment and the potential for active control can be considered.

In this paper, the flow about an oscillating circular cylinder is investigated numerically. Different cases covering a range of high amplitudes of transverse oscillation with various frequencies have been simulated. Some of the consequence of transverse oscillations on vortex shedding are already known, e.g. the capture of the vortex shedding frequency by the oscillation frequency over a range of cylinder oscillation amplitudes. Also, there is a large increase in the spanwise correlation length found in experiments when the vortex shedding frequency coincides with the oscillation frequency. A bluff body forced to oscillate represents a mechanism for synchronising the moment of shedding along the length. With this consideration, 2D numerical simulations should be reliable in terms of analysing flow details and wake structures.

2 Numerical Method

The method used for the simulations is based on the Vortex-in-Cell formulation incorporating viscous diffusion. This approach has been developed by Graham(1) and applied by Arkell et.al.(2). The 2D Navier-Stokes equations are solved in a vorticity(ω)/streamfunction(ψ) formulation. The operator-splitting technique is used, where convection and diffusion of vorticity are treated separately. Force coefficients are calculated by considering the normal gradient of vorticity at the wall. The flow about an oscillating cylinder is simulated by considering a frame fixed to the cylinder. The inertia effects are taken into account by correcting the force coefficient in the oscillation direction.

3 Results

In order to test the computational code, lift (Cl) and drag (Cd) coefficients for the case of a Reynolds number equal to 200 are shown in figure 1a. Vorticity contours are shown in figure 1b and the wake structure in figure 1c. A good agreement is obtained in comparison with simulations using other methods. For the present computations Strouhal number, S = 0.196.

Simulations for different values of forced oscillation frequency have been made. For all calculations Re=200. Cases with an amplitude of transverse oscillation equal to 0.4D were considered. Values of Cd and Cl for frequency of oscillation equal to $0.85f_s$ and $0.975f_s$ (f_s is the shedding frequency for the case without any oscillation) for an amplitude of oscillation 0.4D are presented in figure 2a and 2b, after the steady state has been reached. For these cases, the vortex shedding frequency changes to the oscillation frequency, i.e., lock in occurs. An impulsively started flow was considered in order to study the development of the wake and force coefficients. For the first cycles, as can been noticed from the Cl results, the inertia effects are predominant. In table 1, values of Cd are presented together with the phase angle by which Cl leads body displacement for different values of frequency of oscillation (f_o) or reduced velocities ($Vr=UD/f_o$). As can be seen, the phase angle presents a jump in a relatively small range of f_o. Experimental results by Bearman and Currie(3),for approximately the same value of body oscillation amplitude, show a similar but not quite such a big jump. It must be pointed out that their results were evaluated by considering the suction at a point 90° from the front stagnation point instead of Cl and also their experiments were carried out at a much higher Reynolds number. The results presented here agree reasonably well with results by Sarpkaya(4).

In figures 3a, 3b and 3c, the near wake structure is shown for values of reduced velocity of 0.85fs, 0.975fs and 1.05fs. In all cases the cylinder is at its lowest position. The phase jump is clearly shown: for the case of lower reduced velocities, a vortex is being shed from the upper side of cylinder and for the case of higher reduced velocity is being shed from the lower side of the cylinder. This kind of behaviour has been known to happen and is clearly illustrated in flow visualisations by Ongoren and Rockwell (5).

Now, the question that arises is: what happens during the transient condition when there is a change of phase? To have some insight about the dynamics of this process, a simulation with a frequency of oscillation very close to the resonance value was made. In figure 4a, Cl and Cd time histories and vorticity contours are shown for a frequency of oscillation slightly below the resonance point and a very long transient condition was observed until a steady state was reached. A change of phase occurred before that happened. Five vorticity contour plots are shown, each one when the body was at its lowest position and the phase angle was changing, corresponding to the points indicated in figure 4b. In figure 4c, a vortex was being shed from the upper position. Figure 4d shows vorticity contours when the drag was minimum. As can be noticed, the vortex that was shed had a lower strength. The next plot shows the phase in change. In figure 4f and 4g there is a clear change in phase, after that a steady state has been reached.

4 Concluding Discussion

All these results indicate the potential of this numerical algorithm for future investigations concerning bluff body flow control. A modulation around the shedding frequency could be used in order to obtain a forced and controlled change of phase. Nakano and Rockwell (6) made an experimental investigation using similar ideas but with a modulation with a large frequency deviation from the vortex shedding frequency in order to destabilize the wake. With the present algorithm, the possibility of achieving vortex shedding attenuation by applying frequency modulated oscillation could be studied and optimum parameters of modulation could be chosen.

Another possible means of attenuating vortex shedding is with a closed loop system including a feedback signal to force a change of phase. Berger (7) was the first to try such an approach. However, to what extent vortex shedding reduction can be obtained is not clear. The work of Monkewitz (8) presents a theoretical proof based on the Ginzburg-Landau equation that vortex shedding can be suppressed only for very low Reynolds numbers, slightly above the critical Reynolds number. Above that, suppression of one mode could excite other modes. If these other modes give attenuation or enhancement in relation to the normal shedding is not known. Of course, this analysis does not include any three dimensional effects which are already known to play an important role in vortex shedding attenuation in flows with Re > 150. All the conclusions drawn here are intended to give only some insight into the mechanics of the phenomena at low Reynolds numbers.

References

1 Graham,J.R.M. Num.Meth.FluidDynamics,ed.K.W.Morton,vol.3,p310,Oxford,1988.
2 Arkell R.H., Graham J.M.R. and Zhou C.Y. Proc.6th BOSS Conference, p445, 1992.
3 Bearman P.W. and Currie I.G. J.Fluid Mech. 91, p.661, 1979.
4 Sarpkaya T. J.Waterw., ASCE 104, p275, 1978.
5 Ongoren A. and Rockwell D. J.Fluid Mech., v.191, p197, 1988.
6 Nakano M. and Rockwell D. Phys.Fluid A 3 (5), p723, 1991.
7 Berger E. Phys.Fluid, 10, S191-193, 1967.
8 Monkewitz, P.A. AIAA Paper89-0991, 1988.

Table 1. Amplitude of oscillation equal to 0.4 D.

f_o/f_s	Vr	Phase Angle	Cd
0.85	6.00	177	1.36
0.90	5.66	162	1.46
0.95	5.37	150	1.53
0.975	5.23	138	1.60
0.995	5.13	105	1.71
1.05	4.86	-4	1.35

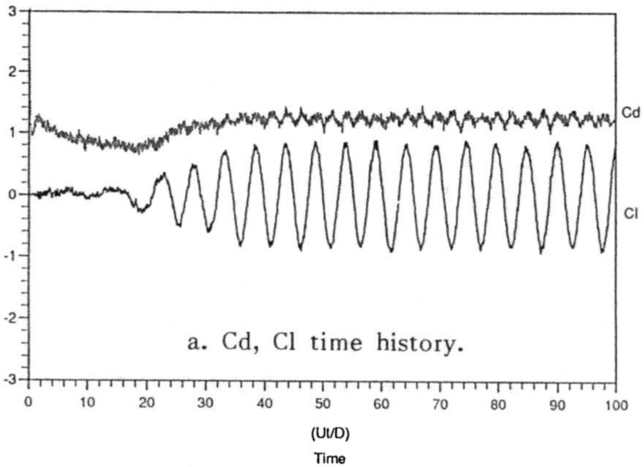

a. Cd, Cl time history.

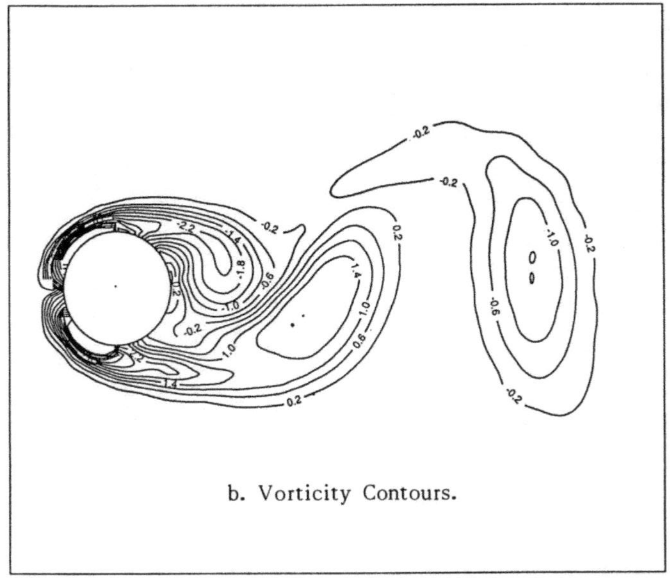

b. Vorticity Contours.

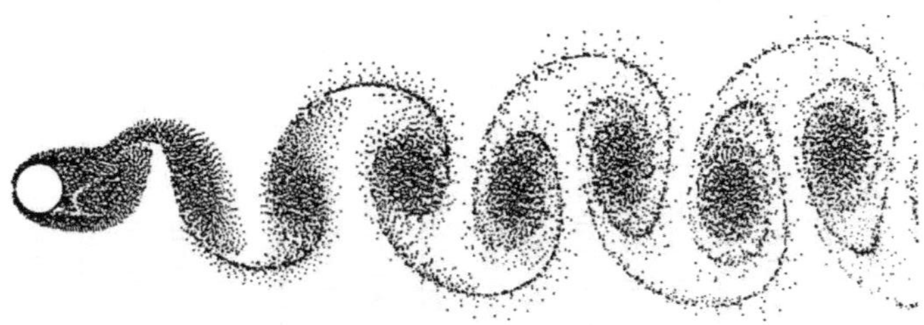

c. Wake structure.

Figure 1. Re = 200.

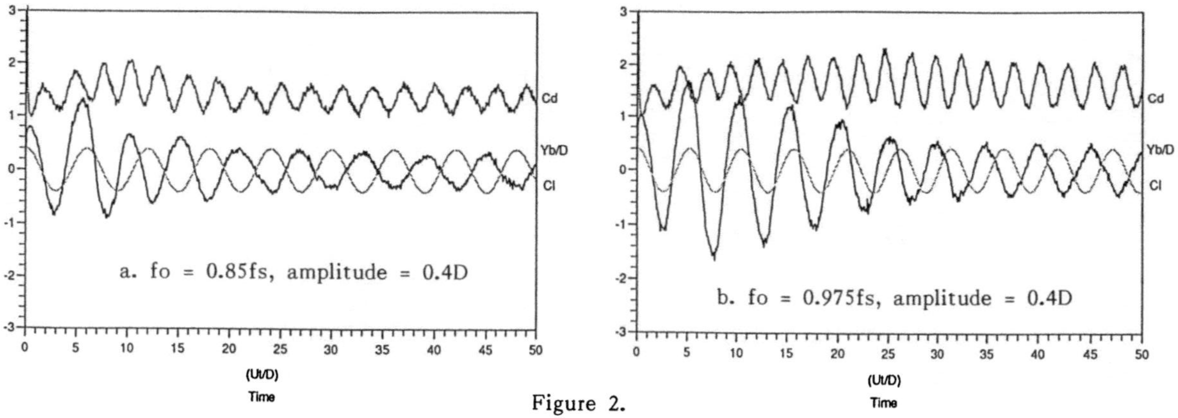

Figure 2.

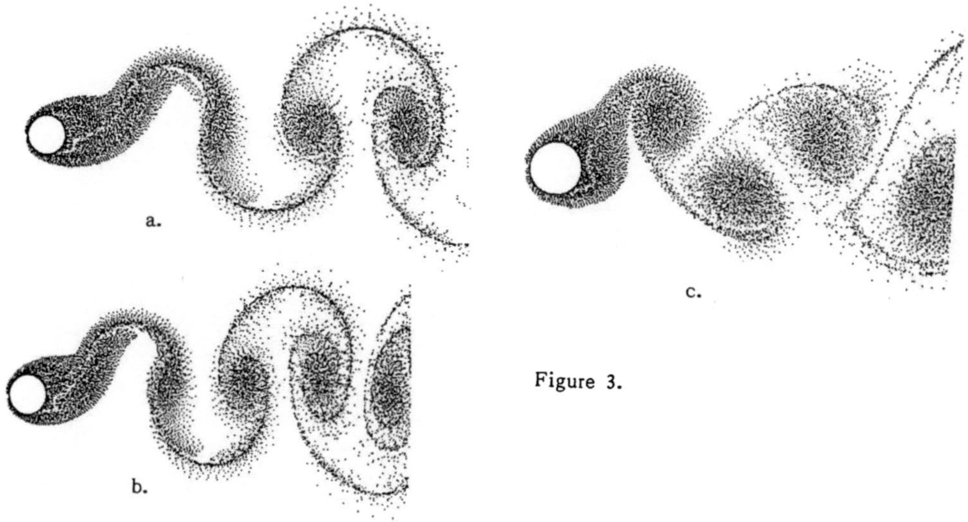

Figure 3.

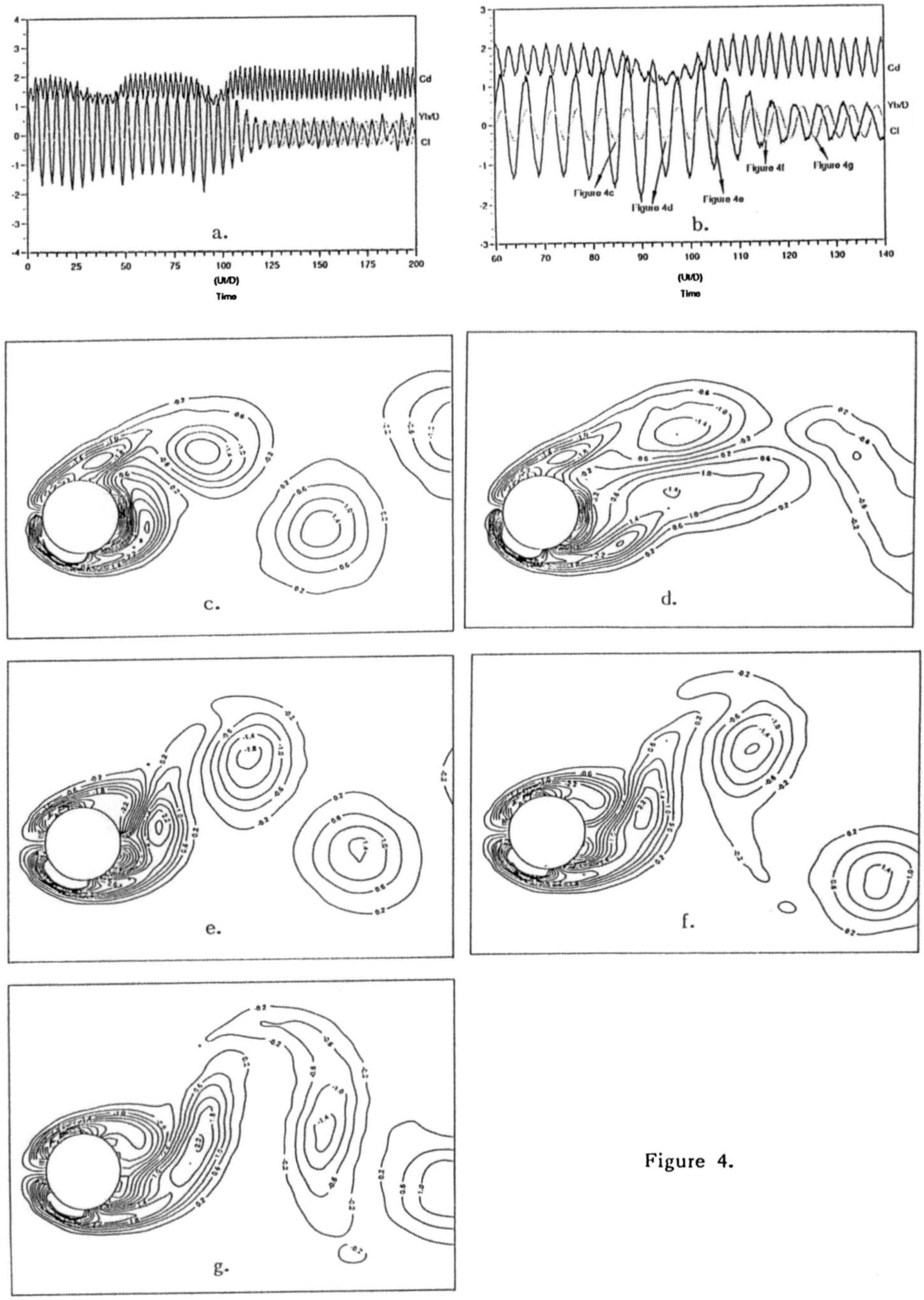
Cd
Yb/D
Cl
a.
(Ut/D)
Time
Figure 4c
Figure 4d
Figure 4e
Figure 4f
Figure 4g
b.
c.
d.
e.
f.
g.

Figure 4.

Quantitative Visualization of Bluff-Body Wakes Via Particle Image Velocimetry

D. ROCKWELL

Department of Mechanical Engineering and Mechanics
Lehigh University
Bethlehem, Pennsylvania

Summary

The instantaneous structure of generic types of wakes from cylinders and trailing-edges is characterized using high-image-density particle image velocimetry. Quasi-two-dimensional and three-dimensional features of these flows are described by patterns of sectional steamlines and vorticity distributions.

Introduction

Wakes formed from bluff bodies are representative of a broader class of unsteady separated flows, which are particularly well-suited to characterization by techniques of high-image-density particle image velocimetry, as described by Rockwell, Magness, Robinson, Towfighi, Akin, Gu, and Corcoran [1]. Use of very small particles and particle displacements provides the basis for quantitative interpretation of the instantaneous structure of these flows. Adrian [2, 3] reviews the concepts of both high- and low-image-density techniques, their optimization, and sources of uncertainty. The ultimate goal, of course, is to represent the flow in terms of the instantaneous velocity field and its derivatives, including vorticity, as well as the corresponding streamline patterns. With this approach, it is possible to avoid possible misinterpretations associated with qualitative flow visualization, including the effects of disturbance amplification, viscous diffusion, and degree of concentration of vorticity, all of which have an influence on the apparent versus actual flow, as described by Gursul, Lusseyran, and Rockwell [4] and Gursul and Rockwell [5].

Knowledge of the vorticity field surrounding a bluff body is central to determining the loading on the body as well as the sound radiation from it and from downstream regions of the wake. The instantaneous force acting on the surface of the body can be related to the time rate of change of the moment of vorticity throughout the volume surrounding the body, as described by Wu [6]. The source of sound radiation from regions immediately adjacent to, and far from, the solid surface of the cylinder is represented by different classes of formulations; each, however, involves the curl of the local velocity and vorticity, as addressed by Howe [7, 8]. Of course, there are a number of approaches to determining experimentally the vorticity field, each with its own advantages, as assessed by Foss and Wallace [9].

The present overview describes selected features of the instantaneous vorticity fields of representative bluff-body wakes. A full account of the experimental approaches is given by Rockwell, Magness, Robinson, Towfighi, Akin, Gu, and Corcoran [1] and Rockwell, Magness, Towfighi, Akin, and Corcoran [10].

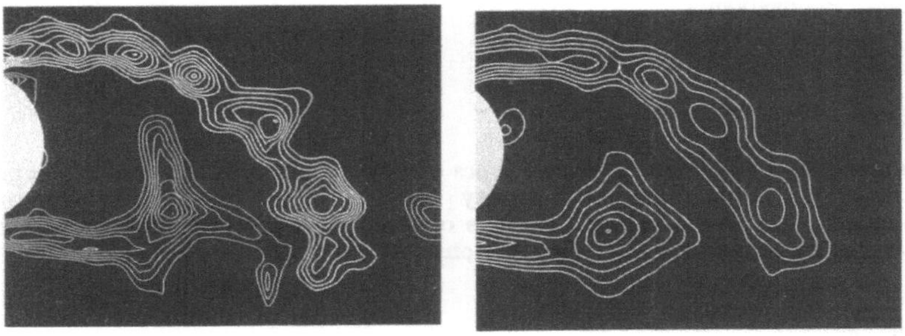

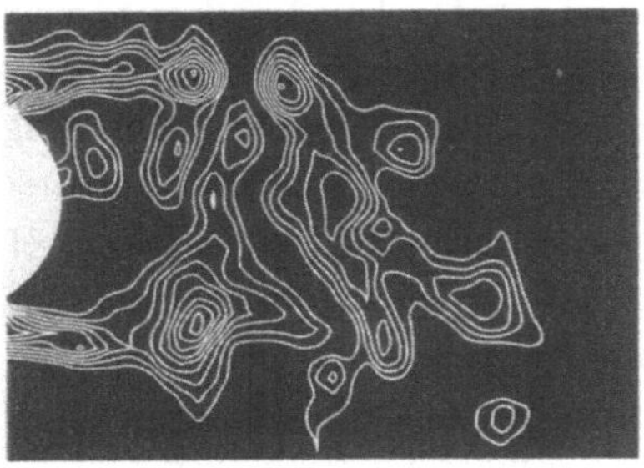

Fig. 1. Instantaneous (left column) and ensemble-averaged (right column) images of vortex formation from an oscillating cylinder at Re = 5,000 (Chyu and Rockwell [11]).

Wake from an Oscillating Uniform Cylinder

The wake from an oscillating cylinder is represented by the series of images of Figure 1. The cylinder is oscillated in the cross-stream direction with an amplitude $A/D = 0.18$ at a forcing frequency f_e relative to the inherent Kármán frequency f_o^* of $f_e/f_o^* = 0.85$. All images were acquired at the maximum- positive (upward) displacement of the cylinder. The two images in the left column are instantaneous representations of the flow structure, while that in the right column is the ensemble-average of five images. In all images, coherent, small-scale vortices arising from the Kelvin-Helmholtz instability of the separating shear layer are evident. They correspond to the instability detected by Bloor [12] and visualized by Gerrard [13] and Wei and Smith [14]. The acceleration of the cylinder in the upward direction apparently enhances the coherent nature of the small-scale vortices, which are particularly pronounced in the upper shear layer. On the other hand, these small-scale structures are barely discernible in the vorticity layer from the lower side of the cylinder. Moreover, as indicated in the image at the lower left, when the lower vorticity layer intrudes across the base region to intercept the upper layer, there is a drastic rearrangement of the small-scale concentrations of vorticity within the upper shear layer. It is evident that the instantaneous shapes and patterns of the large- and small-scale concentrations of vorticity, as well as the degree of concentration of vorticity, are not adequately represented in the ensemble-averaged image at the upper right.

Wake from Stationary Uniform Cylinder

The spanwise structure of the wake from a stationary cylinder is represented in Figure 2. The configuration and flow parameters are the same of those of Figure 1; in this case, however, the laser sheet is parallel to the axis of the cylinder and intersects its shoulder. This orientation of the sheet shows projections of the three-dimensional vorticity field in the y direction, i.e. ω_y, orthogonal to the z axis of the Kármán vortices. The white and gray contours represent respectively negative and positive vorticity ω_y. The spanwise wavelength of this three-dimensional structure is of the order of the cylinder diameter; it apparently has its origin in the instability mechanism described in conjunction with the quasi-two-dimensional visualization of Figure 1. Patterns of both ω_y and ω_x vorticity concentrations relative to the ω_y concentrations are currently being pursued.

Fig. 2. Spanwise structure of wake from a stationary cylinder at Re = 5,000 (Chow and Rockwell [15]).

Wake from an Oscillating Cylinder with Distributed Nonuniformity

Ordered patterns of three-dimensional wakes can be generated at relatively low Reynolds number $Re = 150$ by employing a mildly nonuniform cylinder. The possible locked-in and period-doubled states due to external forcing are described by Nuzzi, Magness, and Rockwell [16]. The structure of period-doubled vortex formation is shown in Figure 3. In this case, the laser sheet is parallel to, and intersects, the axis of the cylinder.

The left plot of Figure 3 shows the instantaneous, biased velocity field and the right plot the instantaneous vorticity contours, with solid and dashed lines representing respectively positive and negative concentrations of vorticity. Although the velocity field suggests concentrations of out-of-plane ω_y vorticity at certain locations, comparison of the instantaneous contours of vorticity with the velocity field shows that the velocity field does not adequately represent the concentrations of ω_y vorticity. The vorticity-free region in the upper part of the right image corresponds to the central portion of a vortex splitting process of the large-scale Kármán vortices. It is evident that this gross distortion of the Kármán vortices is associated with very substantial concentrations of vorticity oriented orthogonally to them.

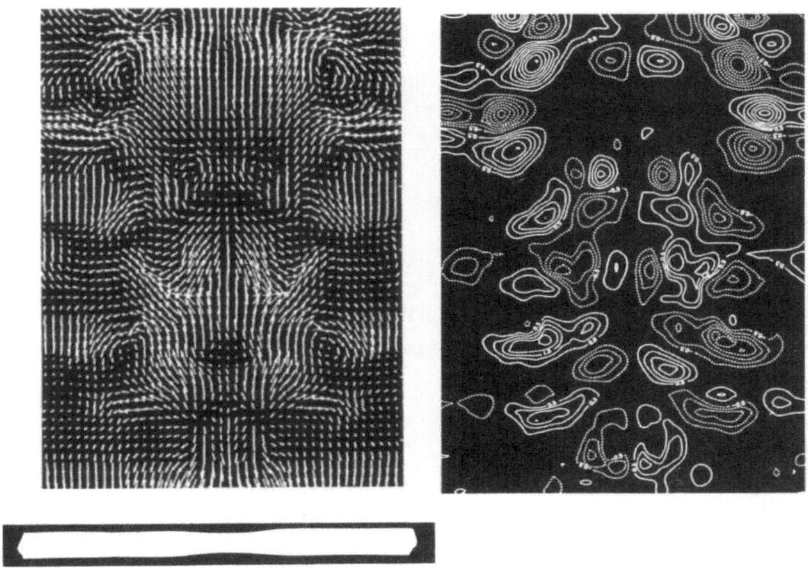

Fig. 3. Ordered patterns of velocity and vorticity fields of a wake from an oscillating, nonuniform cylinder at $Re = 150$ (Towfighi and Rockwell [16]).

Wake from an Oscillating Cylinder with Localized Nonuniformity

In contrast to the highly distributed, geometrical nonuniformity represented in Figure 3, a localized nonuniformity can be imposed on the surface of the cylinder by bleeding from a narrow slit, indicated on the cylinder at the left of Figure 4. Pronounced concentrations of out-of-plane ω_y are generated immediately downstream of the cylinder, which is forced to oscillate at the natural Kármán frequency and at an amplitude $A/D = 0.38$. The dimensionless bleed coefficient $C_\mu = (V_j/U_\infty)^2 A_j/LD = 3.9 \times 10^{-5}$ is orders of magnitude lower than that employed in typical aerodynamic applications; it corresponds to a ratio of the jet bleed velocity V_j to the freestream velocity U_∞ of $V_j/U_\infty = 0.15$. The streamlines superposed on the vorticity concentrations correspond to a reference frame moving at half the average velocity of the image. It is evident that the spanwise extent of the ω_y concentrations is substantially wider than the length of the slit. Moreover, their characteristic circulation $\Gamma/U_\infty D$ exceeds that of the inherent Kármán vortices.

Fig. 4. Ordered patterns of vorticity and sectional streamlines from an oscillating cylinder having a localized nonuniformity due to bleed from a slit. Re = 160. (Gu and Rockwell [18])

Wake from a Rotating, Blunt Trailing-Edge

A rotating impeller in a radial flow machine generates a particularly complex wake. When the system is operated in an off-design condition, the occurrence of stall can substantially modify the structure of the near-wake, as illustrated in Figure 4. In the absolute reference frame, represented by the upper left image, little information on the flow structure is conveyed, and it is necessary to impose a bias on the instantaneous velocity field in order to view the streamline pattern in the relative frame, shown in the image at the upper right. This pattern suggests occurrence of a large-scale vortex just upstream of separation from the trailing-edge. The effect of this large-scale vortex is to distort the near-wake in such a fashion that the developing shear layer from the trailing-edge of the impeller involves a high (streamwise) velocity from the right side of the impeller extending into the wake region, and a much lower (streamwise) velocity on the other side of the wake. As a consequence, one expects coherent instabilities to form, similar to those occurring in mixing-layer flows. The gray (positive) contours of instantaneous vorticity, shown in the image at the bottom left, suggest that this is indeed the case. On the other hand, the white (negative) contours, originating from the large-scale stall region immediately upstream of separation, show lower levels of concentration. By comparing the images at the upper right and bottom left, it is evident that sectional streamlines do not adequately describe the underlying vorticity field.

Fig. 5. Wake from a stalled, rotating impeller blade at $Re_C = 1.2 \times 10^4$ (Akin and Rockwell [19]).

Acknowledgements

The authors are pleased to acknowledge support of the Office of Naval Research (Grants N00014-89-J-1799 and N00014-90-J-1510) monitored by Drs. Steven Ramberg and Edwin Rood and the National Science Foundation (Grant CTS 8922095) monitored by Dr. Stephen Traugott. Moreover, support of the Air Force Office of Scientific Research (Grant AFOSR-91-0055) monitored by Major Daniel Fant contributed to development of related instrumentation and techniques.

References

1. Rockwell, D., Magness, C., Robinson, O., Towfighi, J., Akin, O., Gu, W., and Corcoran, T.: Instantaneous structure of unsteady separated flows via particle image velocimetry. Report PI-1, Fluid Mechanics Laboratories, Lehigh University, Bethlehem, Pennsylvania (1992).

2. Adrian, R. J.: Particle-imaging techniques for experimental fluid mechanics. Ann. Rev. Fluid Mech. 23 (1991) 261-304.

3. Adrian, R. J.: Multi-point optical measurements of simultaneous vectors in unsteady flows - review. Int. J. Heat & Fluid Flow 7 (1986) 127-145.

4. Gursul, I., Lusseyran, D. and Rockwell, D.: On interpretation of flow visualization of unsteady shear flow. Exp. Fluids 9 (1990) 257-266.

5. Gursul, I. and Rockwell, D.: Effect of concentration of vortices on streakline patterns. Exp. Fluids 10 (1991) 294-296.

6. Wu, J. C.: The theory of aerodynamic force and moment in viscous flows. AIAA J. 19 (1981) 432-441.

7. Howe, M. S.: Contributions to the theory of aerodynamic sound, with application to excess jet noise and the theory of the flute. J. Fluid Mech. 71 (1975) 625-673.

8. Howe, M. S.: On unsteady surface forces and sound produced by normal chopping of a rectilinear vortex. J. Fluid Mech. 6 (1989) 131-153.

9. Foss, J. F. and Wallace, J. M.: The measurement of vorticity in transitional and fully developed turbulent flows. In Advances in Fluid Mechanics Measurements, M. Gad-el-Hak (ed.). New York: Springer-Verlag (1989) 45, 264-321.

10. Rockwell, D. Magness, C., Towfighi, J. , Akin, O. and Corcoran, T.: High image-density particle image velocimetry using laser scanning techniques. Exp. Fluids (1992) (in press).

11. Chyu, C. and Rockwell, D. (1992) Manuscript in preparation.

12. Bloor, M. S.: The transition to turbulence in the wake of a circular cylinder. J. Fluid Mech. 19 (1964) 290-304.

13. Gerrard, J.: The wakes of cylindrical bluff-bodies at low Reynolds number. Proc. Roy. Soc. Lond. A 288 (1978) 352-382.

14. Wei, T. and Smith, C. R.: Secondary vortices in the wake of circular cylinders. J. Fluid Mech. 169 (1986) 513-533.

15. Chow, G. and Rockwell, D. (1992) Manuscript in preparation.

16. Nuzzi, F., Magness, C., and Rockwell, D.: Three-dimensional vortex formation from an oscillating, non-uniform cylinder. J. Fluid Mech. 238 (1992) 31-54.

17. Towfighi, J. and Rockwell, D. (1992) Manuscript in preparation.

18. Gu, W. and Rockwell, D. (1992) Manuscript in preparation.

19. Akin, O. and Rockwell, D. (1992) Manuscript in preparation.

A Reflection on Two Modes of Eddy Shedding at Re = 180 ~ 300

M. M. Zdravkovich
Department of Aeronautical and Mechanical Engineering,
University of Salford, Salford M5 4WT, England

Summary

The word 'reflection' means to review something casually and often inconclusively, according to the Webster Dictionary. This has been attempted in this paper where the two modes of eddy shedding are briefly reviewed. It is argued that in the transition in wake state of flow, there is an overlapping range where both modes of eddy shedding, low-speed and high-speed operate at the same time at different frequencies.

Two Modes of Eddy Shedding

Kovasznay[1] noted that laminar eddies are not shed from the cylinder but are formed as they are carried downstream by the laminar wake instability. Figure 1 shows a typical sinusoidal trail at $Re=140$ photographed by Taneda[2]. The formation of eddies is accomplished by the gradual roll-up of free shear layers at crests and troughs along the wake. Note that the trail-streakline emerges from the middle of the nearwake and remains distinct up to the end in Figure 1.

Gerrard[3] postulated that turbulent eddies grow and develop almost in a fixed position for one half of the shedding period until they are strong enough to draw the other shear layer across the wake, so that the subsequent eddy is cut-off from a further supply of the circulation and shed downstream. Figure 2 illustrates the high-speed mode of shedding at $Re=300$ (Freymuth et al[4]). Large eddies are formed immediately behind the cylinder and the nearwake appears stationary in comparison with that in Figure 1. The eddies appear disconnected further downstream and the trail-streakline is not visible.

The Overlapping of Two Modes of Eddy Shedding

Recent measurements of St in TrW by Williamson[5] are shown in Figure 3. Two distinct discontinuities occur at the start and end of $TrW1$ regime. The first discontinuity shows a small histeretic effect. The frequency spectra (a), also included in Figure 3, shows a narrow-band peak at $Re=172$ when Re increases and a wide-band peak when Re decreases at a lower frequency.

There is a wide overlapping region, $230 < Re < 250$, where two values of St are found. The

*The term eddy is used to distinguish a real viscous vortex

frequency spectra (*b*) show a gradual changeover from a low to a high value of St. Williamson[5] suggested that two peaks are produced by an intermittent changeover of the two modes of eddy shedding which alternate in time but do not coexist simultaneously.

Homann[6] photographed the wake at $Re = 225$ in an oil tunnel fitted with heat exchanger. Figure 4 exhibits three streaklines behind the cylinder. The upper and lower one correspond to a small part of the invisible shear layers while the trail-streakline, the typical feature of the low-speed mode, is in the middle. The nearwake appears similar to that in the high-speed mode while the rest of the wake resembles the low-speed mode. The peculiar elongated structure on the upper side of the cylinder draws back the trail-streakline, see the thickened part of the streakline in Figure 4. A similar fully elongated structure is seen on the other side of the wake accompanied by the start of the eddy roll-up in a low-speed mode manner. The impression is that the laminar wake instability interferes with the free shear layers at an incipient phase of roll-up.

An alternative interpretation to the overlapping St curves in Figure 3 may be offered based on Homann's[6] Figure 4. If the elongated structures in Figure 4 are the incipient phase of the high-speed mode of eddy shedding, then the frequency of their formation and 'shedding' should be close to the frequency of the high-speed mode. The elongated structures appear weak at first, hence the wide-band peak in Figure 3. They eventually grow to a full size eddy roll-up immediately behind the cylinder. At the same time, the laminar wake instability is weakened by the shortening of the nearwake and its frequency peak widens until it disappears at the upper end of $TrW1$ regime, see Figure 3b.

Closing Comments

The transition-in-wake, TrW, state of flow may be conveniently divided into:

$TrW1$, *Lower Transition Regime*: $160\text{–}180 < Re < 210\text{–}250$

- Random appearance of fingers in increasing numbers.
- Three-dimensional and irregular wake.
- Shortening of nearwake with rising Re.
- Increasing pressure drag with rising Re.
- Laminar eddies of irregular shape.
- Low-speed mode of eddy shedding, St increases with rising Re.
- Broad-band frequency peak and 'unstable' St range.

***Figure* 1:** Laminar eddy street at $Re = 140$, Taneda[2]; an example of low-speed mode of eddy shedding.

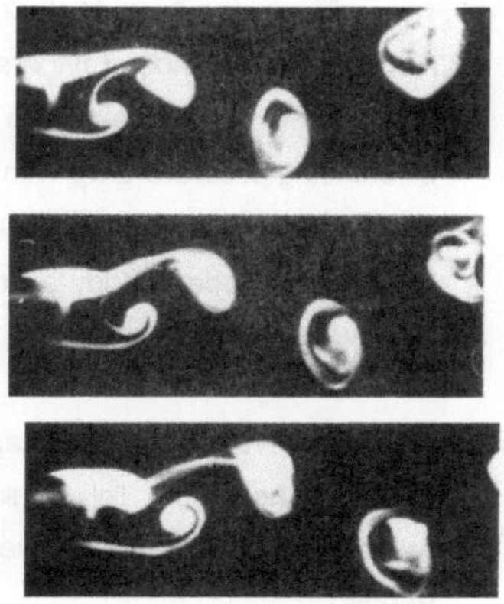

***Figure* 2:** Transitional eddy street at $Re = 300$, Freymuth et al[4]; an example of high-speed mode of eddy shedding.

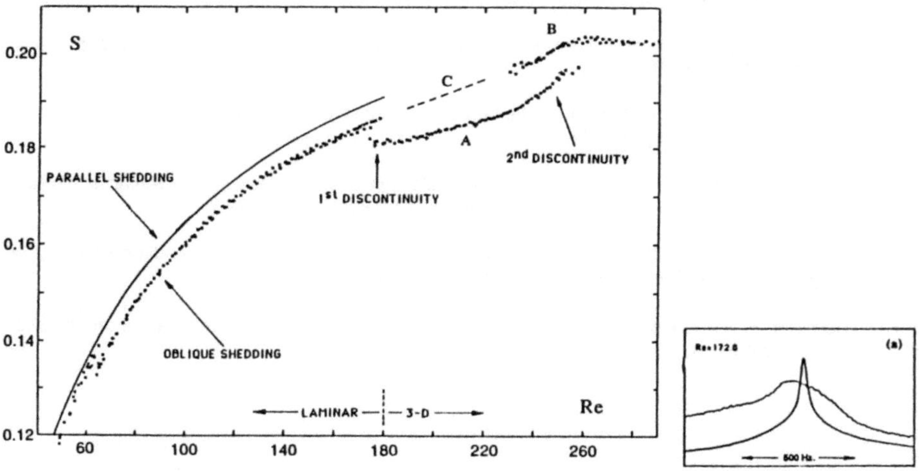

***Figure* 3:** Variation of Strouhal number in terms of Reynolds number; frequency spectra at: (*a*) first discontinuity, (*b*) second discontinuity, Williamson[5].

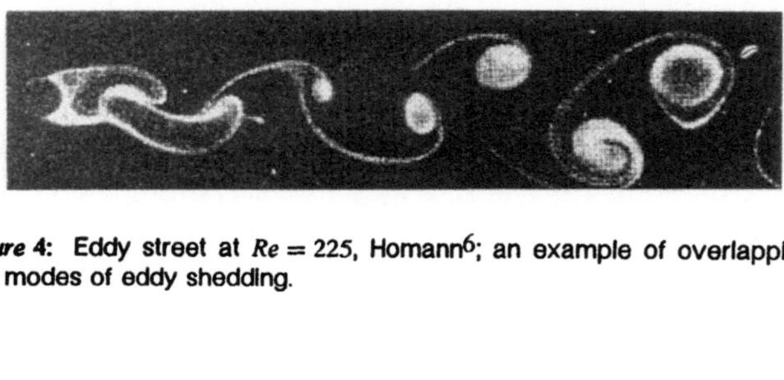

***Figure* 4:** Eddy street at $Re = 225$, Homann[6]; an example of overlapping of two modes of eddy shedding.

TrW2, Upper Transition Regime: $210-250 < Re < 400-500$

Random appearance of fingers in a decreasing number

Lengthening of nearwake with rising *Re*.

Decreasing pressure drag with rising *Re*.

Transitional eddies with laminar cores and turbulent periphery.

High-speed mode of eddy shedding, $St \simeq const.$

Narrow-band frequency peak, 'irregular' *St* range.

TrW1–TrW2: The well-defined overlapping region $230 < Re < 260$ of low- and high-speed modes of eddy shedding. This region can be postponed to higher Reynolds numbers if the two-dimensional flow is preserved in the wake. The topology of this region has to be verified.

References

1. Korvasznay, L. S. G.: Hot wire investigations of the wake behind cylinders at low Reynolds numbers. Proc. Roy. Soc. A 198 (1949) 174-190.

2. Taneda, S.: In Album of Fluid Motion, assembled by M. van Dyke. The Parabolic Press, Stanford (1982) 56, Figure 94.

3. Gerrard, J. H.: The mechanics of the formation region of vortices behind bluff bodies. J. Fluid Mech. 25 (1966) 401-413.

4. Freymuth, P., Finaish, F., and Bank, W.: Visualisation of the vortex street behind a circular cylinder at low Reynolds number. Phys. Fluids. 29 (1986) 1321-1323.

5. Williamson, C. H. K.: The existence of two stages in the transition to three-dimensionality of a cylinder wake. Phys. Fluids. 31 (1988) 3165-3168.

6. Homann, F.: Influence of high viscosity on flow around cylinder (in German). Forsch. Geb. Ing. 7 (1936) 1-10.

Pressure Forces on a Circular Cylinder in Cross Flow

C. NORBERG

Department of Thermo- and Fluid Dynamics
Chalmers University of Technology
S-412 96, Gothenburg, Sweden

Summary

Mean pressure forces have been measured at Reynolds numbers from about 100 to 3×10^5 (fluctuating pressure forces from about 650). In the range $Re = 10^3 - 10^4$ there was a change in the sectional r.m.s. lift coefficient by about one order of magnitude while the corresponding change in the mean pressure drag coefficient was 20%.

Introduction and Experimental Details

Despite the enormous amount of experiments on the circular cylinder in cross flow there is still a need for accurate data on pressure forces, especially at Reynolds numbers less than 10^4. The main objective of the present investigation was to provide reliable measurements of both mean and fluctuating pressure forces over a large range of Re in smooth flow. Another objective was to give further evidence of a transitional behaviour at around $Re = 5 \times 10^3$ [1].

The free stream turbulence intensity was 0.06%. In the measurements of fluctuating pressures either Endevco 8507C miniature pressure transducers (D = 4 mm) or Brüel & Kjær 4135 condenser microphones (D = 10, 40 and 120 mm) were employed. The dynamic pressures were corrected for blockage effects [3] at blockages greater than 0.5% i.e. at $D > 6$ mm. Sectional r.m.s. lift coefficients were estimated from distributions of r.m.s. pressure coefficients together with the cross-correlation coefficients of fluctuating pressures between two diametrically opposite points on the cylinder surface ($\varphi = \pm 90^\circ$ with respect to the free stream). Further experimental details can be found in [2].

Results and Discussion

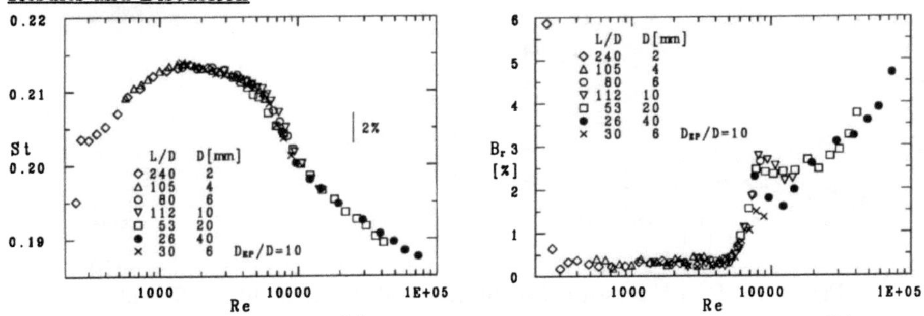

(a) (b)

Fig. 1. (a) Strouhal number and (b) relative bandwidth (D_{EP} – end plate diameter).

Fig. 1 shows some details of the Strouhal number St and the associated relative bandwidth B_r ($-3dB$), respectively. The most prominent feature was the transitional behaviour of B_r at around $Re = 5 \times 10^3$. In the beginning of the subcritical regime, B_r remained at a constant level of about 0.3%. This high peak sharpness continued up to $Re \simeq 5 \times 10^3$ where there was an approx. tenfold increase in B_r up to a local maximum at $Re \simeq 8 \times 10^3$ – followed by a recovery up to $Re \simeq 10^4$ where B_r again showed an increase with increasing Re. As shown in [1] the behaviour of B_r at even higher Re is very sensitive to free stream turbulence. In order to fully capture the details of B_r at Re from about 4×10^3 to 10^4 the aspect ratio L/D had to be large – at least greater than 50 [2]. However, as shown in Fig. 1b, the steep increase in B_r at around $Re = 5 \times 10^3$ was captured at aspect ratios as small as $L/D = 30$ (the phenomenon also occurred for $L/D = 15$ but was not as distinct as for larger L/D). The fact that rather large aspect ratios were required in this range is believed to be a consequence of the maximum in axial correlation at $Re = 5 \times 10^3$, with a correlation length of about 15 diameters [1].

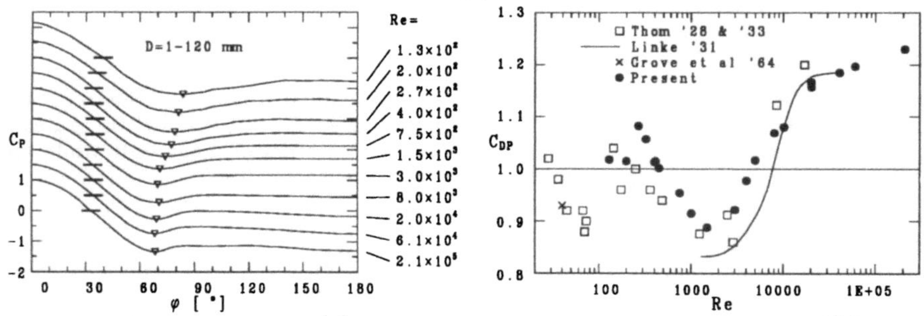

(a) (b)

Fig. 2. (a) Pressure distributions. Fat lines: $C_P = 0$; markers: minimum C_P.
(b) Pressure drag coefficient.

In Fig. 2a some mean pressure distributions are presented. The angles φ were corrected due to finite angle resolution (modified form of Linke [4]). Pressure drag coefficients (C_{DP}) are plotted in Fig. 2b. Since C_{DP} is closely related to the base suction coefficient [2, 8], it can be inferred that local minimum values in C_{DP} occur at the start of the laminar shedding regime ($Re \simeq 48$), at $Re \simeq 190$ and at $Re \simeq 1500$ whereas local maxima occur at the end of the laminar shedding regime ($Re \simeq 160$) and at the start and end of the subcritical regime ($Re \simeq 260$, $Re \simeq 2 \times 10^5$), respectively.

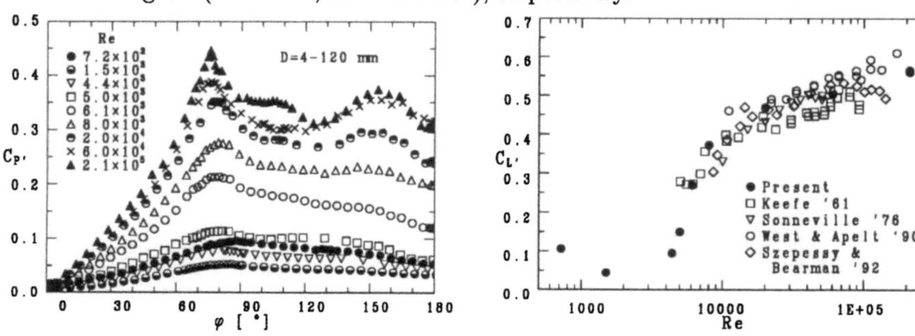

(a) (b)

Fig. 3. (a) R.m.s. pressure distributions and (b) r.m.s. lift coefficient.

In Fig. 3a some r.m.s. pressure distributions are depicted. The corresponding r.m.s. sectional lift coefficients, calculated using the formula of Ribeiro [9], are shown in Fig. 3b. Of particular interest are the indeed low fluctuation levels at around $Re = 1500$ and the dramatic changes when passing through $Re \simeq 5 \times 10^3$. In addition, it is worth noting that the absolute value of the correlation coefficient (used in the calculation of $C_{L'}$) was as low as 0.1 at $Re \simeq 1500$.

$Re = 3 \times 10^3$ $\quad$ $Re = 5 \times 10^3$ $\quad$ $Re = 8 \times 10^3$

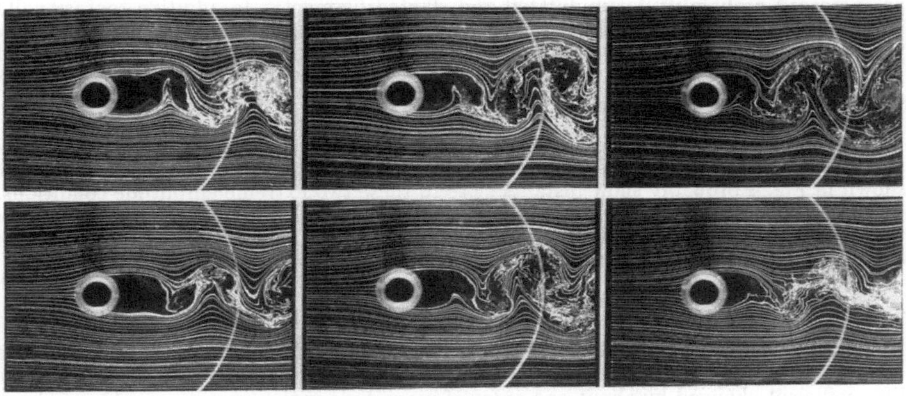

Fig. 4. Side views; $D = 20$ mm, $L/D = 15$.

(a) $Re = 3.0 \times 10^3$ $\quad$ (b) $Re = 5.5 \times 10^3$

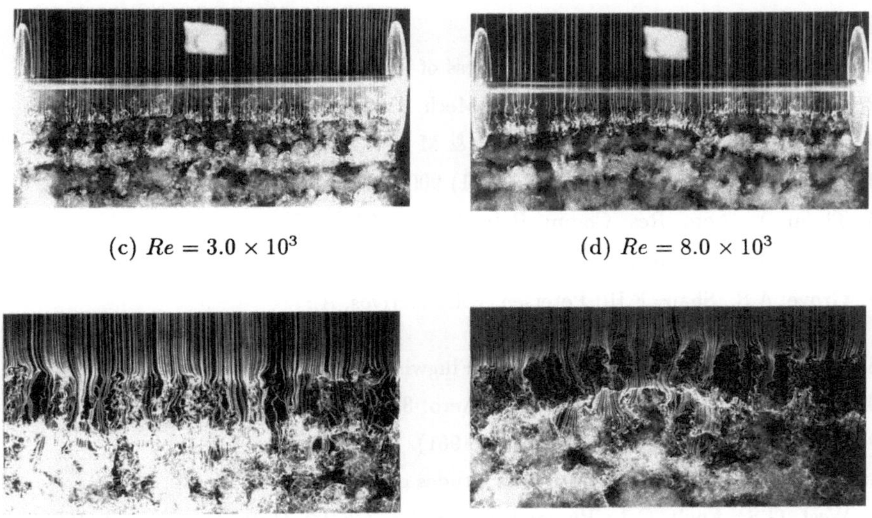

Fig. 5. Spanwise views; (a) and (b) $D = 10$ mm, $L/D = 30$, (c) and (d) $D = 16$ mm, $L/D = 65$.

From the side views of Fig. 4 we notice a substantial decrease in formation length when increasing Re from 3×10^3 to 8×10^3, in agreement with previous hot wire measurements [1]. When considering the phase of shedding, it was notable that all side view pictures taken at $Re < 5 \times 10^3$ were similar in character (Fig. 4, left). At $Re > 5 \times 10^3$, however, the flow exhibited large amplitude modulations ("bursts") and this similarity was lost (Fig. 4, right). Some typical spanwise views are shown in Fig. 5. At $Re < 5 \times 10^3$ the near-wake vortex filaments usually were surprisingly straight and aligned with the cylinder axis – in accordance with the low values of B_r (Fig. 1b). At higher Re, there was an increased probability for undulated and slanted filaments. In addition, occasionally and at seemingly random positions along the span, Y-shaped connections between vortex filaments were present, see e.g. Fig. 5b. In similarity with the flow around cylinders joined with a step in the diameter [14], these connections probably occur between cells having a slight difference in vortex shedding frequency. At these Re also, the shedding frequency has a substantial variation with time [1, 15] – in accordance with the high values of B_r. Conceivably, the vortex shedding is weak and highly disordered at the connections whereas strong and highly correlated shedding occurs otherwise. The side views at $Re = 8 \times 10^3$ shown in Fig. 4 probably show instants close to these two extreme amplitude conditions. As shown in [1] the high amplitudes are associated with a shedding frequency lower than the mean value. Evidently, the fundamental changes in the near-wake shedding process at around $Re = 5 \times 10^3$ are of great importance for the pressure forces on the cylinder.

Conclusions

At Reynolds numbers from about 100 to 3×10^5 the mean pressure drag coefficient was confined within the interval 1.05 ± 0.2. Within $Re = 650$ to 3×10^5 the sectional r.m.s. lift coefficient changed by about one order of magnitude with a minimum value of about 0.05 at $Re = 1500$. The dramatic alterations at around $Re = 5 \times 10^3$ in e.g. the r.m.s. coefficients and the relative bandwidth of the vortex shedding frequency, respectively, were shown to be associated with a transitional change in the three-dimensionality of the near-wake vortex shedding.

References

1. Norberg, C.: Proc. 4th Asian Congress of Fluid Mech. (1989), Suppl. Vol.
2. Norberg, C.: Submitted to J. Fluid Mech. (1992).
3. Maskell, E.C.: Aero. Res. Counc. R & M No. 3400 (1963).
4. Linke, W.: Physik. Zeitschr. 22 (1931) 900-914.
5. Thom, A.: Aero. Res. Comm. R & M No. 1194 (1928).
6. Thom, A.: Proc. Roy. Soc. London, Ser. A 141 (1933) 651-669.
7. Grove, A.S.; Shair, F.H.; Petersen, E.E.; Acrivos, A.: J. Fluid Mech. 19 (1964) 60-80.
8. Williamson, C.H.K.; Roshko, A.: Z. Flugwiss. Weltraumforsch. 14 (1990) 38-46.
9. Ribeiro, J.L.D.: J. Wind Eng. Ind. Aero. 37 (1991) 311-326.
10. Keefe, R.T.: UTIA Report No. 76. (1961).
11. Sonneville, P.: Bull. de la Dir. des Etudes et Rech. Ser. A, No. 3. (1976).
12. West, G.S.; Apelt, C.J.: Research Report No. CE109, Univ. Queensland (1990).
13. Szepessy, S.; Bearman, P.W.: J. Fluid Mech. 234 (1992) 191-217.
14. Norberg, C.: Proc. 11th Australasian Fluid Mech. Conf. (1992) 507-510
15. Blevins, R.D.: J. Fluid Mech. 161 (1985) 217-237.

Prediction of Certain Transition Characteristics in the Wake of a Circular Cylinder in Free and Forced Flow

M. BRAZA[1] , H. PERSILLON[2] & F. SERS[3]

Research group: Transferts en Ecoulements Laminaires et Turbulents,Institut de Mécanique des Fluides de Toulouse, URA D0005 CNRS, Avenue du Professeur Camille Soula, 31400 TOULOUSE, FRANCE.

ABSTRACT: The objective of the present study is to predict several 2-D phenomena of the transition towards turbulence for the flow around a circular cylinder by using directly the Navier-Stokes equations without turbulence modelling. The Reynolds numbers examined are 20,000 and 30,000. This study allows to predict not only the Von Kàrmàn instability leading to the alternating vortices, but also the frequency of the Kelvin-Helmholtz instability, appearing in the separated shear layers and leading to mixing layer vortices. The interactions among different classes of vortices is studied in their details in the near wake region. Especially the influence of a forcing frequency applied at a point next to the wall, to the above transition features is analysed. Moreover a criterion of localizing the onset of transition in the separated shear layers past the cylinder is suggested.

I . INTRODUCTION

The structure of the flow around a circular cylinder has been studied over long years, due to important transition steps that are developed in this kind of flow as the Reynolds number increases, phenomena which have a great importance in naval and aeronautical applications. Detailed presentation of these phenomena can be found in the experimental works and reviews of Roshko [1], Gerrard [2] & [3] and Bearman [4], among others. In addition to the well known phenomenon of alternating Von Kàrmàn vortices, several experimental studies detect a 2-D transition wave frequency in the separated shear layer past the cylinder (Bloor [5], Gerrard [3], Kourta *& al* [6], Wei & Smith [7], Unal & Rockwell [8]) in the Reynolds number range of 1,000 - 50,000. For Reynolds numbers higher than 2,000 a new type of vortices appears in the separated shear layer, the Kelvin-Helmholtz eddies, first visualized by Crausse [9]. Moreover, recent studies of Williamson [10] & [11] have shown experimentally the onset of 3-D instabilities and a vortex dislocation pattern beyond Reynolds number of 180.

Concerning the forced flow around a circular cylinder, studies of Bearman [12], Williamson & Roshko [13] among others, give a detailed measurement and computation of the unsteady body forces and of the flow pattern for the oscillating case. Furthermore, Strykowski *& al* [14] studied the effects of a punctual permanent disturbance in the wake by the means of a control body.

The majority of numerical simulations does not predict, to our knowledge, the 2-D shear-layer transition mentioned above at relatively high values of the Reynolds number in the transition to turbulence regime. There is also a poor knowledge of the way of amplification of the transition wave frequencies in the separated shear layers and whether the flow pattern is affected by a forcing applied in the vicinity of the base pressure point. Hence, this study will examine the ability of the Navier-Stokes system to predict these features at the Reynolds numbers 20,000 and 30,000 in a two-dimensional approach.

II . THE NUMERICAL METHOD

The numerical method is based upon a predictor - corrector pressure scheme, presented in it's details in [19] and based on Amsden & Harlow [16]. The Navier-Stokes equations are written in logarithmic - polar coordinates system (figure 1), in order to provide a sufficiently fine grid near the wall and in dimensionless variables (x,y,t, P, u and v) normalized by the cylinder's radius a and the uniform stream velocity U_∞. The time-dependent Navier-Stokes equations for an incompressible fluid are written in conservative form with respect to the above orthogonal coordinate system.

---

[1] Chargé de recherche CNRS

[2] Research student

[3] Research student

A semi - implicit second order time and space accurate scheme is used. The equations are integrated over finite volumes. The numerical method (code CERCLE2) has been developed by Braza [15]&[25].

III . BOUNDARY CONDITIONS

The boundary conditions are those of impermeability and non-slip for the cylinder's surface. At infinity (external boundary at R_∞ distance from cylinder) the boundary conditions are deduced from the Oseen solution in order to not confine the present external flow.
Hence,

$$V(x,y)_{y\text{-}y_\infty} \rightarrow U_\infty \quad \text{outside the wake,} \qquad \left(\frac{\partial V}{\partial y}\right)_{y\text{-}y_\infty} \rightarrow 0 \quad \text{in the wake region.}$$

The number of grid points, time step and size of the computational domain are carefully studied in order to provide physically valid results, as shown in next paragraph. In particular, the independence of the solution, relative to two large values of the external radius and of grid spacings was tested for Reynolds number 5,000 (Braza & *al* [25]). The relative difference obtained for the global parameters (mean drag, Strouhal number) has been found within the range of the space accuracy of the numerical method (second order accurate) and in agreement with experimental results, as it is discussed in the following paragraph. This allows us to suppose that the values of the external radius adopted in this study, associated with the Oseen boundary conditions are sufficiently large in order to respect the physics of the flow. The numerical parameters are shown in table 1. A careful choice of the numerical parameters is also done for Re=20,000 and 30,000. The optimum values chosen provide a good agreement with the physical experiment for the same Reynolds numbers.

IV . RESULTS

A. Global parameters

The computed global parameters of the free flow (without applied any forcing frequency) are shown in figure 2a (Re=20,000). It is obtained that the flow pattern is governed by a mainly periodic frequency, where there are superimposed non-linearly other frequencies. The mean values of C_D, C_{DV} are $C_D = 0.972$ and $C_{DV} = 0.0081$ (Re=20,000), $C_D = 0.952$ and $C_{DV} = 0.0045$ (Re=30,000). A good agreement is obtained with experimental results (figures 3a & 3b). However the present simulation overpredicts the amplitudes of the lift coefficient.

The Strouhal number is evaluated by taking the Fast Fourier Transform of the time-dependant evolution of the u velocity component (figure 4) and of the lift coefficient. The Strouhal number has the value of 0.195 for Re=20,000 and Re=30,000. A good agreement with various experimental studies is obtained (figure 5).

The forcing is achieved in this study by imposing an oscillation at the U velocity component at $\Delta x/2$ distance from the rear stagnation point. The choice of this forcing is guided by two reasons:

- This forcing, by the means of the velocity variation, modifies directly the base pressure region through the present numerical elliptic code for an incompressible fluid.
- The forcing at this point and at a suitable frequency can readily be used in industrial applications for attenuating vibrations due to vortex shedding.

The forcing applied is $U = U_\infty \sin(2\pi f_S / 2)$.

The global parameters are presented in this case in figure 2b. Beyond the dimensionless time value 84, a phase lag is produced which increases as a function of time and leads to a phase opposition beyond t = 130. However the main periodic character of the flow is not completely attenuated.

B. Transition wave frequency and mixing layer eddies

Figure 8 shows clearly through the unsteady pressure and velocity fields the existence of different kinds of organized structures in the near wake.

It is clearly pointed out that apart from the fundamental frequency the spectrum shows a multitude of well distinct frequencies among which a peak at a frequency f_B which appears systematically on all the spectra calculated along the separated shear layer (figure 4a). It is obtained that the present frequency satisfies the law $f_B / f_S \sim Re^{1/2}$ (figure 4b). Hence, at the present Reynolds number range, the above solver of Navier-Stokes equations, having a sufficiently refined grid and time step, is capable to generate naturally this transition characteristic, confirmed by a wide number of spectra in the separated shear layer. This feature has mainly a 2-D origin, as it is also confirmed by the physical experiments. Owing to this instability, which develops in the shear layer, there is formation of mixing layer vortices as it is shown in figure 8, showing the history of the flow in a period. These vortices grow downstream and form alternating eddies after successive pairings. The growth of the wake is clearly shown through the instantaneous iso-pressure fields.

The amplitude of the above transition wave frequency is plotted versus x/D along the upper shear layer at y/D=0.433 (figure 6) for Re=20,000 and Re=30,000. It is clearly obtained that the amplitude of frequency f_B becomes maximum at x/D=1.662 for Re=20,000 and at x/D=0.474 for Re=30,000. This is a pronounced characteristic of the onset of 2-D transition features in the separated shear layers and it can be used as a criterion of appearance of transition at the present Reynolds number.

The forcing case corresponds to a reduced growth of the wake. The mixing layer eddies persist in the forced case despite the forcing, but the alternating character seems to be attenuated.

ACKNOWLEDGEMENTS: The computational CPU time is supplied partly by the joined program CNUSC-IBM C3NI (Centre de Compétence en calcul Numérique Intensif) and partly by the CCVR (Centre de Calcul Vectoriel de Recherche). The authors thank MM Chassaing, HaMinh and Boisson for valuable discussions. Part of this work is encouraged by the NATO Grant n° CRG920176 where C. Williamson and M. Braza are co project investigators.

Références

1 - ROSHKO A. 1954. On the drag and shedding frequency of two-dimensional bluff bodies. N.A.C.A. Tech. Note n°3169.
2 - GERRARD J.H. 1966. The three-dimensionnal structure of the wake of a circular cylinder. J. F. M. **25**, 1-143.
3 - GERRARD J.H. 1978. The wake of cylindrical bluff bodies at low Reynolds numbers. Philos. Trans. R. Soc. London Ser. A 288, 358.
4 - BEARMAN P.W. 1967. On vortex streets wakes. J. F. M. **28**, 625.
5 - BLOOR M.S. 1964 The transition to turbulence in the wake of a circular cylinder. J. F. M. **19**, 290.
6- KOURTA A., BOISSON H.C., CHASSAING P., HAMINH H. 1987. Non linear interaction and the transition to turbulence in the wake of a circular cylinder. J. F. M. **181**, 141-161.
7 - WEI T. & SMITH C.R. 1986. Secondary vortices in the wake of circular cylinders. J. F. M.**169**, 513-533.
8 - UNAL M. F. & ROCKWELL J. 1988. On vortex formation from a cylinder. Part 1. The initial instability. J. F. M. **190**, 491.
9 - CRAUSSE E. 1936. Thèse de Docteur ès Sciences, Université de Toulouse. Contribution expérimentale à l'étude des phénomènes transitoires et périodiques se produisant dans les lignes en mouvement.
10 - WILLIAMSON C.H.K. 1988. The existence of two stages in the transition to three-dimensionality of a cylinder's wake. Phys.Fluids **31** 2742.
11 - WILLIAMSON C.H.K. 1989. Oblique and parallel modes of vortex shedding of a circular cylinder at low Reynolds numbers. J. F. M. **206**, 576.
12 - BEARMAN P.W. 1984. Ann. Rev. Fluid Mech. **16**, 195-222.
13 - WILLIAMSON C.H.K. & ROSHKO A. 1990. Zeitschrift fur Flugwissenschaften und Weltraumforschung **14**,38-46.
14 - STRYKOWSKI P.J. & SREENIVASAN K. R. 1990. On the formation and suppression of vortex "shedding" at low Reynolds numbers. J. F. M. **218**, 71-107.
15 - BRAZA M., CHASSAING P., HAMINH H. 1986. Numerical study and analysis of the pressure and velocity fields in the near wake of a cylinder. J. F. M. **165**, 79-130.
16 - AMSDEN A.A. & HARLOW F.H. 1970. The SMAC method: a numerical technique for calculating incompressible fluid flows. Los Alamos Scientific Laboratory Report. L.A.-4370.
17 - HARLOW F.H. & WELCH E.J. 1965. Numerical calculation of time-dependent viscous incompressible flow of fluid with free surface. Phys. Fluids **78**, N° 12, 819.
18 - BRAZA M.,CHASSAING P. & HAMINH H. 1990. Prediction of large-scale transition features in the wake of a circular cylinder. J. Phys. Fluids .A., Vol. **2** N° 8, 1461-1471.

Reynolds Number	Grids	External radius R∞	Number of unknowns	Time step	Total physical time	CPU (sec) per time step
20000*	222x139	204.08	123,432	0.01	174	22.2
30000**	222x139	204.08	123,432	0.01	133	13.35

Table 1 The numerical parameters.
* IBM 3090-600-VF Centre National Universitaire Sud de Calcul (CNUSC).
**CRAY-2 Centre de Calcul Vectoriel de Recherche (CCVR)

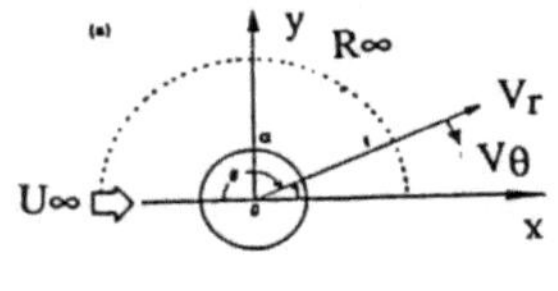

Figure 1. Flow configuration and coordinates system.

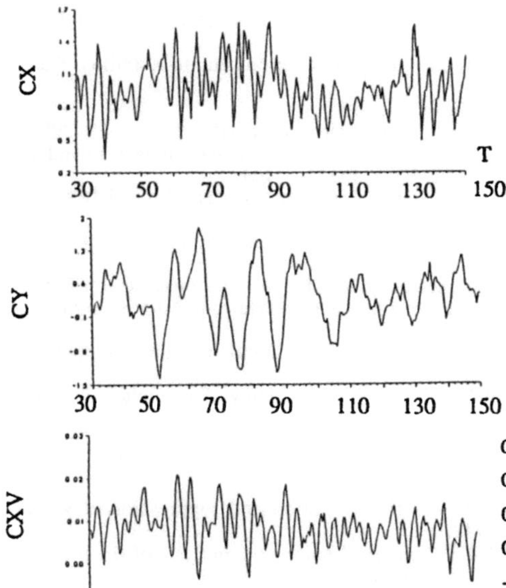

Figure 2a. Time-dependent drag, lift and viscous drag coefficients, Re=20,000 Non-forced case.

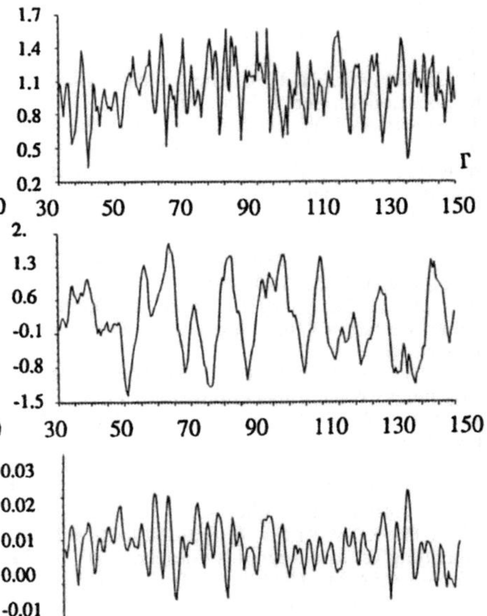

Figure 2b. Time-dependent drag, lift and viscous drag coefficients, Re=20,000 Forced case.

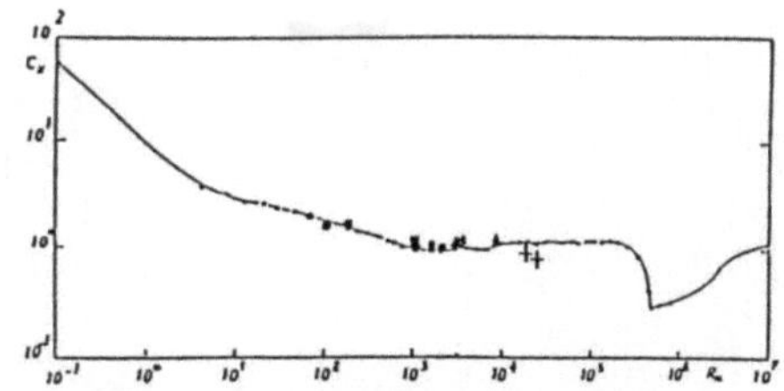

Figure 3a. Mean drag coefficient versus Re. ••• Collected experimental data of Wieselsberger [1921], Roshko [1953], & [1961], Tritton [1959]. •, *, ▲ Braza et al [26]; + this study.

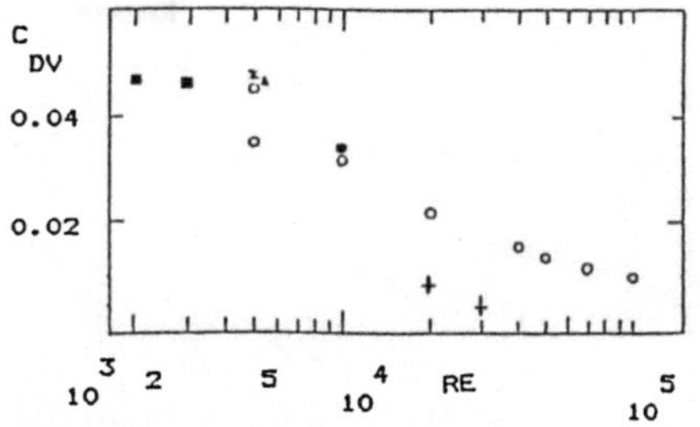

Figure 3b. Mean viscous drag versus Re. O Experimental data of Son & Hanratty (1969). Numerical results of Braza [24]■Re=2000 and Re=3000. Re=5000, case d; ▲ ● Re=10,000. + this study.

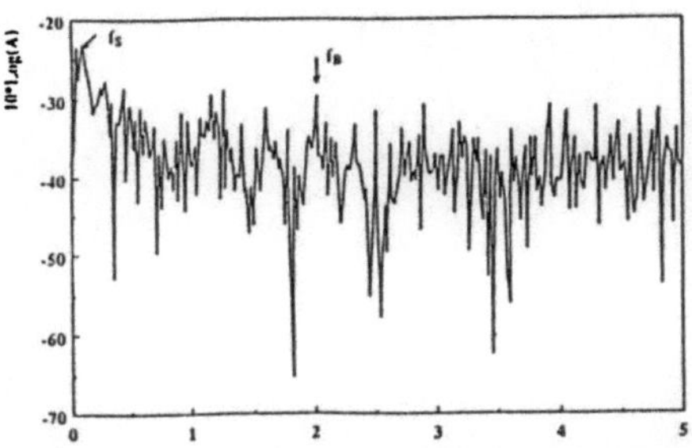

Figure 4a. U component spectrum. re=30,000, x/D=0.474, y/D= 0.455.

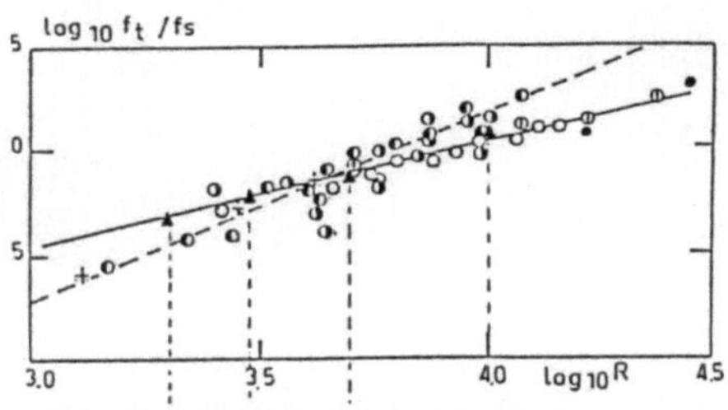

Figure 4b.Ratio of the transition wave frequency over Strouhal number versus Reynolds number. O, + : Bloor [6]; Gerrard [3]; Kourta et al [7]; ◐, ◑ : Wei & Smith [8].▲Braza et al [26]. ● : This study

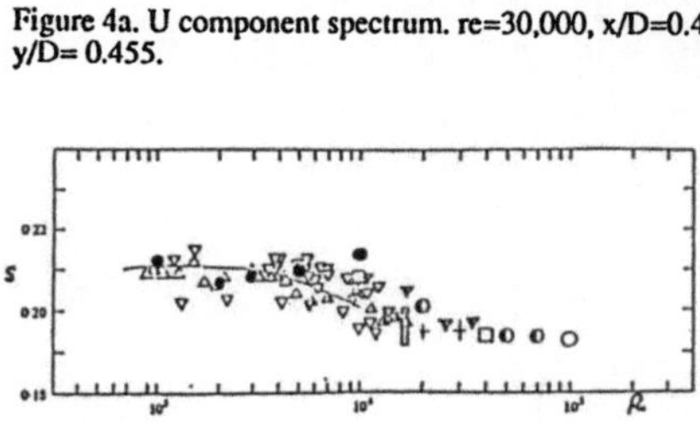

Figure 5 Strouhal number versus Reynolds number : Son & Hanratty [1969]:□ d/h=0.062; O, d/h=0.125. ◐:0.083 ___, Roshko [1], Gerrard [2], :∇ d/h=0.012; ▼ , d/h=0.05;▯ d/h= 0.05, represents 22 points. O Braza et al [26], case c. ● Braza et al [26], cases a,b,d,e. +: this study

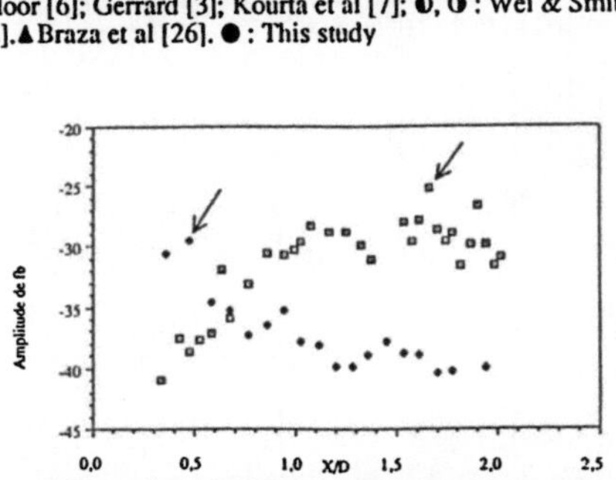

Figure 6. Spectral amplitude of the transition wave frequency versus x/D at y/D=0.433.⊡ Re=20,000. ♦ Re=30,000.

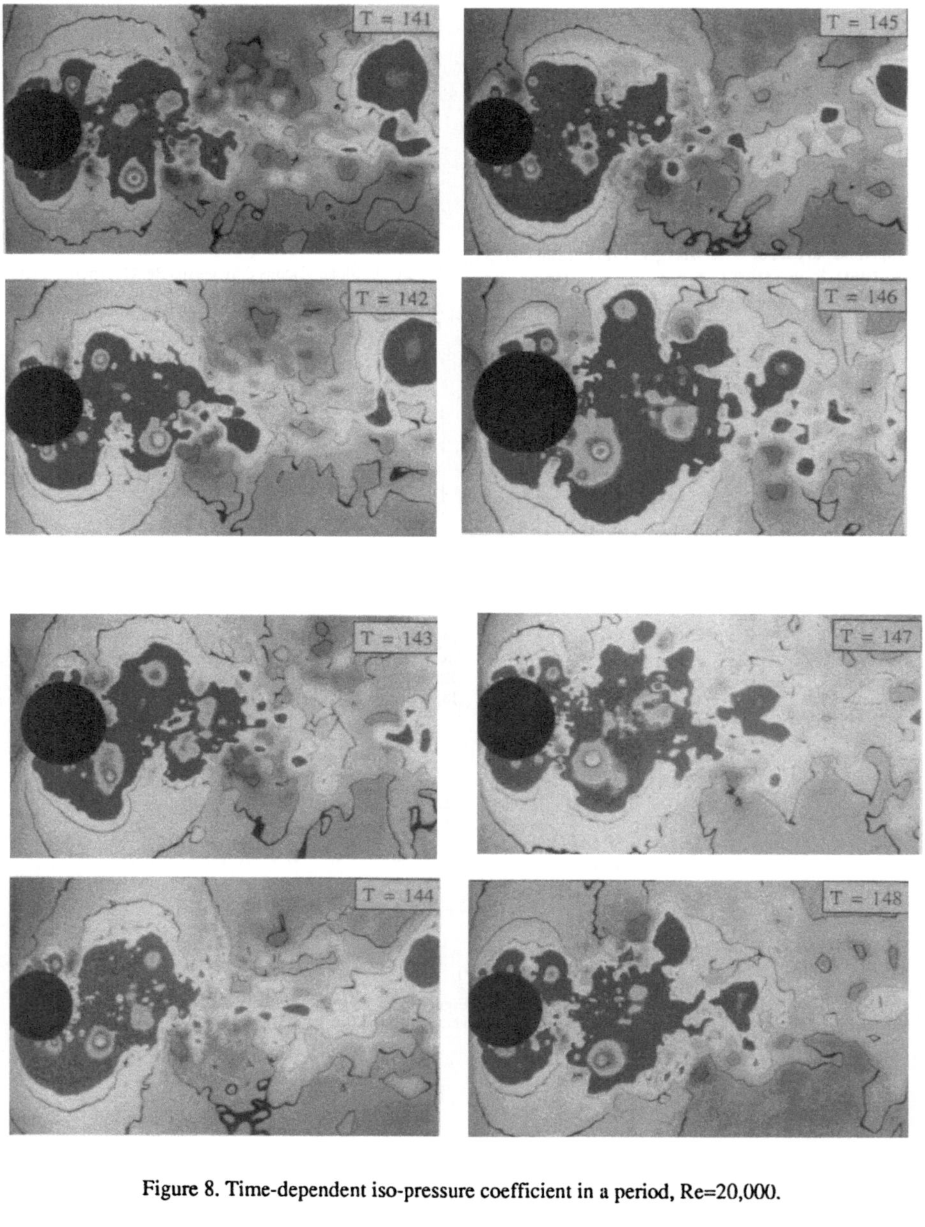

Figure 8. Time-dependent iso-pressure coefficient in a period, Re=20,000.

Shedding Process of the Initial Vortices from Impulsively Started Cylinders at Re = 1000: End and Body Geometry Effects

G. POLIDORI , G. PINEAU , K. ABED MERAIM and M. COUTANCEAU
Laboratoire de Mécanique des Fluides (URA CNRS 191)
Université de POITIERS - FRANCE

Summary

The 3D effects, induced by two symmetrical endplates upon the flow around variously shaped and impulsively started cylinders at Re=1000, are evidenced by means of three complementary experimental methods of visualization. Comparative data, such as the vorticity propagation velocity along the spanwise direction, or the evolution of topological and kinematical parameters of the near wake in a specific cross-section, have been deduced. From streamline and streakline patterns, the change from two-to three-dimensional flow configuration has also been estimated, showing the influence of the body shape upon the 3D effect manifestation.

Introduction

According to recent studies, it appears that boundary conditions, especially end and body geometry may have an important influence on the vortex shedding process and consequently on the wake structure. However, it is recognized that these phenemena are much too often ignored and that there is still a lack of data in this domain [1 to 7]. This work is devoted to this problem by considering the initial phase of the time-development of the flow generated by the impulsive start from rest of short bodies having different shapes but the same aspect ratio AL. These bodies are fitted with two symmetrical large endplates; their shape evolves from the circular cylinder to a flat plate, with a semicircular configuration as an intermediate situation. The object of the study is to determine how the 3D effects, induced by the endplates, set up with time , and how the body shape influences the phenomena, especially the vorticity propagation along the spanwise direction. Moreover, a comparative study, based upon a quantitative analysis of both streamline and streakline patterns of the initial vortex shedding process, gives the time-evolution of some significant topological parameters such as centers and saddles, as well as of the velocity distribution along a selected direction in the middle cross-section of flow field. From streakline patterns, the change from two-to three-dimensional flow configuration has been estimated, as well as the influence of the body shape on this phenomenon and on the appearance of secondary instabilities.

Experimental set-up and visualization methods

The visualization experiments are conducted in a tall vertical water tank (80x100x120 cm^3) equipped with a movable carrier. The carrier, to which the body is attached, translates vertically under the gravity, with a constant speed, after an impulsive start ensured by cutting out the current supplying an electro-magnet. The photographs are taken sequentially by an automatic driving camera which accompanies the body in its displacement.
The time-development of the flow behind the cylinders is visualized by choosing either rilsan particles, or dye filaments, or electrolytic tracers according to the type of information required (velocity field or vorticity propagation). A volumic halogen illumination and a static laser sheet have been used for continuous and individualized particle visualizations respectively. Streamline patterns are obtained by a graphic integration of the streak velocity field.

Experimental parameters

The governing dimensionless parameters are $t^* = tVo/D$, $Re = VoD/\nu$ and the aspect ratio AL. In the present investigation, Re has been fixed to 1000, AL to 5, and the explored time limited to $t^*= 6.0$, because of the complex character of the further flow development. Although, in the experiments, the time step was $\Delta t^*=0.5$, the sequence of the frames presented in this paper corresponds to $\Delta t^*=1.0$.

Results

The respective influence of both endplates and body shape upon the flow structure at Re=1000 around different cylinders has been evaluated by means of the complementary visualization techniques described above; fig. 3,4,5 are given as examples. From these pictures, qualitative as well as quantitative data have been deduced, and used as basic elements for a comparative study. The data concern the influence of the body shape on:
- the three-dimensional vorticity propagation coming mainly from the body ends.
- the initial stages of the shedding process in the middle cross-section, evidenced by the appearance of different flow patterns: two-dimensional and symmetrical, two-dimensional and asymmetyrical, and three-dimensional.
- the time-evolution of near wake topological and kinematical parameters (singular points, velocity profiles and maxima); fig. 1,2 are given as examples.
- the birth and propagation of the wake boundary instabilities.within the vortex rolling-up, which appear as undulations of the separated shear layers, visualized by electrolytic filaments emitted from the body surface.
The analysis of these different results shows clearly the preponderant influence of either the downstream geometry side or the upstream one on flow structures. Thus, it appears that an

upstream circular geometry reduces the nearwake development, especially the recirculating zone, and the three-dimensional effect propagation. On the other hand, both an upstream and downstream flat face make the nearwake larger, and consequently increase the velocity values within the near flow; the three-dimensionality propagates faster in such a case. Moreover, we have shown that the wake boundary instabilities appear only for sharp-edged body (flat plate and semicircular cylinder), in the experimental conditions (Re=1000, t*≤6.0).

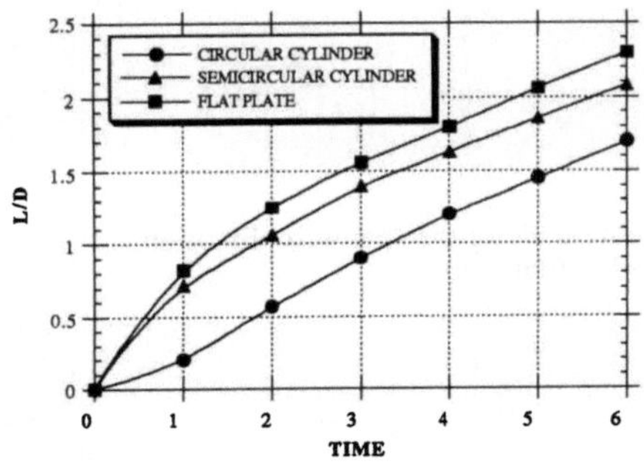

Fig. 1.

Time-evolution of the saddle location corresponding to the near wake length in these cases.

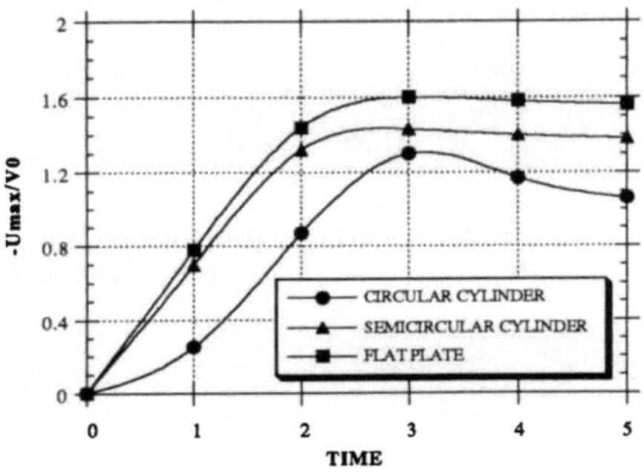

Fig. 2.

Time-evolution of the maximum velocity in the near wake along flow direction

Ackowlegment

A part of this work has been supported by the *Direction des Recherches, Etudes et Techniques*, French Ministry of Defence, under contracts n° 88/047 and 90/174.

References

1. EINSENLOHR & ECKELMAN,Phys. Fluids A 1 (2), 1989
2. GERICH & ECKELMANN, J. Fluid Mech., vol. 122, pp. 109-121, 1982
3. GERICH, Flow Visualization IV, Paris, 1986
4. SLAOUTI & GERRARD, J. Fluid Mech., vol. 112, pp. 297-314, 1981
5. SZEPESSY & BEARMAN, J. Fluid Mech., vol. 234, pp. 191-217, 1992
6. WILLIAMSON, J. Fluid Mech., vol. 206, pp.579-627, 1989
7. WILLIAMSON & ROSHKO, Z. Flugwiss. Weltraumforsch., 14, pp. 38-46, 1990

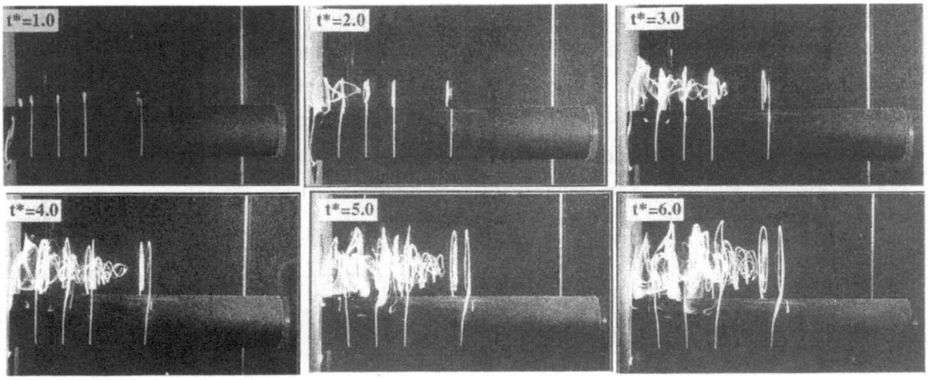

Fig. 3. Vorticity propagation, time-evolution along the spanwise direction (circular cylinder - dye filaments)

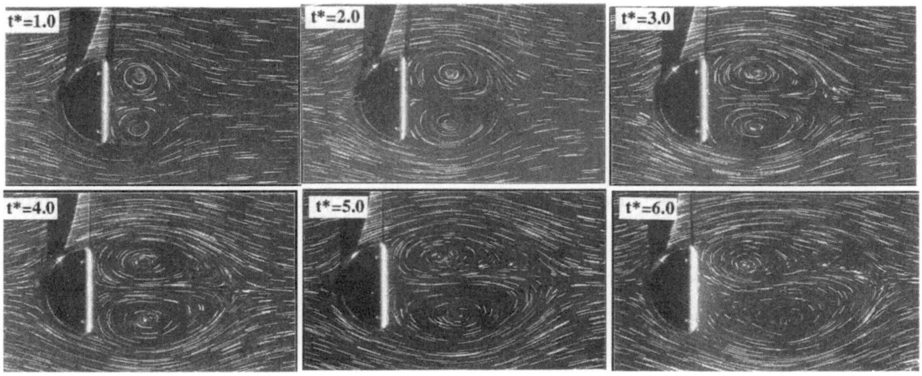

Fig. 4. Velocity field, time-evolution in the middle cross-section (semicircular cylinder - rilsan particles)

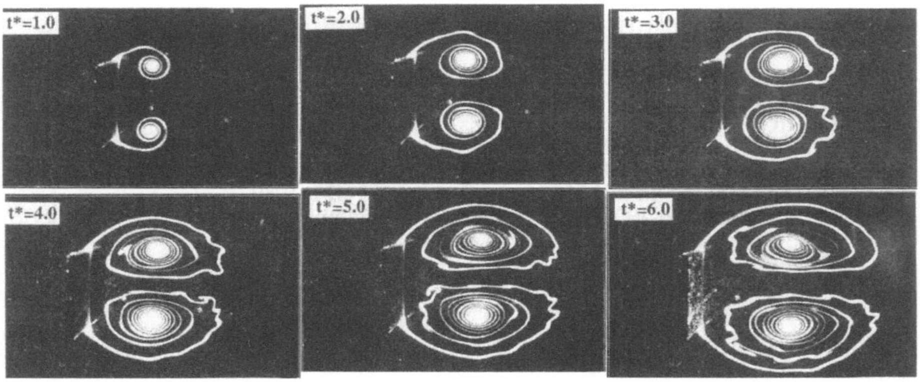

Fig. 5. Wake boundary instabilities, time-evolution in the middle cross-section (flat plate - electrolytic particles)

Control of Wakes by Manipulating Mean Velocity Profiles

Ho, Chih-Ming and Leu, Tzong-Shyng

Mechanical, Aerospace and Nuclear Engineering Department
University of California, Los Angeles
Los Angeles, California 90024

SUMMARY

The free shear layer is known to be sensitive to upstream boundary conditions. Many experiments have shown that low level perturbations can alter the evolution of the shear flow. In this experiment, we demonstrate that a steady suction modifies the mean velocity profile and significantly changes the instability process of a wake. A threshold suction velocity was found. When the suction is higher than the threshold, the wake becomes stable and no vortex shedding exists.

INTRODUCTION

The global development of a free shear layer, mixing layer or wake, depends on the evolution of the coherent structures (see review by Ho and Huerre 1984). The vortices change their arrangement by mutual induction of neighboring structures. Therefore, the flow can be controlled by applying active forcing to 2-D vortices (Ho and Huang 1982, Oster and Wygnanski 1982). The perturbation level usually required to make the difference is very low because the forcing is used only as a seeding energy and the shear flow acts as a spatial amplifier of disturbances injected upstream. This occurs because the mixing layer is convectively unstable with all unstable disturbances growing in the downstream direction. The propagation properties of shear flow instabilities are very sensitive to the shape of the velocity profile (Huerre and Monkewitz 1990).

In some flows, like a wake for example, the velocity profile experiences significant streamwise development and the propagation properties of local instabilities change character along the flow direction. In particular, there may exist a region of significant streamwise extent where local instabilities may propagate in both the upstream and downstream direction (i.e., the flow is absolutely unstable). Of particular interest is the possibility of controlling global flow characteristics through modification of the local velocity profiles.

At present, we are investigating the effect of modified mean velocity distribution on the stability process and the global flow features in a wake-shear layer configuration. Phenomenal changes are observed with a constant suction or blowing at the origin of the flow.

FACILITY

The experiment was performed in an open surface water channel. The stagnation chamber was separated by a vertical splitter plate. Flows in the two sections were provided by two pumps so that the velocity ratio can be easily adjusted. A honeycomb and six layers of screens were used to reduce the mean flow uniformity and turbulence level. Since the splitter plate was vertical in this case, the problem of removing the air bubbles from the screens was not difficult. The splitter plate divided the test section into two equal parts, each 21.3 cm wide and 30.48 cm deep. The maximum velocity was 20 cm/sec. The trailing edge of the splitter plate was a slot one cm wide. A perforated tube inside the slot was connected to a pump. The tube provided either suction or blowing by merely changing the connection to the pump. The flow was visualized by the hydrogen bubble technique and the velocity was measured by laser Doppler velocimetry, particle image velocimetry and hot-film probes.

EXPERIMENTAL FINDINGS

At zero suction velocity, typical wake vortices can be seen by using the hydrogen bubble technique (Fig. 1). The hot-film provides vortex induced velocity fluctuations. The time trace shows an amplitude modulated sinusoidal when no suction is applied. The vortex shedding frequency decreases with increasing suction velocity. The Strouhal number, St, normalized with the width of the slot, D, and the velocity difference, $U - U_C$, has a constant value of 0.12. U_C is the magnitude of the suction speed.

There exists a threshold suction velocity. Above that value, the vortices disappear (Fig. 2). The flow becomes stable for a long distance. The threshold velocity equals about half of the free stream speed of the wake, U. As the suction velocity becomes close to the threshold, the vortices become intermittent. As soon as the threshold velocity is reached, the fluctuation vanishes and the hot-film output is almost constant as a function of time. The wake is stabilized by the suction.

The mean velocity profiles were measured by LDV. Near the trailing edge of the splitter plate, a region of reversed flow is observed. Downstream from this region, a strong velocity deficit is observed in the stable wake (Fig. 3). This phenomenon can not be explained by the quasi-parallel stability results and perhaps needs to be understood from a fully 2D stability analysis (Morzynski and Thiele 1992). A CFD investigation is carried out at USC to study this flow field (Redekopp 1992).

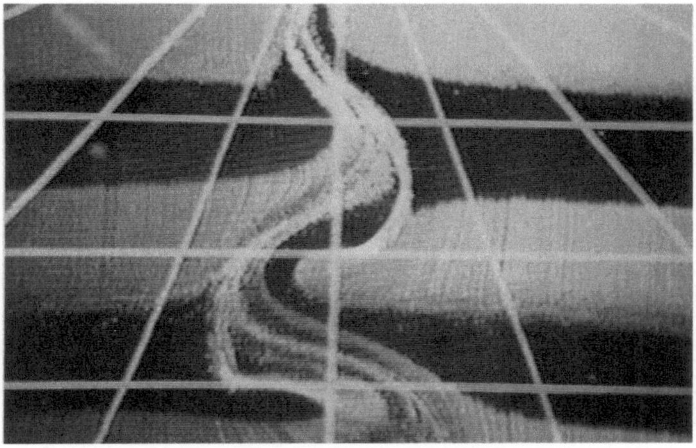

Fig. 1 Wake without suction or blowing

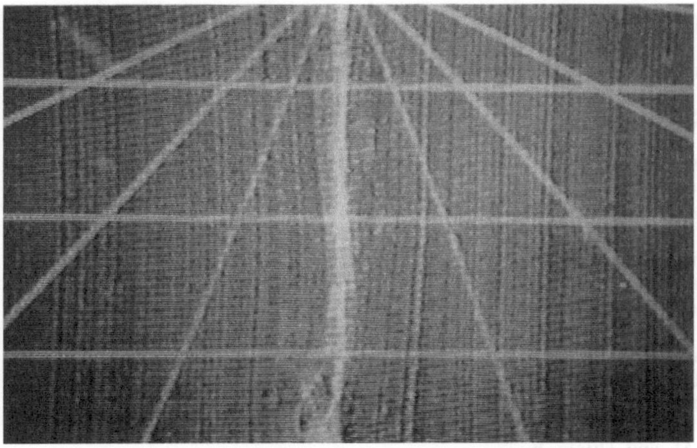

Fig. 2 Stabilized wake by suction

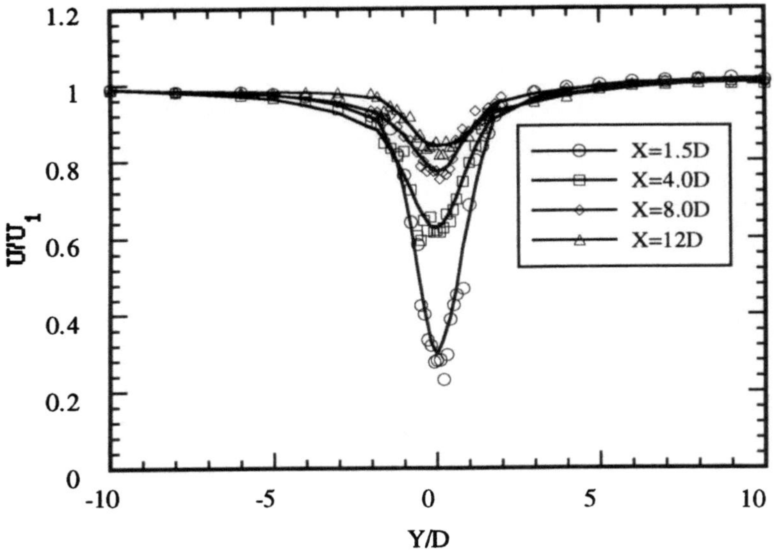

Fig. 3 Time-averaged Streamwise Velocity Profiles

ACKNOWLEDGMENT

The work is supported by Taylor Research Center through ONR and by AFOSR.

REFERENCES

1. Ho, C. M. and Huang, L., Subharmonics and vortex merging in mixing layers, *J. Fluid Mech.*, 1982, **119**:443--473.

2. Ho, C. M. and Huerre, P., Perturbed free shear layer, *Annual Review of Fluid Mechanics*, 1984, **16**:365--424.

3. Huerre, P. and Monkewitz, P., Local and global instabilities in spatially developing flows, *Annual Review of Fluid Mechanics*, 1990, **22**:473--537.

4. Morzynski, M. and Thiele, F., Numerical Investigation of Wake Instabilities, Proceeding of IUTAM symposium on Bluff-Body Wakes, 1992.

5. Oster, D. and Wygnanski, I., The forced mixing layer between parallel streams, *J. Fluid Mech.*, 1982, **123**: 91--130.

6. Redekopp, L., Private communication, 1992.

Unsteady Wakes of Three Dimensional Bodies

E. G. Duell and A.R. George
Cornell University, Ithaca, New York

Abstract

The effect of a splitter plate on the near wake of a three dimensional bluff body in proximity to a ground plane was studied. Static pressure taps on the base of the body were used to measure both the mean and fluctuating pressures acting on the model base. It was determined that the splitter plate raised the spatially averaged mean pressure on the base by 15% when placed one fourth model height below the top of the base. The splitter plate also decreased the spatially averaged fluctuating pressure level between 13% and 31%, but the decrease in fluctuating pressure level was not necessarily accompanied by a decrease in drag. These measurements indicate that the aerodynamic drag of bluff bodies is related to unsteady effects in the wake. The near wake structure behind the bluff body was investigated using flow visualization and two hot wire anemometry probes and cross-correlating the velocity signals. The development of a periodic flow structure that was one half the model height in size was examined.

Introduction

The aerodynamic drag of bluff bodies is predominantly caused by the pressure deficit from the separated near wake acting on the base of the body. The near wake is a highly time dependent flow field which also makes the aerodynamic drag time variant. Much of the current work in bluff body drag consists of making various changes to the body until the average aerodynamic drag is decreased, without understanding how the wake has been affected. To most effectively reduce pressure drag on bluff bodies, the time varying, near wake structure must first be understood and controlled.[1]

Berger et. al.[2] identified three periodic mechanisms in the unsteady near wakes of spheres and discs in a freestream flow. This work extends the understanding of three dimensional bluff body near wakes. To identify large scale periodic structures in the near wake and study their downstream development, cross-correlations were performed on the velocity signals from different points in the near wake.

Modifying the pressure field of the near wake which acts on the base of a bluff body can change the aerodynamic drag of the body. One method of changing the pressure distribution is to add a splitter plate to prevent or modify unsteady vortex formation. Mansingh et. al.[3] achieved a 50% drag reduction on a two dimensional cylinder with the use of a splitter plate. This reduction was accompanied by a change in near wake unsteadiness as indicated by a 60% decrease in Strouhal number. In our preliminary work[4], it was determined that decreasing the drag of a three dimensional body with the insertion of a splitter plate at the model base also resulted in lower pressure fluctuations at some points on the base of the model. In this work, static pressure taps on the model base measure the effect of a splitter plate on both mean and fluctuating base pressure distribution across the entire base.

Experimental Equipment

The experiments were conducted at the Low Noise Wind Tunnel Facility at Cornell University. The tunnel has a nominal cross section of 50 cm high by 74 cm wide. The maximum velocity of the tunnel is 27.5 m/s and is uniform within 1%. The average turbulent intensity is 0.3% in an empty test section at maximum velocity. The experiments were performed at maximum velocity so the Reynolds number based on model length was $7.05 \cdot 10^5$. At frequencies below 500 Hertz, the background tunnel noise levels are at least 5 decibels below typical fluctuating pressure levels measured at the base of the model.

The shape of the model used in this study was chosen to model the near wake structure of a typical ground vehicle. The length was chosen so that the boundary layer on the model will be turbulent when separation occurs at the base. The model dimensions are shown in figure 1. The splitter plates used in the experiments were as wide as the model, 10 cm in length and 3 mm thick. The model was mounted on a stationary ground plane to simulate flow around buildings and other nonmoving bodies. Previous results have shown that this mounting can also generate the basic structure of a ground vehicle body wake.

Hot wire anemometry measurements were taken with a TSI 1050 anemometer and single wire probes mounted on a two component computer-controlled traverse. Pressure measurements were taken with Omega PX160 Series pressure transducers and Scanivalve pressure measurement system mounted inside the model. Thus, a minimum of tubing length was used to minimize damping of higher frequencies resulting in accurate measurements up to 200 Hz. From previous tests with microphones, it is known that most of the energy in the spectrum is concentrated below 200 Hz. The pressure taps were located at 100 points of 1.15 cm increments in a rectilinear pattern across the model base. Data was acquired with a Macintosh data acquisition system.

Results and Discussion

Effects of Splitter Plate on Model Base Pressure Distribution

The mean and fluctuating pressure distribution on the base of the model was measured for three geometries. The base case was the model without a splitter plate, the second geometry included a horizontal splitter plate at z=0 and the third geometry had the plate located at z=H/4.

For the base case of no splitter plate, the mean and rms pressure distributions are shown in figure 2. The spatially averaged mean C_p and rms C_p levels are presented in table 1. Spatially averaged values are simply an arithmetic mean to determine the overall mean C_p and rms C_p. The mean pressure has a suction peak near the center, across the entire base. The higher fluctuating pressure levels are concentrated in the lower half of the base. Flow visualization with helium filled bubbles has shown that there is a large, stable vortex with primary axis along the y direction which causes the suction peak. A secondary, counter-rotating vortex is formed in the corner of the base and ground plane that causes the high fluctuation level in the lower half of the base.

For the cases of the splitter plate at z=H/4 and z=0, another vortex is formed in the upper corner between the base and the splitter plates. Representative changes in the pressure distribution are shown in figures 3 and 4. The vortex produces both high rms pressure fluctuations and strong suction. However, the presence of the plate alters the near wake and the mean pressure on the remainder of the base is raised. For the case of the plate at z=H/4, the overall base pressure is raised 15% because the suction region is confined to one fourth of the base area. In the case of the plate at z=0, the stronger vortex produces higher local suction which acts over one half of the base area and this results in 3% lower overall base pressure. For the two cases, the spatially averaged rms pressure levels are lowered 13% and 31% respectively, which shows how much the unsteadiness in the near wake has been affected. The asymmetry shown in the mean pressure distribution was insensitive to changes in model yaw angle and appeared only when a splitter plate was in place. It is not yet understood what causes this asymmetry and further investigations are planned.

Cross-Correlations

Cross-correlations of velocity signals in the near wake were calculated to identify periodic large scale structures in the flow and their location. The cross-correlation coefficient $\rho_{xy}(T')$ is defined as

$$\rho_{xy}(T') = (u_1(t)u_2(t')) \ / \ ((u_1{}^2)^{1/2} * (u_2{}^2)^{1/2}) \qquad 1)$$

where $u_1(t)$ is the fluctuating velocity at location 1, $u_2(t)$ is measured at location 2 and T'=t-t'.

Two hot wire probes, separated by a distance of h/2, were located at z=0, y= ± H/4 and at a various x locations. The data for the cross-correlation was sampled at 1500 samples per second.

Each velocity signal contains 30000 samples per block and 6 blocks of data were taken at each x location. A cross-correlation was calculated for each block and the six results were averaged to reduce noise. The cross-correlation coefficient with the maximum absolute value was measured at x/H=2.0 and is shown in figure 5. The dominant frequency at this location was 45 Hz. The variation of the maximum value of the cross-correlation coefficient along x is shown in figure 6. This figure shows the development of flow structures with wave length of H/2, the separation distance of the two probes. This size structure reaches a maximum strength beyond the shear layer between x/H=2.0 and x/H=2.5 before its intensity begins to decay.

Conclusions

The effect of a splitter plate on the base pressure distribution was measured and the periodic structure in the unsteady near wake was studied which led to the following conclusions.

1) A splitter plate located in the proper position in the near wake region can increase the spatially averaged mean pressure on the base by 15%. A splitter plate also affects the fluctuating pressure distribution on the base, but a decrease in spatially averaged fluctuating pressure does not necessarily correspond to a decrease in drag.

2) The experimental technique of outfitting the base of the model with pressure taps and examining the mean and fluctuating pressure distribution is very useful in determining the flow structures near the base.

3) Cross-correlations of velocity signals in the near wake have identified periodic structures of size equal to one half the model height in the near wake. The development and decay of these structures was shown.

References

1. Nouzawa, T., Hiasa, K., Nakamura, T., Kawamoto, A., and Sato, H., "Unsteady-Wake Analysis of the Aerodynamic Drag of a Notchback Model with Critical Afterbody Geometry", SAE Paper 920202, 1992.

2. Berger, E., Scholz, D., and Schumm, M., "Coherent Vortex Structures in the Wake of a Sphere and a Circular Disk at Rest and Under Forced Vibrations", *Journal of Fluids and Structures*, 4, pp.231-257, 1990.

3. Mansingh, V., and Oosthuizen, P.H., "Effects of Splitter Plates on the Wake Flow Behind a Bluff Body", *AIAA Journal*, Vol. 28, No. 5, pp. 778-783, 1990.

4. Duell, E.G., and George, A.R.,"Unsteady Near Wake Flows of Ground Vehicle Bodies", AIAA Paper 92-2641, AIAA 10th Applied Aerodynamics Conference, June, 1992.

Acknowledgements

The authors would like to acknowledge the suggestions and assistance of the other members of the Aeroacoustics Research Group at Cornell University especially Gustavo Bodstein, Todd Ringler, and Ben Sim.

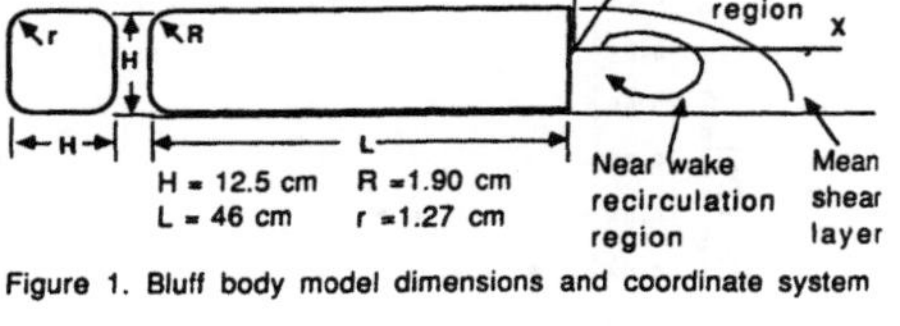

Figure 1. Bluff body model dimensions and coordinate system

Geometry	Spatially averaged mean Cp	% change from base case	Spatially averaged rms Cp	% change from base case
No splitter plate	-0.242	----	0.0218	----
Splitter plate @ z = H/4	-0.207	-15%	0.0189	-13%
Splitter plate @ z= 0.	-0.25	+3%	0.0149	-31%

Table 1.

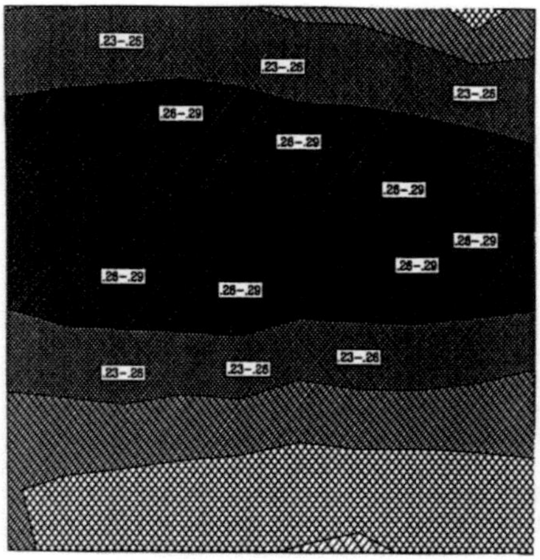

Figure 2a. Mean pressure distribution in terms of - C_p on model base with no splitter plate.

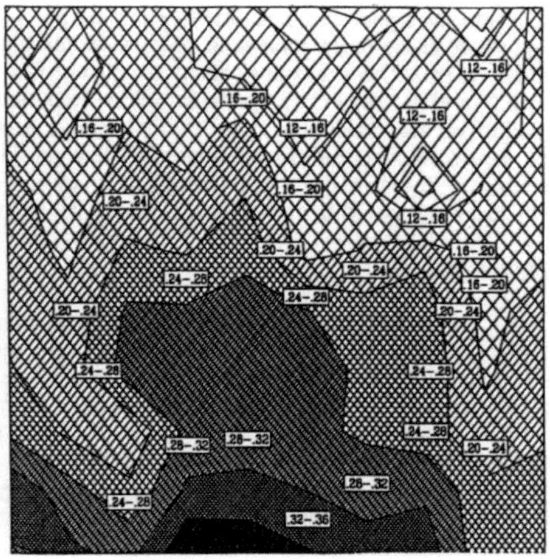

Figure. 2b. Distribution of rms pressure in terms of C_p*10 on model base with no splitter plate.

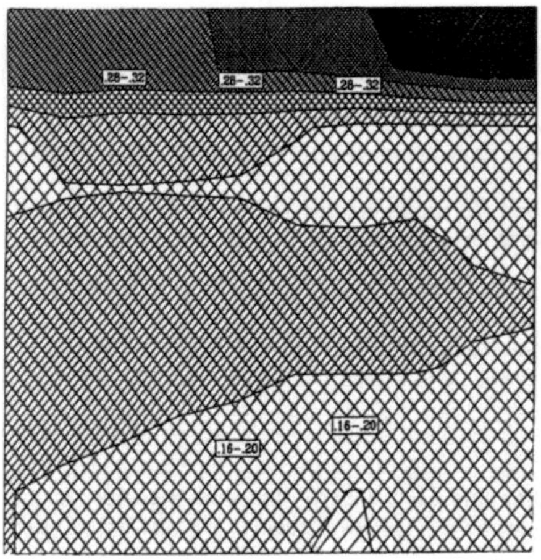

Figure 3. Mean pressure distribution in terms of - C_p on model base with splitter plate at z=H/2.

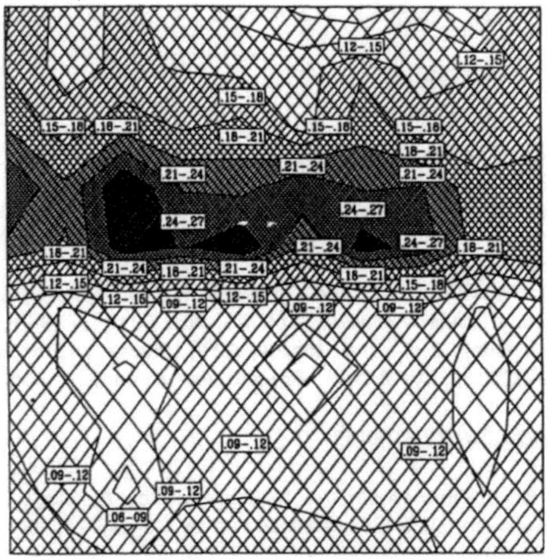

Figure 4. Distribution of rms pressure in terms of C_p*10 on model base with splitter plate at z=0.

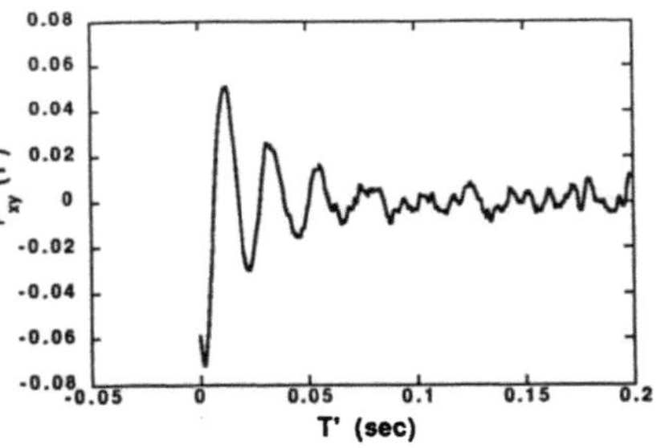

Figure 5. Cross-correlation coefficient for two points separated by H/2 at location x/H=2.0

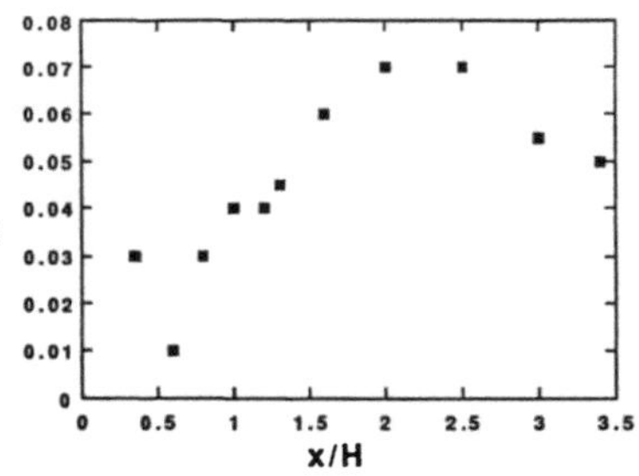

Figure 6. Variation of the maximum value of the cross-correlation coefficient along x-axis

Separated Flows Behind Bluff Bodies Including Ground Effects: Application of The Zonal Method

Heinz-D. Papenfuss and Klaus Gersten

Ruhr-Universität Bochum
Institut für Thermo- und Fluiddynamik
D-4630 Bochum 1, F.R.G.

The flow field past bluff bodies at large Reynolds numbers consists of the inviscid outer flow around the body, the viscous boundary layer close to the wall and the separated flow region behind the body, i.e. the wake. Consequently, the flow field is divided into several *zones*, which interact with each other. For bodies near moving grounds, the boundary layer at the moving ground forms an additional zone. The zonal method takes advantage of this structure of the flow field. It is equivalent to an asymptotic approach to the full time-averaged Navier-Stokes equations for large Reynolds numbers in analogy to the classical boundary-layer theory. The separated region, however, is included in this method. In the present work, only bluff bodies will be considered, where the position of the flow separation is fixed due to sharp corners at the blunt base.

The flow equations for each zone are simplified when large Reynolds numbers are considered. The solutions for the zones must be matched asymptotically to obtain the final solution. The key element of the method is the concept of the "fictitious body" which is introduced to simulate the displacement effect of the viscous zones (primarily the wake) on the outer inviscid flow. Experimental investigations of turbulent flows over backward-facing steps have shown that the displacement effect of the wake on the outer flow can be simulated by a fictitious body whose contour is universal for all step heights and Reynolds numbers when the coordinate parallel to the oncoming flow is scaled using the reattachment length and the normal coordinate using the effective step height. It is assumed that, in analogy to flows with backward-facing steps, the contour of the fictitious body representing the displacement effect of the wake follows this universal function.

The turbulent boundary layer attached to the body is forced to separate at the edges of the blunt base and extends as a free turbulent shear layer whereby a continuous change of the velocities between the outer flow and the reversed flow in the inner part of the separated region is established. From the theory of the turbulent mixing layer it follows that the total pressure is constant along the dividing streamline. Using this fact, a linear relationship between the pressure coefficient at the base, c_{pB}, and the pressure coefficient at the reattachment point, c_{pR}, can be derived. In cases of asymmetrical flow, the additional condition must be satisfied, that the lift of the fictitious body in the wake zone must be zero.

Based on these considerations, a computer code for predicting the flow field past bluff bodies near a ground with separated regions in the wake was developed. The code contains individual routines corresponding to the flow zones mentioned above. The flow fields in all four zones interact with each other. Consequently, the final solution and, particularly, the shape of the fictitious body must be determined iteratively. Integration of the pressure distribution on the given body provides the lift coefficient, c_L, and the pressure drag coefficient, c_{Dp}. The friction drag coefficient, c_{Df}, follows from integration of the shear-stress distribution along the body surface.

In order to check the theoretical results, experimental studies were carried out, mainly with models of automobiles near a ground. In order to simulate the relative motion between the model and the ground, two techniques were employed: 1) a moving belt in a wind tunnel and, 2) the moving-model technique where the model was launched pneumatically along a stationary rail (Papenfuss and Kronast [1]). Pressure distributions on the body contour and the ground and velocity distributions around the body were measured. For the latter, the hot-wire anemometry was applied in the moving-model experiments. In the wind-tunnel experiments, the Laser-Doppler anemometry was applied to measure velocities and turbulent shear stresses, particularly in the wake zone where reversed flow occurs.

Two-dimensional flow past an automobile near a ground

A two-dimensional model with a cross-section typical for a station waggon (squareback) was investigated theoretically and experimentally in a wind tunnel with moving belt and with the moving-model technique. The pressure distributions on the body for a ground clearance $h/L = 0.05$ and a Reynolds number of $Re = 2.5 \cdot 10^5$ are shown in Fig. 1.

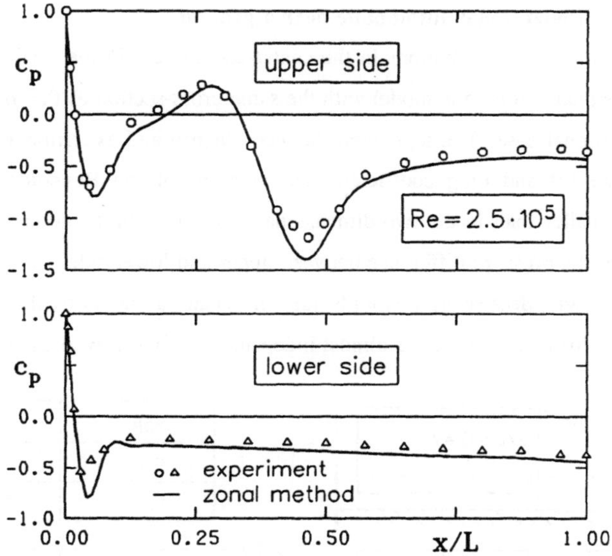

Fig. 1: Pressure distribution on a two-dimensional body near a ground (h/L = 0.05). Cross-section see Fig. 2. Comparison between experiment and zonal method

Re	c_D	c_L	c_{pB}	c_{Df}
∞	0.351	0.177	-0.435	0.000
10^8	0.367	0.311	-0.432	0.015
10^7	0.372	0.382	-0.431	0.021
$2.5 \cdot 10^5$	0.391	0.497	-0.430	0.040

Table 1: Aerodynamic coefficients for the two-dimensional body near the ground

Table 1 contains the lift and drag coefficients calculated with the zonal method. The drag coefficient, c_D, consists of three parts: 1) the friction drag, c_{Df}, 2) the pressure drag, c_{Dp}, which is essentially determined by the base pressure, c_{pB}, and 3) a suction force (c_D - c_{Df} + c_{pB}) resulting from the pressure distribution on the upper and lower side of the body. The results can be summarized as follows:

1) The base pressure which is the main contribution to the pressure drag is practically independent of the Reynolds number.
2) The drag coefficient increases for decreasing Reynolds number. This increase results mainly from the friction drag.
3) The lift coefficient increases considerably with decreasing Reynolds number. This is caused by the blockage effect of. the boundary layers between body and ground resulting in a reduced velocity and, hence, higher pressure in this region.

Flow past a three-dimensional automobile near a ground

Fig. 2 shows the pressure distributions in the center-section and in the section x/L = 0.47 of a three-dimensional automobile model with the same cross-section as the model described in the two-dimensional case. The agreement between theory and experiment is very good. Table 2 gives the lift and drag coefficients as functions of the Reynolds number. The coefficients are smaller than in the two-dimensional case since the additional flow over the side walls reduces the pressure difference between upper and lower side. The increase of the drag coefficient c_D with decreasing Reynolds number is caused mainly by the increase of the friction drag. The lift coefficient c_L is almost independent of the Reynolds number.

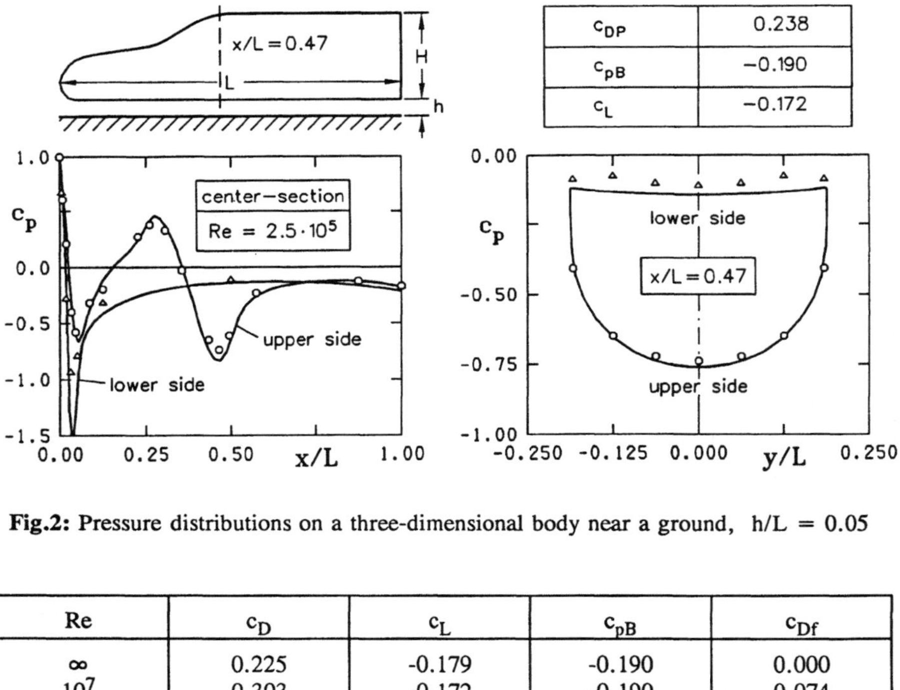

Fig.2: Pressure distributions on a three-dimensional body near a ground, h/L = 0.05

Re	c_D	c_L	c_{pB}	c_{Df}
∞	0.225	-0.179	-0.190	0.000
10^7	0.303	-0.172	-0.190	0.074
10^6	0.329	-0.172	-0.190	0.096
$2.5 \cdot 10^5$	0.349	-0.172	-0.190	0.111

Table 2: Aerodynamic coefficients for the three-dimensional body near the ground

REFERENCES

[1] Papenfuss, H.D.; Kronast, M. (1991): Moving-model technique used in automobile aerodynamics for measurement of ground effects. Experiments in Fluids **11**, 161-166.

Laminar and Turbulent Wakes of Bodies of Arbitrary Shape in Uniform Flow

N.I.Yavorsky

Institute of Thermophysics, Novosibirsk, Russia

1. Laminar wake.

In this section the problem of the laminar wakes of bodies of arbitrary shape in uniform flow is studied on the basis of the complete Navier-Stokes equations. The generalized multipole expansion of the exact solution of boundary hydrodynamic problem is elaborated. At infinity we have the exact solution of the Navier-Stokes equations in the form of constant velocity and pressure $\vec{v}_0$, p_0 (the body is in the rest). We can expand the velocity and pressure fields in the vicinity of the infinite point as

$$\vec{v} = \vec{v}_0 + \vec{w}, \quad p = p_0 + q, \qquad \vec{w} = o(v_0); \quad \vec{v}_0, p_0 = const\,,$$

where $\vec{w}, q$ - defect of velocity and pressure.

The Navier-Stokes equations in terms of $\vec{w}, q$ are

$$\begin{aligned} &\nu\Delta\vec{w} - (\vec{v}_0, \nabla\vec{w}) - \nabla q = (\vec{w}, \nabla\vec{w}), \qquad (\rho = 1) \\ &\text{div } \vec{w} = 0, \qquad \vec{w} = \vec{w}_s, \quad \vec{x} \in \Sigma \end{aligned} \tag{1}$$

Σ - surface of the body in general case having the permeable boundary, $\vec{w}_s$ - velocity on the boundary of the body.

Linearized Navier-Stokes equations that are called Oseen equations (left hand of the first equation) have the fundamental solution (Oseen,1927) in the form of the "tensor of velocity" $E_{ij}(\vec{x})$ and the "vector of pressure" $e_i(\vec{x})$

$$E_{ij}(\vec{x}) = \delta_{ij}\Delta\Phi - \frac{\partial^2\Phi}{\partial x_i \partial x_j}, \qquad e_i(\vec{x}) = -\frac{\partial}{\partial x_i}[\nu\Delta\Phi - (\vec{v}_0, \nabla\Phi)] \tag{2}$$

$$\Phi = -\frac{1}{8\pi\sigma\nu}\int_0^{\sigma s}\frac{1 - e^{-\alpha}}{\alpha}d\alpha, \quad \sigma = \frac{v_0}{2\nu}, \quad s = |\vec{x}| - (\vec{n}_0, \vec{x}), \quad \vec{n}_0 = \frac{\vec{v}_0}{v_0}$$

At infinity the asymptotic of the velocity has the form

$$v_i(\vec{x}) = v_{0i} + a_j E_{ij}(\vec{x}) + b e_i(\vec{x}) + \sigma_i(\vec{x}), \qquad \sigma_i = O(r^{-3/2}\ln r) \tag{3}$$

(Finn,1965; Babenko,1973). The vector $\vec{a}$ and scalar b one can express in the following form

$$a_i = \oint_\Sigma [v_{0i}v_j - v_i v_j - p\delta_{ij} + \nu(\frac{\partial v_j}{\partial x_i} + \frac{\partial v_i}{\partial x_j})] n_j dS, \qquad b = -\oint_\Sigma v_j n_j dS$$

These expressions one can perform as

$$\vec{a} = \vec{v}_0 Q - \vec{J}, \qquad b = -Q, \tag{4}$$

$$\vec{J} = \oint_\Sigma \Pi_{ij} n_j dS, \qquad Q = \oint_\Sigma v_j n_j dS$$

where $\vec{J}$, Q are the flux of impulse and flow rate. So, the main term at infinity is determined by the strict conservation integrals.

With the help of the fundamental solution we can invert the Navier-Stokes equations in the integral form

$$\begin{aligned} w_i(\vec{x}) = &\int_E w_j w_k \frac{\partial}{\partial x_k} E_{ij}(\vec{x}-\vec{y}) d^3y + \\ &+ \oint_\Sigma \{E_{ij}(\vec{x}-\vec{y})(v_{oj}v_k - \Pi_{jk}) + w_j T_{ijk}(\vec{x}-\vec{y})\} n_k dS \end{aligned} \tag{5}$$

$$\begin{aligned} q(\vec{x}) = &\int_E w_j w_k \frac{\partial}{\partial x_k} e_j(\vec{x}-\vec{y}) d^3y + \\ &+ \oint_\Sigma \{e_j(\vec{x}-\vec{y})(v_{oj}v_k - \Pi_{jk}) + w_j T_{jk}(\vec{x}-\vec{y})\} n_k dS \end{aligned} \tag{6}$$

where tensors Π_{ij}, T_{ijk}, T_{ij} are

$$\Pi_{ij} = v_i v_j + p\delta_{ij} - \nu(\frac{\partial v_i}{\partial x_j} + \frac{\partial v_j}{\partial x_i}), \qquad e^* = (\vec{v}_0, \nabla\frac{1}{r})$$

$$T_{ijk} = \nu(\frac{\partial E_{ij}}{\partial x_k} + \frac{\partial E_{ik}}{\partial x_j}) - e_i\delta_{jk}, \qquad T_{ij} = \nu(\frac{\partial e_i}{\partial x_j} + \frac{\partial e_j}{\partial x_k}) - e * \delta_{ij}$$

Similar to the multipole expansion in theory of the Laplace equation we can expand the integral over the surface Σ on the derivatives of the fundamental solution. It is very convenient to explore the multipole approach because our goal is to study the solution of the Navier-Stokes equation in the vicinity of the infinity. Denoting the corresponding integrals over surface in (5) and (6) as $K_i(\vec{x})$, $K(\vec{x})$ we have

$$K_i(\vec{x}) = a_j E_{ij}(\vec{x}) + \sum_{n=1}^{\infty} a_{jk_1 \dots k_n} \frac{\partial^n E_{ij}(\vec{x})}{\partial x_{k_1} \dots \partial x_{k_n}} + b e_i(\vec{x}) + \sum_{n=1}^{\infty} b_{k_1 \dots k_n} \frac{\partial^n e_i(\vec{x})}{\partial x_{k_1} \dots \partial x_{k_n}}$$

$$K(\vec{x}) = a_j e_j(\vec{x}) + \sum_{n=1}^{\infty} a_{jk_1 \dots k_n} \frac{\partial^n e_j(\vec{x})}{\partial x_{k_1} \dots \partial x_{k_n}} + b e * (\vec{x}) + \sum_{n=1}^{\infty} b_{k_1 \dots k_n} \frac{\partial^n e * (\vec{x})}{\partial x_{k_1} \dots \partial x_{k_n}}$$

where $a_j, a_{jk_1 \dots k_n}, b, b_{k_1 \dots k_n}$ are intensities of multipoles. Accepting the point of view of multipole approach we can assume that these intensities are known. So, we can rewrite the integral equations (5), (6) in the following form with known velocity $w_i^0(\vec{x})$

$$w_i(\vec{x}) = w_i^0(\vec{x}) + \int_E w_j w_k \frac{\partial}{\partial x_k} E_{ij}(\vec{x} - \vec{y}) d^3 y, \qquad (7)$$

$$q(\vec{x}) = q^0(\vec{x}) + \int_E w_j w_k \frac{\partial}{\partial x_k} e_j(\vec{x} - \vec{y}) d^3 y$$

and solve these equations. There are the large difficulties for solving (7). Therefore we treat (7) in the Fourier space. We should note that the asymptotics of the field quantities at infinity in the physical space correspond to the limit of vanishing wave numbers in the Fourier space. Fourier image of (7) is

$$W_i(\vec{k}) = W_i^0(\vec{k}) - iA_{im}(\vec{k}) k_n B_{mn}(\vec{k}), \qquad P(\vec{k}) = P^0(\vec{k}) - \frac{k_n k_m}{k^2} B_{nm}(\vec{k}) \qquad (8)$$

$$A_{jl}(\vec{k}) = \frac{1}{(2\pi)^3} \int\limits_{E^3} e^{i\vec{k}\cdot\vec{x}} E_{jl}(\vec{x}) d^3x = \frac{k_i k_j}{vk^2[k^2 - 2i\sigma(\vec{k}\cdot\vec{n}_0)]}$$

$$B_{ij}(\vec{k}) = \frac{1}{(2\pi)^3} \int\limits_{E^3} e^{i\vec{k}\cdot\vec{x}} w_i(\vec{x}) w_j(\vec{x}) d^3x = \int\limits_{K^3} W_i(\vec{p}) W_j(\vec{k} - \vec{p}) d^3p$$

The main difficulty is to calculate the tensor $B_{jl}(\vec{k})$. After very long and tedious mathematical exercises one can obtain the asymptotic of the tensor at vanishing wave numbers

$$B_{ij}(\vec{k}) = -\frac{\pi^2}{4\sigma v^2} \ln \frac{k^2 - 2i\sigma k x_0}{8\sigma^2} G_{ij}(\vec{a}) + \frac{\pi^2}{4\sigma v^2} K_{ij}(\vec{a}) + \tilde{p}_{ij}^0 + p_{ij}^1 k^{1/2} + $$
$$+ O(k \ln^2 k), \quad \vec{k} \to 0$$

where tensors $G_{ij}(\vec{a})$, $K_{ij}(\vec{a})$ are determined by (axis x_3 is oriented along the velocity vector at infinity $\vec{v}_0$)

$$G_{ij} = \begin{bmatrix} \frac{9a_1^2 + a_2^2}{8} & \frac{a_1 a_2}{4} & \frac{a_1 a_3}{2} \\ \frac{a_1 a_2}{4} & \frac{9a_2^2 + a_1^2}{8} & \frac{a_2 a_3}{2} \\ \frac{a_1 a_3}{2} & \frac{a_2 a_3}{2} & a_3^2 \end{bmatrix},$$

$$K_{ij} = \begin{bmatrix} \frac{-11a_1^2 - 3a_2^2 + 4a_3^2}{8} & -5\frac{a_1 a_2}{4} & 0 \\ -5\frac{a_1 a_2}{4} & \frac{-3a_1^2 - 11a_2^2 + 4a_3^2}{8} & 0 \\ 0 & 0 & \frac{a_1^2 + a_2^2 - 2a_3^2}{2} \end{bmatrix}$$

$x_3 \| \vec{v}_0, \quad |\tilde{p}^0_{ij}|, \ |p^1_{ij}| < \infty, \quad \mathrm{Re}\, \tilde{p}^0_{ij} = p^0_{ij} = \int_{\mathcal{E}} (w_i w_j - a_k a_l E_{ik} E_{jl}) d^3x$ - renormalized tensor of defect of kinetic energy of the wake, where $\mathcal{E}$ - fluid region.

The Fourier image for the velocity defect one can express as follows

$$W_l(\vec{k}) = A_{lm}(\vec{k})\{a_m - ik_n G_{mn}(\vec{a})\alpha(\vec{k}) / v^2 -$$
$$- ik_n[a_{mn} + \frac{1}{32\pi\sigma v^2} K_{mn}(\vec{a}) + \tilde{p}^0_{mn}]\} - \frac{ik_l}{k^2}(b - ik_n b_n) + O(k^{1/2})$$

where $\quad \alpha(\vec{k}) = \int_{K^3} d^3p \hat{F}(\vec{p})\hat{F}(\vec{k} - \vec{p}), \quad \hat{F}(\vec{k}) = \dfrac{1}{k^2 - 2i\sigma(\vec{k} \cdot \vec{n}_0)}$

] The final result after inverting the Fourier transformation is

$$w_i(\vec{x}) = a_j E_{ij}(\vec{x}) + \frac{1}{32\pi\sigma v^2} G_{jl}(\vec{a}) \frac{\partial}{\partial x_l} U_{ij}(\vec{x}) + +(b + b_l \frac{\partial}{\partial x_l}) \frac{\partial}{\partial x_i} \frac{1}{4\pi r} +$$
$$[a_{jl} + \frac{1}{32\pi\sigma v^2} K_{jl}(\vec{a}) + p^0_{jl}] \frac{\partial}{\partial x_l} E_{ij}(\vec{x}) + O(\frac{\ln^2 \sigma r}{r^2(1 + \sigma s)^2}) \qquad (9)$$

where

$$U_{ij}(\vec{x}) = \delta_{ij}\Delta U - \frac{\partial^2 U}{\partial x_i \partial x_j}, \quad U(\vec{x}) = \frac{\pi^2}{\sigma}\Phi(s)\ln \sigma r + \sum_{n=0}^{\infty} S_n(s)(\sigma r)^{-n},$$
$$S_0(s) = \int_0^s [\Phi(\alpha) - e^{-\sigma\alpha}\Phi(-\alpha)]d\alpha, \quad S_1(s) = \frac{1 - \sigma s}{8\pi\sigma s} - \sigma s\Phi(s), \ldots$$

$b_i = -\oint_{\Sigma} x_i v_j n_j dS, \qquad a_{ij} = -v_{0i} b_j - \oint_{\Sigma} x_j \Pi_{ik} n_k dS$ - the first moments of flow rate and flux of impulse. The first term of (9) was obtained by Finn

(1965). He proved that this is valid if the velocity and pressure fields are belonging to the so called PR (physical reasonable) class of function. (It means that the functions are vanishing at infinity not slow than $r^{-\alpha}$, $\alpha > 1/2$.) Babenko (1973) has proved that if the functions are from the Dirichlet class this result is correct too under very general condition that surface of the body is belonging to the Lyapunov class. These theorems are applicable in general case, particularly for the other asymptotic terms which have been obtained in our work. The great difficulty was to get correct remainder term. The peculiarity of the asymptotic expansion of the velocity is that it is expressed through exact conservation integrals and their moments. It allows to determine some important characteristics of the body from the observation of the velocity field in the far wake. For example, one can get the drag coefficient (not only the drag force), that is rather important for some practical applications. The obtained formulas are valid not only for the wake problem but for the jets in the uniform stream as well. It could be the vessels moving due to the reactive force and so on.

It is essential that the decay of the vorticity out the wake is exponential which is followed from

$$\omega_i(\vec{x}) = \frac{\sigma}{4\pi v} e_{ijk} \frac{\partial s}{\partial x_j} a_k \frac{e^{\sigma s}}{r} + e_{ijk} \frac{G_{jl}(\vec{a})}{32\pi\sigma v^2} \frac{\partial^2 f}{\partial x_k \partial x_l} -$$

$$+ e_{ijk}[a_{jl} + \frac{G_{jl}(\vec{a})}{32\pi\sigma v^2} + p_{jl}^0] \frac{\partial^2}{\partial x_k \partial x_l} \frac{e^{-\sigma s}}{4\pi v r} +$$

$$+ O(\frac{\ln^2 \sigma r}{r^{5/2}} e^{-\sigma(1-\varepsilon)s}) \qquad 0 < \varepsilon << 1$$

obtained from (9). This result is confirmed by the many experimental data.

One can show that the further terms of the generalized multipole expansion are the fractional derivatives of the basic fundamental tensors like E_{ij}, U_{ij}. The order of the derivatives grows with number of the term. So, one can get the complete expansion of the exact solution of the Navier-Stokes equation for the wake problem.

$$T_0(s) = \frac{1}{\sqrt{2\pi^3}\sigma}, \quad T_0(s) = \frac{\sigma s - 1/4}{5\sqrt{2\pi^3}\sigma}$$

2. Turbulent wake.

In this section we consider the turbulent wake on the basis of the Reynolds equations without any additional hypothesis of closure. Starting from

$$\nu\Delta\vec{u} - (\vec{v}_0, \nabla\vec{u}) - \nabla P = (\vec{u}, \nabla\vec{u}) + \overline{(\vec{w}', \nabla\vec{w}')} \;, \qquad \text{div}\, \vec{u} = 0;$$
$$\vec{u} = \overline{\vec{w}}, \quad P = \bar{q}, \quad \vec{w} = \vec{u} + \vec{w}'$$

similar to (5) we can obtain the integral equation for defect of the averaged velocity by inverting this equation with help of the fundamental solution (2)

$$u_i(\vec{x}) = \int_E [u_j u_k + \overline{w_j' w_k'}] \frac{\partial}{\partial x_k} E_{ij}(\vec{x} - \vec{y}) d^3y +$$
$$+ \oint_\Sigma \{E_{ij}(\vec{x} - \vec{y})[\mathrm{v}_{0j}(\mathrm{u}_k + \mathrm{v}_{0k}) - \overline{\Pi}_{jk}] + u_j T_{ijk}(\vec{x} - \vec{y})\} n_k dS \tag{10}$$

One can prove that if the asymptotic of the Reynolds tensor is

$$\overline{w_i w_j} = u_i u_j + \overline{w_i' w_j'} = O(r^{-\alpha}), \qquad \alpha > 1, \tag{11}$$

then we have exact asymptotic for the mean velocity defect

$$u_i(\vec{x}) = a_j E_{ij}(\vec{x}) + O(r^{-3/2} \ln r) \tag{12}$$

This result is based on the slight modification theorems of Finn (1965).

Using the deep mathematical work of Babenko (1973) we can prove that if the turbulent velocity fields are satisfies the Dirichlet condition

$$\int_E (\frac{\partial w_i}{\partial x_j})^2 d^3x \le C = const < \infty \tag{13}$$

the integral over volume in nonstationary case has the estimation

$$|I_i(x,t)| = \left| \int_E w_j(\vec{y},t) w_k(\vec{y},t) \frac{\partial E_{ij}(\vec{x} - \vec{y})}{\partial x_k} d^3y \right| \le C_1 r,$$
$$C_1 = const < \infty, \quad \alpha \ge 2/3$$

From where the integral in (10) satisfies

$$\left|\bar{I}_i(\bar{x})\right| \le \overline{\left|I_i(\bar{x},t)\right|} \le C_1 r^{-\alpha}, \qquad \alpha \ge 2/3 \tag{14}$$

The integral over surface Σ in (10) one can perform as it was done for surface integrals $K_i(\bar{x})$ *and* $K(\bar{x})$ in (5), (6) in the form of the multipole expansion. So, the asymptotic of the surface integral in (10) at $r \to \infty$ is $O(1/r)$. When the condition (11) is satisfied the main term at infinity is the term corresponding to the surface integral in (10) and (12) is valid. When the condition (11) is not satisfied the asymptotic (12) is not correct. In this case the main term at infinity is integral over the volume and we can not state that the asymptotic is determined by the exact conservation integral as in (12). This case may correspond to the situation of the presence of the wake memory of the shape of the body. This fact was observed experimentally by Bukreev, Vasilyev, Lytkin (1972) for 3D axisymmetric wakes. The opposite to (11) condition may express in the terms of the asymptotic behavior of the turbulent intensity ε

$$\varepsilon = \sqrt{\frac{\overline{w'^2}}{u^2}} \ge \sqrt{\frac{\overline{w_i' w_j'}}{u^2}} \ge C_e r^{1/6} \to \infty \quad at \quad r \to \infty \tag{15}$$

For the jets in the uniform stream it is known from the experiment (Abramovitch et al, 1974) that $\varepsilon(r) \to \infty$ *at* $r \to \infty$. Therefore there not exists the wake memory of the nozzle shape for the jets. As for wakes is concerned we should have the infinite turbulent intensity at infinity. This is rather unusual behavior of the turbulent intensity because it is well known that the turbulence is decaying in the wakes, but in the intermediate region of the self-similar turbulent wake may be some peculiarities of the flow, so that the (15) may take place. This fact one can check experimentally and compare to the conclusion of the experimentallists about existence of the wake memory. It is interesting that for the jets, corresponding to the case of the presence of the source of impulse there is no shape memory. If there exists such shape memory for the wakes the question arises what is the physical meaning of the difference between source (jets) and sink (wakes) of impulse in hydrodynamics. For the first glance there is no reason to have such great difference.

References

1. Oseen C.W. Neuere Methoden und Ergebnisse in der Hydrodynamik. - Leipzig: Akademische Verlagsgesellschaft m.b.H. 1927. 132S.
2. Finn R. Arch. Rational Mech. Anal. 1965. Vol.19. N.5. P. 364-406.
3. Babenko K.I. Math. Coll. Papers. 1973. Vol. 91(133). N1(5). P.3-26. (in Russian).
4. Bukreev V.I., Vasilyev O.F., Lytkin Yu.M. Reports of Acad. Sci. USSR. 1972. Vol. 207. N4. P.804-807.
5. Abramovitch G.N., Krasheninnikov S.Yu., Sekundov A.N., Smirnova I.P. Turbulent mixing of gaseous jets. Moscow, Nauka. 1974.

Three-Dimensional Structure and Dynamics of the Turbulent Far Wake

Y. G. Guezennec and T. J. Gieseke

The Ohio State University
Mechanical Engineering Department
206 W. 18th Avenue
Columbus, Ohio 43210, U.S.A.

The stochastic estimation technique, originally proposed by Adrian[1] and applied by Adrian[2], Guezennec[3], and others, is a method by which conditional turbulent velocity fields can be estimated. Velocity "conditions" are imposed at a few locations within a reconstruction space and velocities around the conditions are computed using simple functions of the conditions in space and time. The conditions can be set, to evaluate typical flow structures associated with a specific phenomenon, or they can be measured experimentally, to estimate physically realized flow fields. The mathematical basis for the procedure lies in multiple regressions analysis which requires detailed space-time correlation measurements between all involved variables, usually the perturbation velocities, $\vec{u}(\vec{r},t)$.

Measurements of this correlation tensor were performed in the turbulent wake of a circular cylinder at Re_d=5000. All nine components of the correlation tensor were simultaneously measured on a 17 by 17 grid at x/D=100 using two subminiature triple hot-wire probes. The grid was perpendicular to the flow direction, spanning the entire width of the wake. All possible probe separations within the y-z plane were considered and the appropriate symmetries used. Taylor's hypothesis, after proper validation, was used to estimate the streamwise spatial correlation from the temporal correlation data.

Structures associated with large contributions to the Reynolds stress play an important role in turbulent transport processes. For this reason, conditions that are characteristic of these structures have been selected to reconstruct velocity fields. Conditions are chosen at specific locations using joint probability distribution functions of all three velocity components at one and two points in space. The locations of most interest are points along the y axis where the average $\overline{uv}$ has its maximum, minimum, and zero value. These locations are noted in Figure 1 as points A, B, and C respectively. Figure 2a shows the distribution of u and v weighted by uv measured at point B. Quadrant events are defined as realizations of u and v in specific quadrants of $u-v$ space. Events corresponding to maxima in the uv weighted joint probability distributions make the most significant contributions to $\overline{uv}$, and therefore are of most interest. The spanwise component of velocity associated with these events can be inferred from the distribution of uv and w (Figure 3). It was found that the spanwise component was approximately normally distributed around 0.0 for any value of uv. The high probability of having a non-zero w at the time of strong uv, indicates that the turbulent structures have no preferred spanwise symmetry.

1 Supported by Office of Naval Research under Grant N00014-90-J-1499.

To get a description of the turbulent structures as they span the entire wake width, multi-point stochastic estimates were generated by selecting conditions at both points A and B. The condition at point B was selected based on the second and fourth quadrant events at that location. The condition at point A was selected by looking at the conditional distribution function of u and v at that location, considering only the realizations when the velocities at point B are in a prescribed quadrant. Such a distribution, conditioned on a fourth quadrant event, is shown in Figure 2b. This procedure enables the determination of the largest contributor to the Reynolds stress based on events on both sides of the wake.

The conditions for a two-point stochastic estimation were fixed as a fourth quadrant condition at location B and a first quadrant condition at location A. The resulting flow structure, which can be considered the most probable velocity field given the noted conditions, is shown in Figures 4a and 4b. The 3-D topology is illustrated using an iso-surface of constant perturbation enstrophy, represented as a solid object, superposed on a selected planar cut through the field showing the velocity vectors. For clarity, only every five points in x (or t) is shown. The structure suggested by this analysis is a pair of counter-rotating eddies, inclined approximately 45° in the direction of the mean shear. A similar structure was found by considering a second quadrant condition at location B and a third quadrant condition at location A. However, the distribution at point A did not show as strong a favoritism to a specific quadrant given the second quadrant event at point B as it did for the fourth quadrant event. This indicates that entrainment rollers or the *sweep* events are more likely to be symmetric about the wake center than the *ejection* rollers. These inclined eddy structures are very similar to the ones proposed by Grant[4] and computed using Proper Orthogonal Décomposition (Payne and Lumley[5], Aubry[6]).

To gain a better understanding of the instantaneous dynamics of these dominant structures, the formalism of the stochastic estimation was extended to the *Pseudo-Dynamic Reconstruction Technique* with which the most likely evolution of a three-dimensional velocity field is computed. Using simultaneously measured conditions at points A, B, and C, three-dimensional velocity fields were computed representing the most likely field state given short-time velocity histories at each of the three physical condition locations. By slowly advancing the time segments through a long data record, a realistic animation of the flow evolution during an actual experiment was computed.

Results of such animations confirm the large degree of asymmetry present instantaneously in the flow and illustrate that the dominant vortical structures tend to inhibit one another, rearranging themselves in a loosely staggered array in all directions. In addition, the structures appear in groups separated by less organized motions. This characteristic is consistent with the results of Townsend's [7] where similar groupings of eddies were observed. One possible explanation for the sporadic appearance of eddy groups is the streamwise change in linear stability characteristics of the wake mean profile (Cimbala [8], and others). It is a matter of ongoing research as to whether modes resulting from linear instability can account for the eddy motions seen in these experiments.

Figure 5 is one instant of the reconstruction, illustrating the complex distribution of eddy structures. At the far left of the figure, a set of double roller eddies are observed which are nearly identical to those shown in Figure 4. The slight asymmetry in these rollers is a result of asymmetry in the experimentally realized conditions. Immediately to the right are a combination of several eddies of much smaller amplitude in a staggered configuration. Both entraining and ejecting eddies are present with intense shear layers created between them. This picture is typical of the major Reynolds stress contributing events seen in the animation. One or two strong pairs of sweeping or ejecting eddies are followed by smaller amplitude motions. The strongest motions have v components that are anti-symmetric about the wake center, i.e. sweeps (ejections) on one side of the wake are accompanied by sweeps (ejections) on the other side. As observed from the probability density functions, there is a probability of having any cross-wake combination of events. This is seen in the small amplitude motions between larger events, where the asymmetry is regularly observed.

Using the multi-point stochastic estimation, a picture has been developed describing eddy structures in the turbulent wake. The two point estimates are consistent with published measurements and observations and the reconstructions using experimental conditions provide a level of detail that could not be otherwise attained experimentally without of a very large number of sensors (in this case, 17x17 triple hot-wires!). The pseudo-dynamic reconstruction as proven to be a viable method to study the evolution of the instantaneous three-dimensional structures and has provided insight into possible mechanisms by which far-wake turbulence is sustained.

References

1. Adrian, R.J.,; On the role of conditional averages in turbulence theory. Symposium on Turbulence, Rolla, MO (1973).
2. Adrian, R.J.; Conditional eddies in isotropic turbulence. Physics of Fluids 22 (1979).
3. Guezennec, Y.G.; Stochastic estimation of coherent structures in turbulent boundary layers. Physics of Fluids A. 1 No. 6 (1989).
4. Grant, H.L.; The large eddies of turbulent motion. J. Fluid Mech. 93 No. 3 (1957).
5. Payne, F.R.; Lumley, J.L.; Large eddy structure of the turbulent wake behind a circular cylinder. Physics of Fluids Supplement (1967).
6. Aubry, N.; The structure, dynamics, and control of turbent wakes. The Third ONR Workshop on Vortex Shedding and Vortex Wakes: Dynamics, Instabilities, and Modifications, Columbus, OH (1992).
7. Townsend, A.A.; Flow patterns in a wake and in a boundary layer. J. Fluid Mech. 95 (1979).
8. Cimbala, J.M.; Nagib, H.M.; Roshko, A.; Large structure in far wakes of two-dimensional bluff bodies. J. Fluid Mech. 190 (1988).

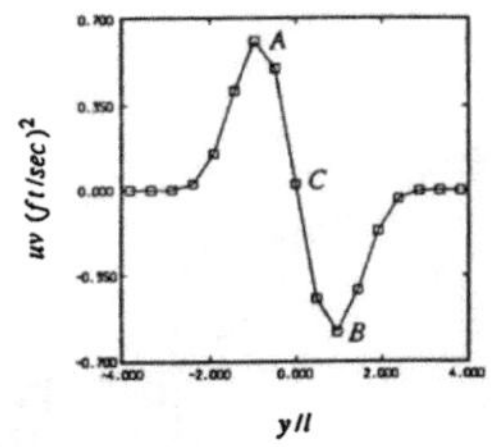

Figure 1. Mean profile of $\overline{uv}$.

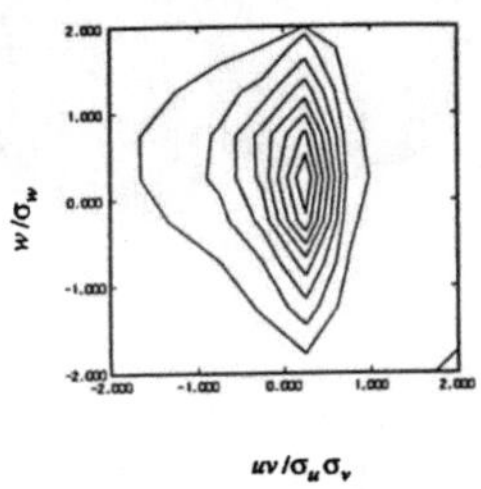

Figure 3. Joint probability density functions of uv and w weighted by uv at location B:

a)

b)

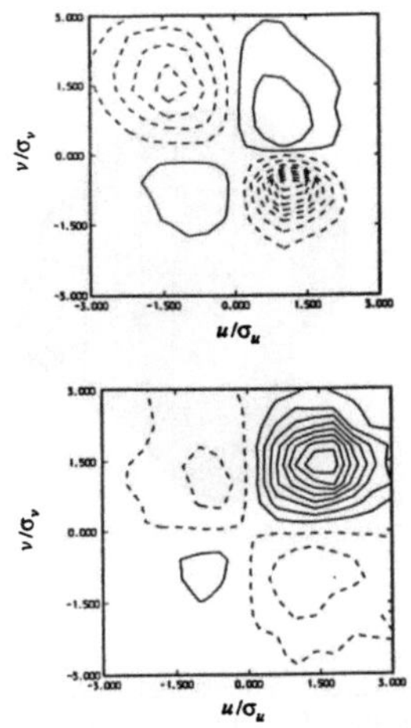

Figure 2. Joint probability density functions of u and v weighted by uv: a) location B and b) location A conditioned on fourth quadrant events at location B.

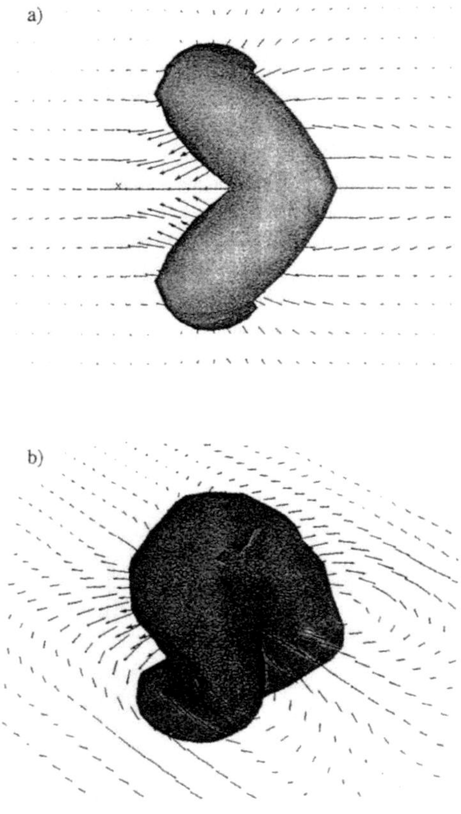

Figure 4. Three dimensional two-point conditional structure for most typical conditions at locations A and B: a) side view and b) oblique view.

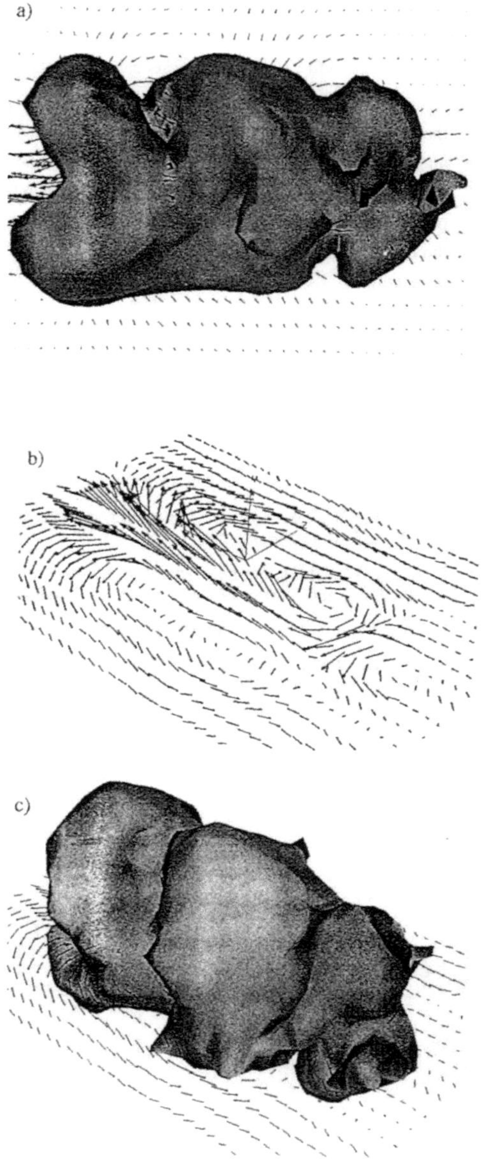

Figure 5. Instantaneous realization of three dimensional conditional flow field for 15 conditions, five in x at each location A, B, and C: a) side view, b) oblique view of velocity vectors at mid-plane and c) oblique view of iso-enstrophy surface.

The Ginzburg-Landau Equation as a Model of the Three-Dimensional Circular Cylinder Wake at Low Reynolds Numbers

Pierre Albarède, Thomas Leweke, Michel Provansal

Laboratoire de Recherche en Combustion, CNRS Marseille, France

1. Introduction

The circular cylinder wake is a basic problem in fluid dynamics. At low Reynolds numbers, there is a transition from a steady to an oscillatory wake, which has been previously identified (Provansal *et al.*) as a Landau-Hopf bifurcation by local measurements. However this model is unsufficient to account for the spatial effects along the axis of the cylinder which turned out to be of crude importance. Many observations of spatial structures have been done: inclined vortex shedding (Berger & Wille, Gerrard), cells of different frequencies (Gerich & Eckelmann), influence of end effects and pressure conditions on parallel or oblique vortex shedding (Williamson, Hammache & Gharib).

In this paper, we want to extend the Landau model to take into account these three-dimensional effects. Since the main experimental effects are along the cylinder axis (z-direction), and in order to keep the symmetry $z \to -z$, we add a diffusive term to the Landau equation and get a Ginzburg-Landau equation. For simplicity, we do not use the direction of the mean flow, x, where the region of absolute instability acts as a wavemaker. The importance of the z-direction is further emphasized by the presence of the physical boundaries of the flow which impose a zero condition on the amplitude of oscillation.

From the Ginzburg-Landau equation, we construct a standard length of the system which depends on both the aspect ratio and the Reynolds number. Increasing the standard length, we increase the number of excited modes and the flow evolves from a single mode to a many-mode regime characterized by chevron-shaped vortices. Measurements confirm the main features of the model.

2. Experimental set-up

Experiments have been performed in an open wind tunnel of square test section (side length 100 mm). The upstream flow is regulated by a sonic nozzle and calibrated by Laser Doppler anemometry. The metallic circular cylinders are solidly held at their ends to avoid any vibration effects. Mobile end plates have been added in order to reduce the thickness of the boundary layer. This configuration permits a continuous variation of the aspect ratio L/d for a fixed cylinder diameter d and Reynolds number Re without important blockage effects. Usually the wake is characterized by the Strouhal/Roshko number-Reynolds number curves exhibiting the influence of the control parameter upon the frequency of oscillation. The diversity of experimental phenomena requires a more detailed characterization of the flow. Here we try to get a more quantitative description by measuring both the amplitude and the phase variation of the oscillation along the axis of the cylinder. Visualizations by the smoke-wire method provide a spatial information along the span direction. The smoke-wire has been positioned 30 mm upstream of the cylinder on one side of the vortex street. An halogen light illuminates the flow and video recordings were made at low exposure time.

3. Results and discussion

The extension of the Landau to the Ginzburg-Landau equation needs the knowledge of two coefficients due to the diffusive term. Near the threshold of instability, an approximate solution of the Ginzburg-Landau equation has been proposed by Albarède & Monkewitz: amplitude and phase vary sinusoidally along the span direction. Experimental curves confirm these predictions and allow to determine the effective length of the cylinder, lower than the geometrical spacing due to the influence of boundary layers. Moreover, the addition of a linear spatial term to the Landau model provides an explanation of the variation of the critical Reynolds number with the aspect ratio. Collected experimental data confirm the predictions of the model (fig. 1). The real part of the diffusive term of the Ginzburg-Landau equation, μ_r, is deduced from these measurements. Other measurements (Lee & Budwig) have been analyzed in the same way, yielding very similar values of μ_r. The typical spatial scale along z is based on this value and gives the standard length of the cylinder as function of both, the aspect ratio and the Reynolds number.

In the single mode regime, the model predicts a variation of the frequency of oscillation with the square of the wave number $(\pi d/L)^2$ at constant Reynolds number. Keeping constant the Reynolds number, while varying the aspect ratio by moving the end plates, we plot the frequency versus the wave number. The linear variation is in good agreement with the prediction of the model. The ratio c_1 of real and imaginary part of the complex diffusive coefficient is deduced from the slope of these curves.

Visualizations exhibit the changes from single mode oscillation to chevrons. On fig. 2, at a higher standard length L, the second mode of oscillation is visible. Chevrons are visible for a large standard length. Video recordings show the dynamics of chevrons: a dissymetry of the upstream flow or of boundary conditions can induce a displacement of the tip of the chevron and lead to an oblique vortex shedding situation.

At large standard length we perform a numerical simulation of the model and its boundary conditions with experimental values. Starting from noise, it leads asymptotically to V-shaped vortices. Our experimental results give the order of magnitude of the coefficients c_1 and c_2. c_1-c_2 is different from zero but we cannot determine the sign of $1+c_1c_2$. We know that the stability of the chevron pattern depends on this sign. The effect of varying these coefficients has been numerically inquired. A slight variation destabilizes the chevron and produces many cells of different frequencies. When increasing the length of the cylinder, the lateral cells keep the same length while the central cell saturates producing a new chevron, itself unstable.

4 Conclusion

These experimental and numerical results show the importance of three-dimensional effects on vortex shedding, even at low Reynolds number. The Ginzburg-Landau model has been successfully tested and the comparison with experiments gives the values of all the coefficients of the equation.

The financial support by the European Community under Grant N°SCI- 0212-C(TT) is gratefully acknowledged.

References

M. Provansal, C. Mathis, L. Boyer 1987, J. Fluid Mech. **182**, 1.

M. Gaster 1971, J. Fluid Mech. **46**, 749.

D.J. Tritton 1971, J. Fluid Mech. **6**, 547.

E. Berger, R. Wille 1972, Ann. Rev. Fluid Mech. **4**, 313.

J.H. Gerrard 1978, Proc. Roy. Soc. Lond. A **288**, 351.

D. Gerich, H. Eckelmann 1982, J Fluid Mech. **122**, 109.

H. Eisenlohr, H. Eckelmann 1989, Phys. Fluids A **1**, 189.

C.H.K. Williamson 1989, J. Fluid Mech. **206**, 579.

M. Hammache, M. Gharib 1991, J. Fluid Mech. **232**, 567.

P. Albarède, P. Monkewitz 1992, Phys. Fluids A **4**, 744.

T. Lee, R. Budwig 1991, Phys. Fluids A **3**, 309.

Figure 1:

Critical Reynolds number Re_1 versus $q_{1R}{}^2 = (\pi d/L)^2$.

d = 1.6 mm, L variable.

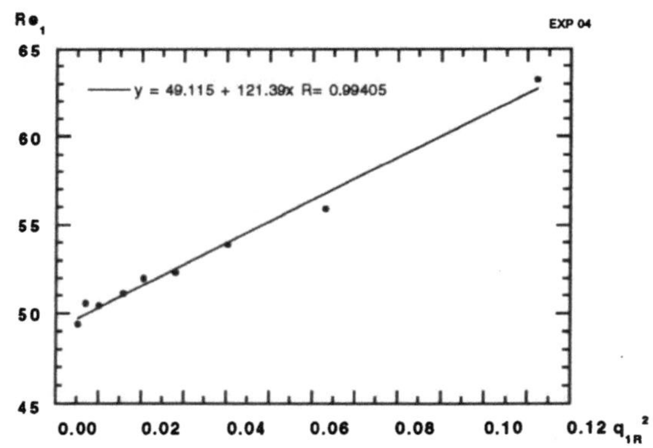

Figure 2:
d=1.3 mm, L/d=16,
L=21mm, Re=95,
L_r = 2.8π,
second mode.

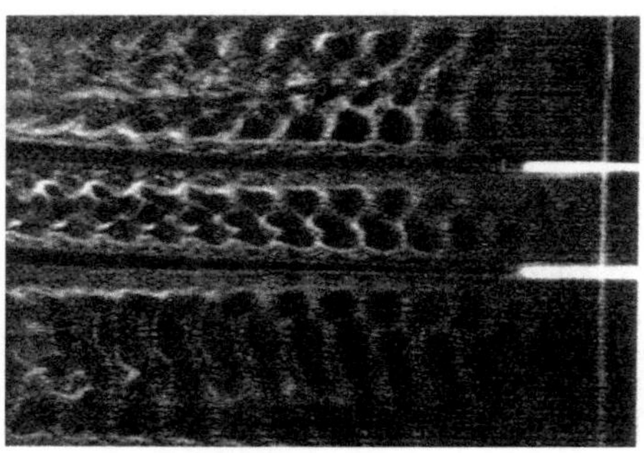

Report on The Session Comparing Computation of Flow Past Circular Cylinders With Experimental Data

J.M.R. Graham
Department of Aeronautics
Imperial College Science, Technology and Medicine
London, SW7 2BY

Summary
All the participants at the symposium were given the following invitation to bring to the meeting computational or experimental data for flow past cylinders. Computations of cross-flow around circular cylinders and in one instance a square section were requested for specific input conditions with the output data in a specified format to aid comparison, as below. This report presents and comments briefly on the global quantities from the data which were submitted.

Data Requested
C.F.D. codes which can simulate unsteady two-dimensional flow fields surrounding a bluff body have been available for some time and computations up to moderately high Reynolds numbers have been undertaken. However the predictions of even the integral quantities such as drag coefficient vary significantly for the same nominal input conditions. We therefore felt that it would be appropriate, in the context of this symposium, to invite participants to submit the results of two-dimensional flow computations for as many of a common set of problems as they were able or wished to do. The intention was, using a common format, to compare the results with each other and with experimental results, in the light of the greater understanding now available on the effects of three-dimensionality in vortex shedding at low Reynolds number. We also invited anyone who had a three-dimensional flow program able to simulate these flows and anyone who had or could generate appropriate experimental data of known two- or three-dimensionality similarly to participate. The main emphasis was to be on the circular cylinder, but one square section case was included as an additional possibility.

The flow cases, in each case a steady incident cross flow past a nominally infinite cylindrical body, were:

1. BOUNDARY CONDITION

Case	(1)	Circular Section	Reynolds number	= 100
	(2)	" "	" "	= 200
	(3)	" "	" "	= 300
	(4)	" "	" "	= 500
	(5)	" "	" "	$= 10^n$, n = 3,4 ...
	(6)	Square Section (Flat side facing flow)	" "	= 250

The following if possible were to be output for each case computed:

2. GLOBAL QUANTITIES

2.1 Strouhal number

2.2 Lift and drag

- $C_D(t)$ for t* = [0,50] (in convective time units Ut/diameter)
- C_D, C'_D
- $C_L(t)$ for t* = [0,50]
- C_L, C'_L (The first quantity is a measure of the accuracy of the time averaging) (diameter = width of face for square)

3. LOCAL QUANTITIES

3.1 Velocity

- Time average and standard deviations of u/U, v/U, w/U (U: oncoming velocity) for specified locations, x/R, y/R) = (1,1), (2,0), (2,1), (2,2) (4,0), (4,4) (8,0), (8,4), (12,0), (12,4) (R: radius = d/2)

- Velocity field for (x/R,y/R) in [−3,10]x[−3,3] in a 13 cm x 6 cm frame at the instance when the upper separation angle was largest.

3.2 Pressure and skin friction

- Pressure distribution, skin friction on the cylinder at the same instant (either the whole field (CFD) or at isolated points (for experiment)).

- The time average and rms-value of the base-pressure might also be of interest.

3.3 Contours of vorticity and streamlines at t* = 50.

In the event 15 groups responded, of which 3 provided experimental data, 11 the results of conventional computational fluid dynamic methods and 1 group a set of results based on the method of cellular automata. This last set of results are not included in the figures shown in this report because the method is not yet proved to converge to a solution of the Navier-Stokes equations and the results were significantly different from the others.

Of the computational methods, one was a fully three-dimensional flow code, but all the rest were limited to two-dimensional flow simulation. The methods can be put into 4 groups: (1) methods based on spectral elements and spectral decomposition, (2) finite element methods, (3) finite difference or volume methods and (4) particle methods based on vortex-in-cell techniques.

The computational results and the experimental data which were presented were all for cases of flow past a circular cylinder. In the event no square section results were submitted.

In the case of a circular cylinder, it has been established that the cause of the discontinuities in the curve of Strouhal number (S) at Reynolds numbers (Re) of about 160 and below is due to the onset of three-dimensional vortex shedding brought on by end effects (Williamson 1988; König, Eisenlohr and Eckelmann 1990). By controlling the end boundary conditions on the cylinder the development of three-dimensionality in the wake may be postponed until a Reynolds number of 200 at least. Therefore because of the restriction to two-dimensional flow in the majority of codes used in the present case it was felt to be important that the comparisons were made with assured two-dimensional data for the lower Reynolds number cases.

At low Reynolds number accurate force measurements are difficult and therefore experimental measurements of transverse force are scarce. More in-line (drag) data is available but in some cases it has been estimated from pressure measurements using semi-empirical correlations. Some force data also lacks the skin friction component which is very significant at low Reynolds numbers. Such data has been corrected using the empirical result $C_{Df} \sim 4.Re^{-1/2}$ and this has been indicated. Similarly comments have been given about other factors which affect the data. In general the Strouhal number is the most accurate data set at low Reynolds number.

At higher Reynolds number the wake, then the separating shear layers and finally, at sufficiently high Reynolds numbers, the boundary layers on the cylinder become turbulent. Several codes extended the calculation range of a 2-D solution to higher Reynolds numbers without incorporating any model for three-dimensionality or turbulence. These may be regarded as '2-D Direct Simulations'. It was clearly hoped that in these computations the small scales of unsteady 2-D motion would be adequate to model the most important components of Reynolds stress. Other codes included a specific turbulence model and this has also been indicated. Eddy viscosity and k-e turbulence models were used in a quasi-steady sense with coefficients based on, but in some cases changed from, their accepted values for turbulence in a steady deterministic flow.

The different computations and experiments are identified by symbols and numbers, given in a reference list at the end of the report.

Results

Figure 1 shows all the results for Strouhal number plotted against Reynolds number. The results for the Strouhal number for low Reynolds number are for 2-dimensional flow and are mainly in reasonable agreement with each other. Figure 2 shows the results for the mean drag coefficient C_D, Figure 3 the fluctuating drag coefficient C_D' and Figure 4 the fluctuating lift coefficient C_L', the latter two plotted as root mean square values. The solid lines in the figures indicate the general trend of other experimental data not requested in this exercise.

Discussion

For Reynolds numbers up to 200 the experimental measurements of Strouhal number (the solid line taken from the uppermost curve of (14) have been shown to correspond to two-dimensional flow. The numerical predictions follow this curve, fairly well. The discrepancies among the computed results in this range are almost certainly due to inadequacies of the meshes or numbers of modes used in the computations. Unfortunately force measurements appear to be inaccurate in this Reynolds number range. Those that are presented have in some cases had the estimates for

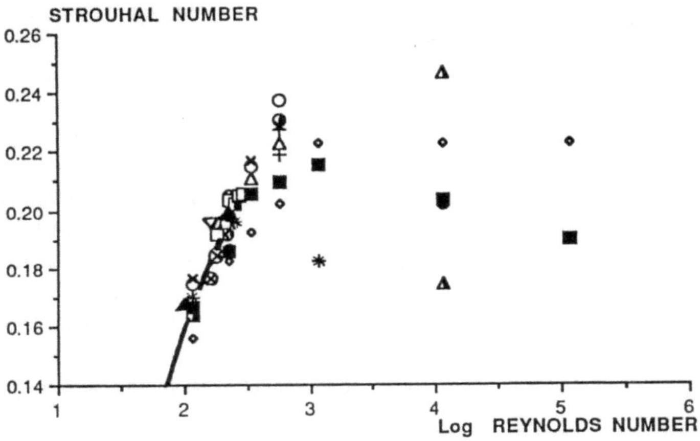

Figure 1 Strouhal number versus Reynolds number.

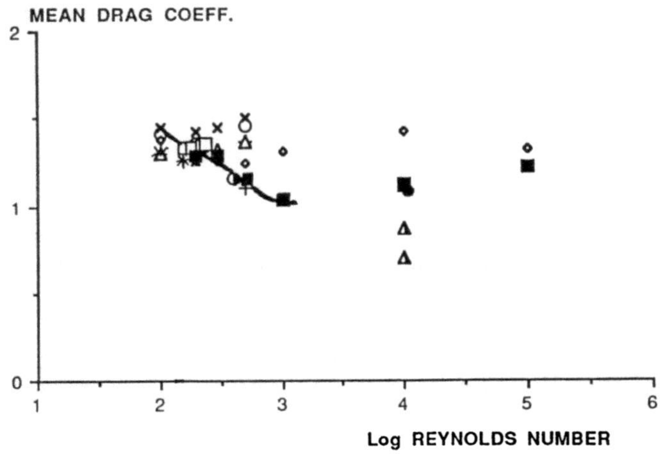

Figure 2 Mean Drag Coefficient versus Reynolds number

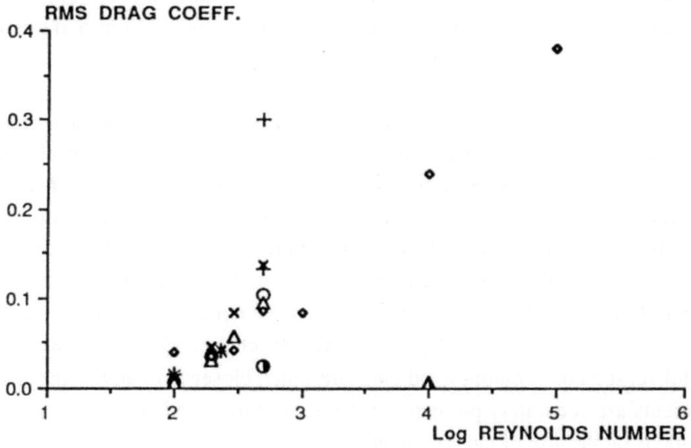

Figure 3 RMS Fluctuating Drag Coefficient versus Reynolds number.

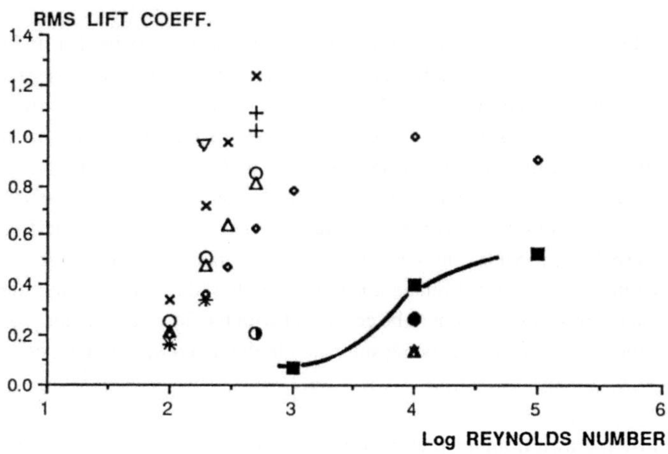

Figure 4. RMS Fluctuating Lift Coefficient versus Reynolds number.

the skin friction component to C_D quoted above added to aid comparison (12). The comparisons indicate larger discrepancies between the different computational predictions and the measured data for C_D, but the experimental data is less reliable here.

For Reynolds numbers above 200 the measured data are increasingly affected by the development of three-dimensionality in the wake. This is not catered for in any of the computations except for the results from the one three-dimensional code (1b) of the set. It is clear from the comparison that in the immediate range above Re = 200 the two-dimensional computations exhibit the wrong trend with Reynolds number for the force coefficients. The forces are seen to be significantly overpredicted when compared with the experimental results. In contrast the results of the three-dimensional code (1b), while not showing perfect agreement in all respects with the measured data, probably due to the restricted number of modes which could be incorporated for reasons of computational cost, are much closer than any of the 2-D results. The force coefficients are accurately predicted, but the Strouhal number estimates may still be affected by limitations on the size of the computations.

At higher values of Reynolds number ~ 10^4-10^5 agreement with measured data is, not surprisingly, less good. But it is notable that some two-dimensional codes without any turbulence model, i.e. the 2-D 'direct simulations' manage to achieve a quite good approximation to the results over the whole sub-critical range. This may be fortuitous and depend on the combination of the numerical algorithms and the grids used and the degree to which the solutions were converged or possibly it may indicate that two-dimensional computations of sufficient refinement capture the majority of the important small scales contributing to the flow.

Attempts to include a turbulence model, Baldwin-Lomax eddy viscosity (7) or k-ε (5,10) indicated a tendency for the standard coefficients for these models predict an excessive level of Reynolds stress leading to damping out of vortex shedding. However variations in the turbulence model (7,10) which suitably reduced the values of eddy viscosity or turbulence intensity were found to permit sustained vortex shedding and give fair but not very accurate prediction by the numerical codes. Again such results for higher Reynolds numbers, where the wake is known to contain significant large scale three-dimensionality, must be treated with caution, since the turbulence models cannot make any claim to model the large scale components in the third dimension which are not simulated. But the need to reduce significantly the level of turbulence energy in the model does seem to be consistent with the fact that a major proportion of the contribution to the Reynolds stress is being simulated in the two-dimensional motions.

Conclusions

The general conclusion drawn from the presentation of this data was that two-dimensional codes can be used to provide a time-accurate simulation of the flow around a circular cylinder in the low Reynolds number regime for which the flow can be made to remain two-dimensional, that is up to about 200. Above this value two-dimensional codes significantly overpredict the forces. The three-dimensional code whose data was presented gave much better predictions here but such codes may be limited at present by computational cost to fairly moderate Reynolds numbers. The use of turbulence models in two-dimensional computations at higher Reynolds numbers generates some realistic results but cannot be regarded as a validated approach as yet.

High Reynolds number two-dimensional 'direct simulations' have similarly been used to generate realistic results, but these are probably the result of the numerical effects of a grid of insufficient resolution acting as an artificial diffusion.

References

1. Williamson, C.H.K. 1988 Phys. Fluids 31, p2742
2. König, M., Eisenlohr, H. and Eckelmann, H. 1990 Phys. Fluids A2 (9), p1607

Key to Data

Symbol		Originator	Type	Comments (variables, grid)
○ ◐	1a	(2-D), 1b(3-D). Karniadakis G.	(1)	Velocity, pressure, irregular grid
Δ	2.	Chaplin J.R.	(1)	Vorticity, s-function, polar grid
◨	3.	Noack B.	(2)	Irregular grid
◇	4.	Gushchin V.	(3)	Polar grid
Δ	5.	Bosch G.	(3)	As (1), k-ε turbulence model
×	6.	Younis B.	(3)	As (1)
◻	7.	Braza M.	(3)	Velocity, pressure, eddy viscosity
∇	8.	Kaykayoglu C.R.	(4)	As (2)
⊗	9.	Meneghini J. and Bearman P.W.	(4)	As (2)
*	10.	Arkell R.H. and Graham J.M.R.	(4)	As (2), irreg. grid, (k-ε)
+	11.	Unal M.F.	(4)	As (2)
■	12.	Norberg C.	Experiment	Skin friction added as estimate
•	13.	Szepessy S.	Experiment	
▲	14.	Williamson C.H.K.	Experiment	

The Three-Dimensional Evolution of Oblique Waves in Plane Wakes

Eckart Meiburg

Department of Aerospace Engineering
University of Southern California
Los Angeles, CA 90089

Summary

We investigate the two- and three-dimensional evolution of an oblique wavy perturbation to a plane vorticity field. We first discuss the case of a plane mixing layer, which, if perturbed by an oblique wave only, forms an oblique Kelvin-Helmholtz roller with swirl that contains streamwise vorticity of the opposite sign from the connecting braid vorticity layer. If an additional spanwise perturbation is introduced, the braid region becomes subject to a Kelvin-Helmholtz instability of its own, which leads to the formation of streamwise braid vortices all of the same sign. In addition, a wavy modulation of the oblique roller's core is observed which resembles a vortex breakdown situation. For the plane wake, an oblique wave only leads to the evolution of an oblique Karman vortex street, in which the streamwise vorticity is again of opposite sign in the core and in the braid. As for the mixing layer, an additional spanwise perturbation wave leads to the formation of concentrated braid vortices, which are all of the same sign on the same side of the wake, but of opposite sign on opposite sides.

1. Introduction

Although the phenomenon of organized periodic two-dimensional structures in plane wakes has been known for a long time, recent developments have led to a renewed interest in these flows. The analysis of plane wake flows in terms of the concepts of absolute and convective instabilities (Koch [1], Huerre and Monkewitz [2], Oertel [3], and references therein) has shed new light on the formation of the two-dimensional Karman vortex street. Furthermore, the experimental work by Williamson, e.g. [4], as well as by the groups around Eckelmann, e.g. [5], and Gharib, e.g. [6], has clarified some discrepancies between earlier measurements of shedding frequencies in wakes behind circular cylinders. The observation and documentation of oblique shedding regimes, e.g. Williamson [7] and Hammache and Gharib [8], demonstrates the richness of the canonical cylinder wake flow, while giving rise to many questions regarding the downstream evolution of the vortical structures and potential strategies for their control. As far as the three-dimensional structure of incompressible plane wakes is concerned, our knowledge is still relatively limited. For the far wake, Meiburg and Lasheras [9] as well as Lasheras and Meiburg [10] demonstrated, both experimentally and numerically, the existence of two distinct three-dimensional vorticity modes. These modes occur if, in addition to a wavy streamwise perturbation, spanwise perturbation waves are present that introduce transverse or streamwise vorticity into the flow, respectively. The evolution of concentrated counterrotating streamwise braid vortex structures is then observed to lead to closed vortex loops forming either a staggered or an aligned array. A three-dimensional far wake structure has also been observed experimentally by Cimbala, Nagib and Roshko [11]. More recently, Karniadakis and Triantafyllou [12] numerically described the existence of wavy perturbations to the vortex lines in the near wake of a circular cylinder as well. Corke, Krull and Ghassemi [13] investigated the extent to which resonance interactions between the two-dimensional wave and oblique waves are responsible for the three-dimensional evolution of the far wake. A spanwise cellular structure of the near wake has been observed experimentally by König, Eisenlohr and Eckelmann [14] as well as by Piccirillo and Van Atta [15], see also Szepessy and Bearman [16]. The transition region between such cells appears to involve vortex dislocation or defects of the kind studied by Nassef and Browand [17] in the plane mixing layer. Progress towards modeling some of the above oblique shedding phenomena has been reported by Albarede and Monkewitz [18] as well as by Park and Redekopp [19]. The above findings demonstrate the importance of oblique waves for the evolution of the wake. We hence aim, in the present study, at shedding some light on their two- and three-dimensional growth by means of numerical simulations. Section 2 will give a brief description of the numerical technique. In order to gain some first insight into the effect of an oblique wave onto a plane vorticity layer, we will first discuss some results for the mixing layer in section 3, before proceding to the plane wake in section 4.

2. Numerical technique

Our intention is to numerically simulate incompressible flowfields of which only a relatively small fraction is rotational. Since it is our goal to obtain insight into the concentration, reorientation, and stretching of vorticity caused by oblique waves, it thus appears convenient to formulate the problem in terms of the vorticity variable. Its definition along with the condition of a nondivergent velocity field allows for a complete description of the *kinematics* of the flow in the form of the Biot-Savart law (see e.g. Batchelor [20]). In the case of an inviscid flow, the dynamics can then be properly accounted for by employing Helmholtz's theorem. For general reviews of the above concepts, see Leonard [21], [22]. For the representation of the vorticity field, we employ vortex filaments with a finite core radius σ. The diameter of the filaments at the start of the simulation represents a measure for the thickness of the rotational layer and was taken to be the unit length, which means that the initial core radius σ comes to 0.5. The two vorticity layers are initially centered at y=+/-0.65. The velocity jump of the sides of the wake is taken as the reference velocity, from which the definition of the unit time follows. In order to obtain a numerically tractable problem, we limit ourselves to the analysis of the temporally amplified problem. Our flowfield is thus considered to be periodic both in the spanwise and the streamwise direction. For more details and a more complete discussion of the numerical scheme, see Ashurst and Meiburg [23].

3. Plane mixing layer

Figure 1 shows the evolution of a plane mixing layer under the influence of an oblique wave only. Here x represents the streamwise, y the cross-stream, and z the spanwise direction, respectively. Shown are top and side views of the vortex filament centerlines, as well as streamwise vorticity contours in the plane x=4.19. Initially, the vortex filaments are all aligned in the z-direction. At time t=0, an oblique wave is superimposed, which slightly displaces the filaments' centerlines in the y-direction, thereby introducing cross-stream perturbation vorticity. The overall velocity field immediately tilts this y-vorticity component into the x-direction, thus generating streamwise vorticity as well. Figure 1a (time 13.44) shows that in the evolving oblique Kelvin-Helmholtz roller the streamwise vorticity begins to orient itself in the upstream direction, i.e. it is of negative sign. The assumption of inviscid dynamics of course implies that the vortex filaments have to remain connected. Consequently, the filament has to become oriented in the downstream direction in the emerging braid region between two consecutive rollers. This development has become more pronounced by time 15.31 (figure 1b), as a significant amount of vorticity has become concentrated in the oblique KH roller. The connecting braid region assumes the form of an elongated layer of positive streamwise vorticity, which in its central part is of nearly uniform strength. The extensional strain field generated by the emerging KH rollers tends to intensify the braid vorticity component normal to the oblique rollers, thereby producing additional streamwise vorticity. By time 17.66 (figure 1c), the braid vorticity layer has begun to wrap around the KH vortex, so that we now observe an oblique roller of negative streamwise vorticity surrounded by braids in which the streamwise vorticity is of positive sign. It is important to note that the shape of the large-scale vortex structure, i.e., the oblique KH roller, is different from that of the individual vortex lines. Each vortex line forms part of the roller at one location, and it represents a part of the braid region at a different spanwise location. In addition, we observe that a vortex line, from the location where it first begins to wind into the oblique roller until it emerges from it again, has components both in the direction of the roller's axis as well as in the circumferential direction. Hence we can conclude that

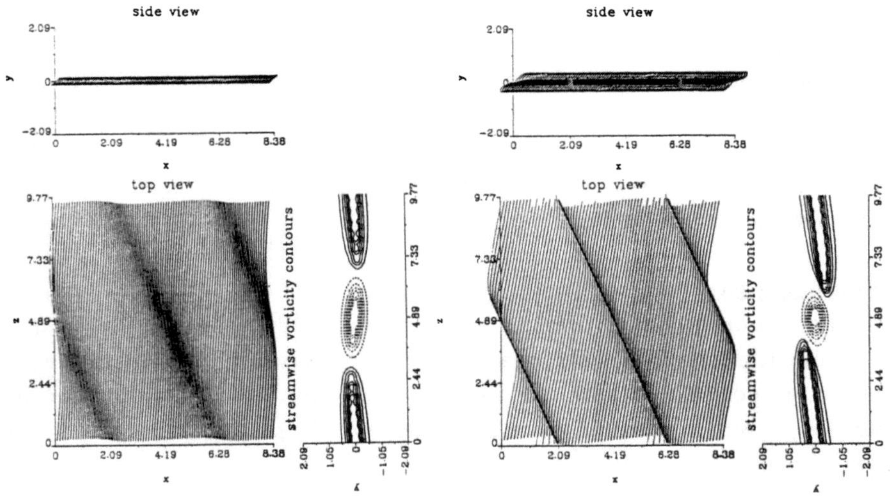

a

b

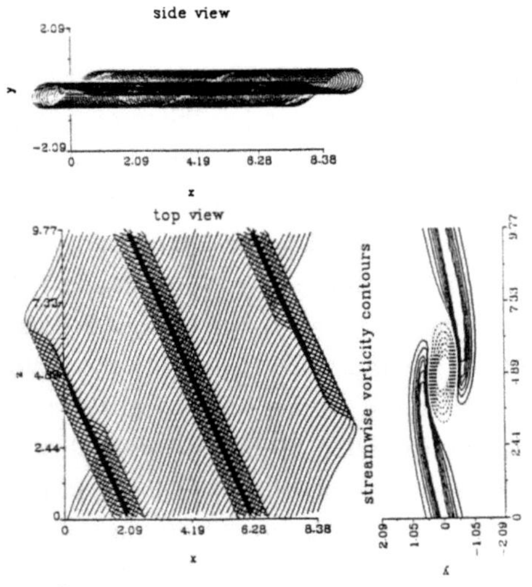

c

Figure 1: Evolution of a plane mixing layer perturbed by a single oblique wave. Shown are top and side views of the vortex filament centerlines as well as streamwise vorticity contours at x=4.19 for times (a) 13.44, (b) 15.31, and (c) 17.66. Observe that the oblique wave leads to the formation of an oblique Kelvin-Helmholtz roller which contains streamwise vorticity of the opposite sign from the braid region. As the braid vorticity wraps around the KH roller, it forms an elongated, nearly uniform vorticity layer of one sign.

the present oblique rollers, in contrast to strictly spanwise KH vortices, contain swirl. This fact opens up new routes for potential core instabilities of oblique rollers. Furthermore, the existence of nearly uniform braid vorticity layers raises questions regarding the stability of these regions of the flowfield as well.

In order to address the above issues, we now consider a flow field in which we introduced a spanwise perturbation wave in addition to the oblique wave. The spanwise wave is again taken to affect the cross-stream position of the filament centerline and mimics the experimental situation of a wavy trailing edge modulation. Figure 2a shows the resulting flow field at time t=14.69. Again a top and a side view of the vortex filaments are shown, along with contours of the streamwise vorticity at x=4.19. We recognize that the spanwise wave has triggered a Kelvin-Helmholtz-like instability in the braid region, thereby causing the braid vorticity to collapse into concentrated streamwise vortices. Since the case of an oblique wave only had resulted in the formation of streamwise braid vorticity of one sign, it does not come as a surprise that the braid region is now dominated by streamwise vortices which are all of the same sign. This is in marked contrast to the vorticity field obtained from a perturbation consisting of a strictly streamwise and a spanwise wave, which shows

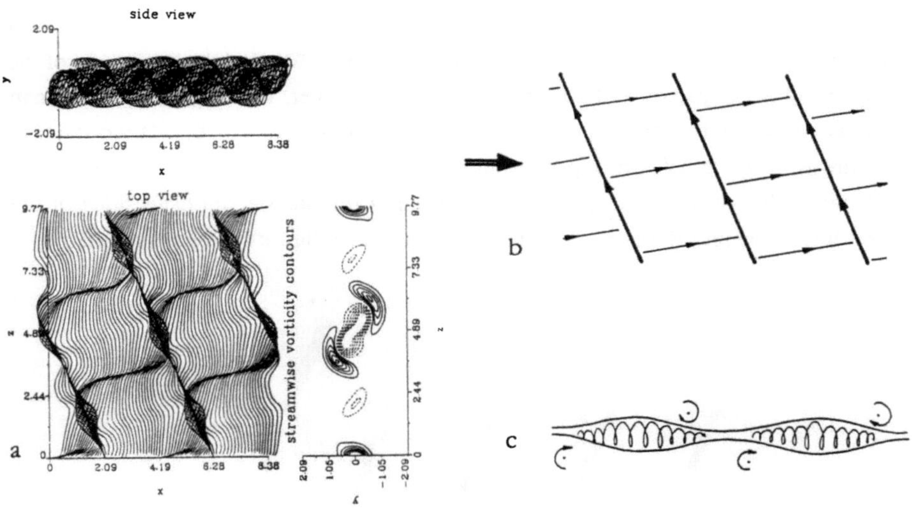

Figure 2: a) The mixing layer perturbed by an oblique and a spanwise wave. Shown are a top view and a side view of the vortex filaments as well as streamwise vorticity contours at x=4.19. b) Schematic representation of the vorticity field with concentrated braid vortices of a single sign. c) The staggered formation of the concentrated streamwise braid vortices leads to the convection of fluid from certain vortex core regions to neighboring ones, thereby creating 'vortex core bulges'.

the formation of counterrotating streamwise braid vortex pairs, e.g. Ashurst and Meiburg [23] as well as Lasheras and Choi [24]. In the terminology of Lin and Corcos [25], the oblique wave leads to the formation of an infinite aspect ratio braid vorticity layer, which then collapses into streamwise structures all of the same sign. A corresponding situation is found in the three-dimensional evolution of nominally axisymmetric jets perturbed by axial and helical waves, respectively (Martin and Meiburg [26], [27). We furthermore observe a strong wavy modulation along the cores of the oblique KH vortices. The creation of these 'vortex core bulges' can best be understood on the basis of the schematic representation of the vorticity field in figure 2b and 2c. We recognize that the slightly staggered formation of the streamwise braid vortices leads to a situation in which fluid is convected from certain axial locations to neighboring ones, thereby causing a local thickening of the vortex core. These regions of local reverse flow, in combination with the swirl generated by the vortex lines' winding into and out of the core creates a local velocity field that resembles vortex breakdown situations, a connection that will have to be explored in more depth.

4. Plane wake

On the basis of the above discussion of the plane mixing layer, the plane wake can now be understood in a straightforward manner. Figure 3 shows a top view of the vortex filaments, as well as streamwise vorticity contours at $x=4.19$, for the situation in which the perturbation consists of an oblique wave only. We find that each side of the wake develops approximately as the above mixing layer, forming oblique Karman vortices with swirl and streamwise vorticity, connected by braids containing streamwise vorticity of the opposite sign. We furthermore see that the streamwise vorticity is also of opposite sign in corresponding regions on opposite sides of the wake. If we now introduce an additional spanwise wave (figure 4), we again find the

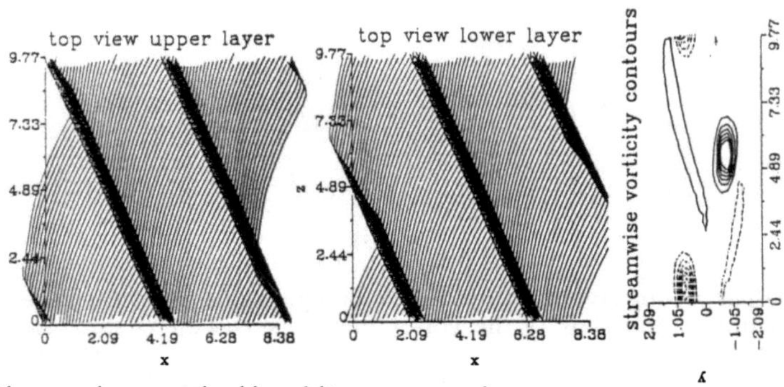

Figure 3: Plane wake perturbed by oblique wave only.

braid regions undergoing a Kelvin-Helmholtz-type instability leading to the emergence of concentrated streamwise braid vortices all of the same sign on the same side of the wake, but of opposite sign on opposite sides. The oblique Karman vortices furthermore display a wavy modulation similar to the one observed for the mixing layer vortices. It should be noted that the streamwise braid vortices on both sides form a staggered pattern of their own, i.e., a 'braid Karman vortex street'.

The above findings so far are of a qualitative nature, and their dependence on quantities such as the oblique wave angle will have to be studied in more detail. Furthermore, a combined careful experimental and numerical investigation appears necessary in order to establish the connection between the oblique waves described here, which emerge from a parallel flow instability, and the near wake vortices generated by oblique shedding.

6. Acknowledgments

Helpful discussions with J.C. Lasheras are gratefully acknowledged. This work has been supported by the National Science Foundation and by the San Diego Supercomputer Center.

References

1. Koch, W.: Local instability characteristics and frequency determination of self-excited wake flows. J. Sound Vib. 99, p. 53 (1985).
2. Huerre, P. and Monkewitz, P.A.: Local and global instabilities in spatially developing flows. Annu. Rev. Fluid Mech. 22, p. 473 (1990).
3. Oertel Jr., H.: Wakes behind blunt bodies. Annu. Rev. Fluid Mech. 22, p. 539 (1990).
4. Williamson, C.H.K.: On defining a universal and continuous Strouhal-Reynolds number relationship for the laminar vortex shedding of a circular cylinder. Phys. Fluids 31, p. 2742 (1988).

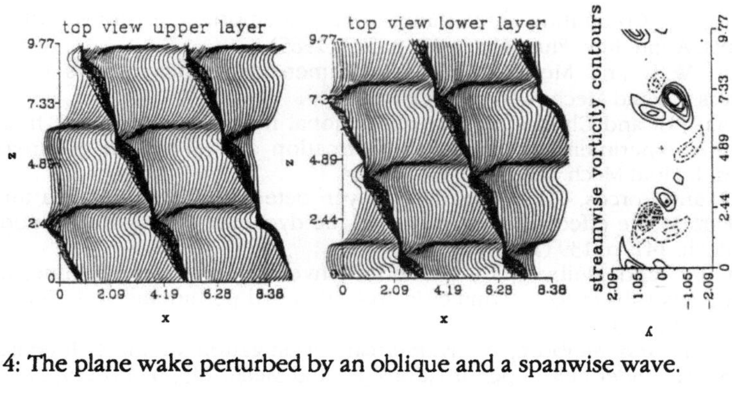

Figure 4: The plane wake perturbed by an oblique and a spanwise wave.

5. König, M., Eisenlohr, H., and Eckelmann, H.: The fine structure in the Strouhal-Reynolds number relationship of the laminar wake of a circular cylinder. Phys. Fluids A 2, p. 1607 (1990).
6. Hammache, M. and Gharib, M.: A novel method to promote parallel vortex shedding in the wake of circular cylinders. Phys. Fluids A 1, p. 1611 (1989).
7. Williamson, C.H.K.: Oblique and parallel modes of vortex shedding in the wake of a circular cylinder at low Reynolds number. J. Fluid Mech. 206, p. 579 (1989).
8. Hammache, M. and Gharib, M.: An experimental study of the parallel and oblique vortex shedding from circular cylinders. J. Fluid Mech. 232, p. 567 (1991).
9. Meiburg, E. and Lasheras, J.C.: Experimental and numerical investigation of the three-dimensional transition in plane wakes. J. Fluid Mech. 190, p.1 (1988).
10. Lasheras, J.C. and Meiburg, E.: Three-dimensional vorticity modes in the wake of a flat plate. Phys. Fluids A 2, p. 371 (1990).
11. Cimbala, J.M., Nagib, H.M., and Roshko, A.: Large structures in the far wakes of two-dimensional bluff bodies. J. Fluid Mech 190, p. 265 (1988).
12. Karniadakis, G.E. and Triantafyllou, G.S.: Three-dimensional dynamics and transition to turbulence in the wake of bluff bodies. J. Fluid Mech. 238, p. 1 (1992).
13. Corke, T.C., Krull, J.D., and Ghassemi, M.: Three-dimensional-mode resonance in far wakes. J. Fluid Mech. 239, p. 99 (1992).
14. König, M., Eisenlohr, H., and Eckelmann, H.: Visualization of the spanwise cellular structure of the laminar wake of wall-bounded circular cylinders. Phys. Fluids A 4, p. 869 (1992).
15. Piccirillo, P.S. and Van Atta, C.W.: An experimental study of vortex shedding behind linearly tapered cylinders at low Reynolds numbers. To appear in J. Fluid Mech.
16. Szepessy, S. and Bearman, P.W.: Aspect ratio and end plate effects on vortex shedding from a circular dylinder. J. Fluid Mech. 234, p. 191 (1992).
17. Nassef, H. and Browand, F.K.: On the formation of a defect structure in a two-dimensional mixing layer. Submitted to Phys. Fluids A.
18. Albarede, P. and Monkewitz, P.A.: A model for the formation of oblique shedding and chevron patterns in cylinder wakes. Phys. Fluids A 4, p.744 (1992).
19. Park, D.S. and Redekopp, L.G.: A model for pattern selection in wake flows. Phys. Fluids A 4, p. 1697 (1992).
20. Batchelor, G.K.: An introduction to fluid dynamics. Cambridge University Press (1967).
21. Leonard, A.: Vortex methods for flow simulation. J. Comp. Phys. 37, p. 289 (1980).
22. Leonard, A.: Computing three-dimensional incompressible flows with vortex elements. Annu. Rev. Fluid Mech. 17, p. 523 (1985).
23. Ashurst, W.T. and Meiburg, E.: Three-dimensional shear layers via vortex dynamics. J. Fluid Mech. 189, p. 87 (1988).
24. Lasheras, J.C. and Choi, H.: Three-dimensional instability of a plane free shear layer. An experimental study of the formation and evolution of streamwise vortices. J. Fluid Mech. 189, p. 53 (1988).
25. Lin, S.J. and Corcos, G.M.: The mixing layer: deterministic models of a turbulent flow. Part 3. The effect of plane strain on the dynamics of streamwise vortices. J Fluid Mech. 141, p. 139 (1984).
26. Martin, J.E. and Meiburg, E.: Numerical investigation of three-dimensionally evolving jets subject to axisymmetric and azimuthal perturbations. J. Fluid Mech. 230, p. 271 (1991).
27. Martin, J.E. and Meiburg, E.: Numerical investigation of three-dimensionally evolving jets under helical perturbations. J. Fluid Mech. 243, p. 457 (1992).

Three-Dimensional Phenomena in Bluff Body Wakes: Part 1: 3-D Phase Dynamics Part 2: Wave Interactions in the Far Wake

C.H.K. Williamson

Mechanical and Aerospace Engineering, Cornell University, Ithaca, NY 14853, U.S.A.

A number of recent investigations have been concerned with three-dimensional aspects of cylinder wakes, in both the laminar and transition regimes. In the laminar regime, it has been found that the end boundary conditions of long cylinders are important in determining the flow over the whole span, and for inducing various parallel and oblique vortex patterns. These near wake patterns have implications to the far wake structure.

Part 1: 3-D Phase Dynamics

In Part 1 of the present work, we have devised a method to control end conditions of a cylinder span by the use of end suction. This work follows from the discovery in 1988 that oblique shedding and "chevron" patterns of vortex shedding are caused by the end conditions (Williamson, 1988; Eisenlohr & Eckelmann, 1989; Konig, Eisenlohr & Eckelmann, 1990; Hammache & Gharib 1989, 1991) and that, with suitable novel manipulation, the shedding can be induced to be parallel to the body. Investigations have also been conducted from rather fruitful and useful analyses of wake patterns using equations such as the Guinzburg-Landau and Van der Pol equations (Albarede, Provansal &Boyer, 1990; Albarede & Monkewitz, 1991; Noack, Ohle & Eckelmann, 1991; Park & Redekopp, 1992).

Our present control allows a continuous variation of the end boundary conditions, and admits transient or impulsive control. We have shown that the laminar shedding regime exists up to a Reynolds number (Re) of 205, which is beyond the typical value of Re=150-160. Using the suction end control, we have controlled steady-state patterns such as the "chevron" pattern, and indeed parallel shedding. We have also found that the shedding angle may be determined solely by the end conditions, and that the wake is receptive to a whole range of oblique angles at a given Re, rather than a single oblique angle. As a part of this study, from a very sensitive set of experiments, we have found excellent agreement with the "cosθ" formula for collapsing oblique-shedding data onto a single "universal" curve (Williamson, 1988).

The suction method has also allowed us to undertake, for the first time, transient experiments whereby impulsive changes in end conditions send "phase shocks" across the span. An example of such a phase shock is demonstrated from Figure 1, where we can see an impulsive change of end conditions at the top of the span sends a phase shock travelling downwards. Conditions may be simply deduced for which phase shocks, and other phenomena, will occur.

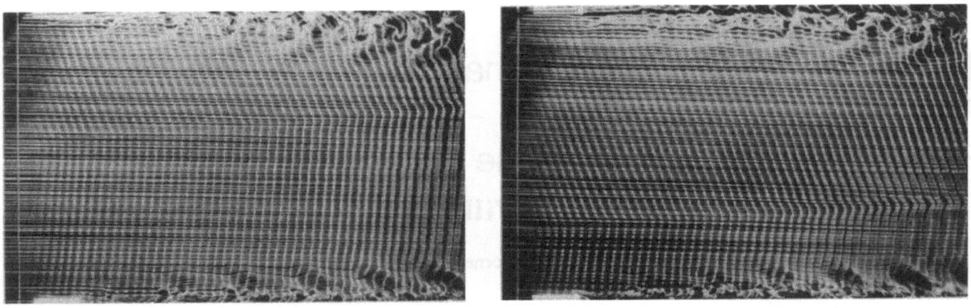

Figure 1. "Oblique shock" travelling from the top of the span downwards, after an impulsive increase in oblique shedding angle at the top end. The initial vortex pattern was the "chevron" pattern.

Part 2: Wave interactions in the far wake.

In part 2 of the present work, we have obtained some results pertaining to wave interactions in the far wake, and which were originally presented at the ONR Workshop on "Wake Vortex Dynamics" in March 1991. At that time, it was generally believed that the far wake structure was independant of both the scale and frequency of the near wake, and that it was therefore "unconnected" with the near wake. It was suggested by Cimbala, Nagib & Roshko (1986) that the evolution of a secondary vortex street structure in the far wake is by hydrodynamic instability, as distinct from vortex pairing, suggested earlier by Matsui & Okude (1983). Cimbala et al. further discovered a 3-D "honeycomb"-like pattern in the far wake, and suggested that it could be related to a parametric subharmonic instability, similar to other shear flows. This work has triggered further studies (Lasheras & Meiburg, 1990; Corke et al. 1991).

As a preliminary result of the present investigation, we have found that the "honeycomb" pattern only appears when there is oblique vortex shedding, as shown in Figure 2. The honeycomb pattern is formed by the interaction between the decaying oblique shedding waves from upstream, and large-scale parallel (2-D) waves which grow in the far wake. In Figure 3, we show that the honeycomb pattern of Cimbala et al. (1988) is indeed caused by the same wave interaction phenomenon. Such a pattern is not observed when there is parallel shedding, suggesting that the honeycomb pattern is not a manifestation of a parametric subharmonic instability. The scale of the honeycomb pattern is directly influenced by the angle of oblique shedding. In particular, with this kind of wave interaction, the spanwise wavelength must, by definition, be the same for the oblique shedding vortices and the honeycomb pattern. Important results for the far wake, obtained recently by Hammache & Gharib (ONR Workshop, June 1992; APS Meeting Nov. 91) confirm, from careful measurements, that the wavelength of a spanwise waviness on their far-wake parallel vortices is equal to that for the oblique shedding vortices, which is consistent with the present work. *There is therefore a "connection" between the near and far wakes, under certain conditions.*

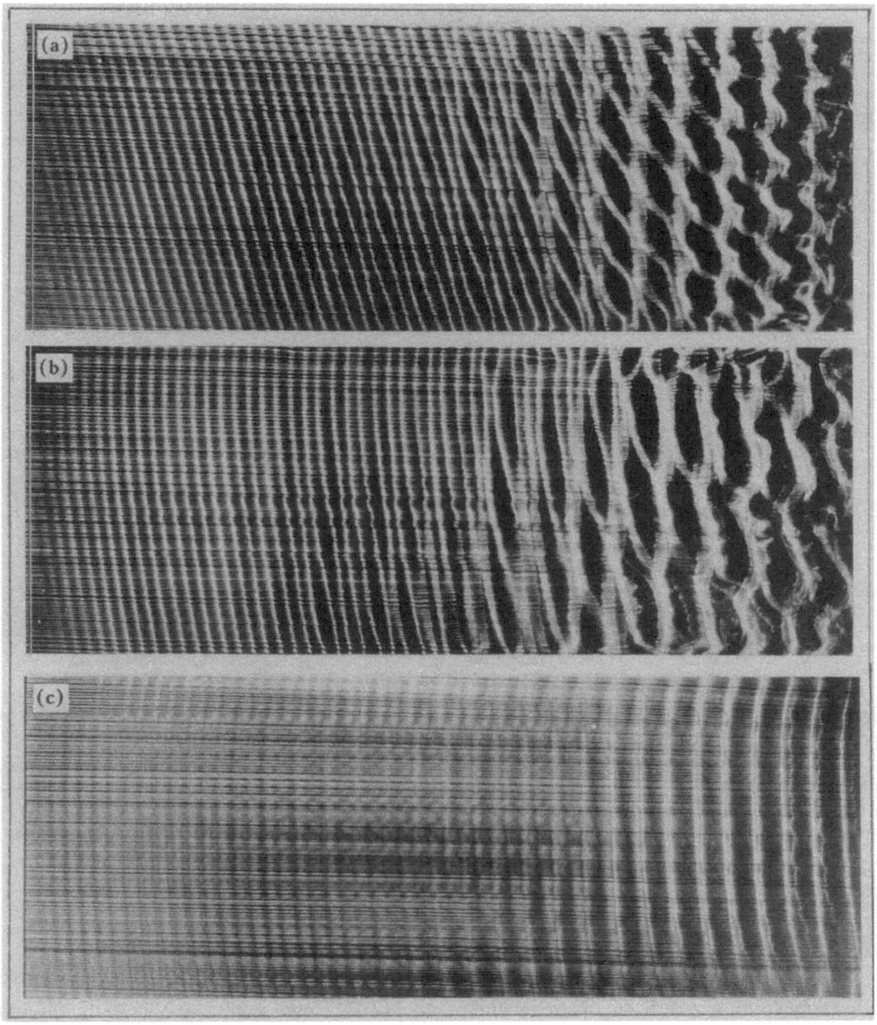

Figure 2. *Demonstration of a direct link between the "honeycomb"-like pattern and the existence of oblique vortex shedding.*

The far wake honeycomb pattern is caused by the interaction of oblique shedding waves with large-scale parallel (2-D) waves in the far wake. The angle of oblique shedding influences the structure of the honeycomb pattern as shown in (a) and (b), most obviously the spanwise wavelength, which is precisely equal to the spanwise wavelength of the oblique vortices. If there is parallel shedding in (c), then there is no honeycomb pattern in the far wake.

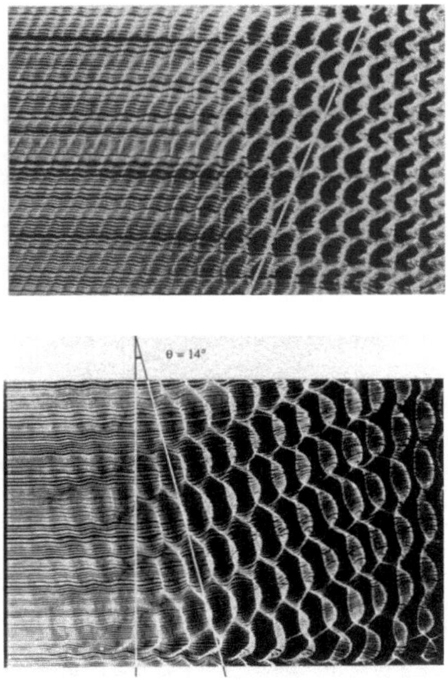

Figure 3. *Demonstration that the honeycomb pattern of Cimbala, Nagib & Roshko (1988) is caused by oblique shedding waves interacting with far wake parallel waves.*

Oblique vortices may be seen in the upper photograph being deformed by parallel waves in the far wake, as they travel to the right. The oblique shedding angle is drawn as the white line in the "honeycomb" pattern, and it can be seen that it represents a line of symmetry in this pattern. The lower photograph shows the honeycomb pattern observed by Cimbala et al. (1988), and demonstrates that their oblique shedding angle (14°) represents the same line of symmetry in their pattern.

In conclusion, the honeycomb pattern of the far wake is caused by an interaction between oblique shedding waves from upstream and parallel (2-D) waves, which evolve in the far wake. We present evidence that the honeycomb pattern of Cimbala et al. (1988) is indeed the same phenomenon. The far wake waves are caused by hydrodynamic instability, rather than vortex pairing, principally because the parallel waves and oblique shedding waves are not parallel to each other. The scale of the honeycomb pattern is directly dependant on the oblique shedding geometry, and thus the far wake is "connected" to the near wake, under certain conditions.

Acknowledgements: The very fine work of Anil Prasad and Greg Miller is acknowledged. This work has been supported by the U.S. Office of Naval Research, Contract # N00014-90-J-1686, as part of the O.N.R. "Bluff Body Wake Vortex Dynamics" Accelerated Research Initiative.

Symmetry of Two- and Three Dimensional Modes at a Vortex Shedding Interface in a Cylinder Wake

D. R. Williams, H. Mansy and P.-M. Yang

Fluid Dynamics Research Center
Illinois Institute of Technology
Chicago, Illinois USA 60616

Summary
The wave interactions which produce the low-frequency disturbances that dominate the far-wake spectrum of a circular cylinder have been studied. The spatial symmetry of interacting 2-D and 3-D waves that occur in the near wake region under natural and forced conditions are shown to follow a simple set of symmetry rules. The spanwise phase distribution of the low-frequency wave is predicted assuming a simple three-wave interaction.

Introduction

Experiments by Mansy and Williams (1991) studied the interaction of two-dimensional waves in the near wake region of a circular cylinder. It was shown that the spatial symmetry of the two-dimensional modes could be identified as being *even* modes (symmetric in streamwise velocity about the x-axis) or *odd* modes (antisymmetric) using a modal decomposition technique introduced by Marasli, et al. (1988). The symmetry of two-dimensional modes created by the interaction of two other modes, i.e., a triad interaction was found to follow a simple set of rules: $o * o \rightarrow e$ $e * e \rightarrow e$ $o * e \rightarrow o$, where "o" represents an odd mode and "e" represents an even mode. For example, the antisymmetric Karman vortex shedding mode must produce a symmetric harmonic, $o * o \rightarrow e$. Chow (1992) showed analytically that the interaction coefficients for weakly nonlinear wave interactions will be nonzero only for wave triads that follow the same set of rules described above. Thus, in order for energy to be transferred to a difference mode, the symmetry of the modes must follow the constraints described above, which rules out certain types of wave interactions, such as $o * o \rightarrow o$. Furthermore, it shows that knowledge of the mode frequencies alone is not sufficient to describe the wave interaction or to predict the resulting wave structure. The structure of a forced wake will depend on the symmetry of the forcing function.

The previous studies have been limited to two-dimensional waves. It is well known since the work by Gaster(1969,1971) that a circular cylinder may have different shedding frequencies at different locations along its span. This may be caused by nonuniformity in the cylinder diameter, mean flow or end conditions. At the interface between the regions of different vortex shedding frequencies, one finds a variety of combination modes of oscillation resulting from the nonlinear interaction. In this experiment the symmetry of the various modes produced at the region of interaction between the two types of shedding is examined to determine if the symmetry rules still apply to wave triads involving three-dimensional waves. The development of the low-frequency mode in the streamwise direction is examined.

Experimental Technique

The experiments were conducted in a recirculating water channel with a test section 20 cm x 30 cm and 242 cm long. The mean flow was uniform to within 0.5%, and the free stream turbulence level was less than 0.3%. The cylinder diameter was D=0.954 cm, and the Reynolds number based on the diameter was varied from 140 to 1,000. A scanning laser anemometer was used to obtain instantaneous velocity profile data. Scanning was at 10 scans/sec in directions either parallel to the cylinder axis (z-scans), or perpendicular to the cylinder axis and flow (y-scans).

Two techniques were used to generate the waves necessary to study the interaction process. The first involved misalignment of the end plates on the cylinder to produce a region of oblique vortices and parallel shedding vortices (Gerich and Eckelmann, 1982). Typical oblique wave angles were from $\theta = 10^o$ to $\theta = 30^o$ to the cylinder axis. The frequency ratio between the oblique (f_l) and parallel (f_h) shedding closely followed the relation $f_l/f_h = \cos(\theta)$ described by Williamson (1988). This type of "forcing" can produce only a limited variation in the difference frequencies, i.e. due to the limitation in θ the ratio was typically $f_l/f_h \geq 0.87$. The oblique vortex shedding was dominant over the lower half of the cylinder, and the parallel shedding was found in the upper half. The second method was the unsteady bleed technique developed by Williams, Mansy and Amato(1992), which produced symmetric disturbances at frequency $2f_e$. These were introduced into the boundary layer on the surface of the cylinder to generate an antisymmetric vortex shedding mode at half of the excitation frequency, f_e. The unsteady bleed forcing was applied on only one half the span of the cylinder, with the other half shedding vortices at the natural frequency. In this way the interaction region was fixed in space at the interface between the forced and unforced regions, and a larger ratio $f_l/f_h < 0.6$ was achieved.

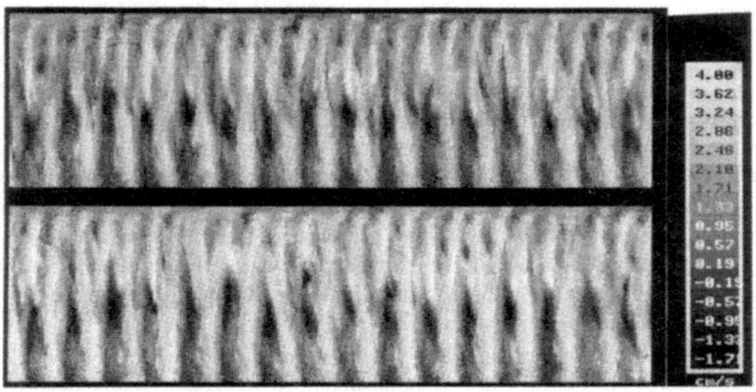

Fig. 1. Scanning LDV data at y/D=0.5, x/D=3 for Re=320 showing the appearance of vortex dislocations connecting the Karman vortices across the interface.

Results

Figure 1 is an example of "raw" z-scan data acquired at y/D = 0.5, x/D = 3 for Re = 320. Each strip represents 46.5 seconds in time. The z-scans cover 4.4 D across the span. The gray level represents the streamwise velocity corresponding to the scale on the right side of the figure. Forcing produces vortex shedding at f_e = 0.35 Hz on the bottom (z < 0) of each time strip, and the natural vortices on the top (z > 0) are shedding at f_o = 0.57 Hz. The requirement of constant circulation produces vortex dislocations between the two sides, which is clearly seen in the data. The region of interaction near z=0 produces a spectrum with many interaction modes. The dominant peaks are listed in Table 1 along with their symmetry, origin and the applicable rule that produced the combination mode.

Table 1 Combination Modes at Location (3,0.5,0) Re = 320

frequency	.14	.20	.35	.57	.68	.88	1.03	1.23	1.38
symmetry	o	e	o	o	e	e	o	o	e
origin	2fe-fo	fo-fe	fe	fo	2fe	2fe+(fo-fe)	fe+2fe	3fe+(fo-fe)	4fe
interaction	eo→o	oo→e				ee→e	oe→o	oe→o	ee→e

The mode interactions shown in the table verify that the spatial symmetry of the modes comprising the region of interaction obey the same rules of symmetry as seen in two-dimensional wave interaction. As a result, the low-frequency difference mode shown by Lewis and Gharib(1992) to be associated with the passage of dislocations is an even mode and must be symmetric about the x-axis. It is the low-frequency mode that persists farther downstream as the wake develops. Ultimately it becomes the dominant mode in the spectrum in the far wake.

Similar types of wave interactions occur in the unforced wake. End effects produce oblique vortex shedding at a slightly lower frequency than the parallel shedding vortices. Fig. 2 shows the amplitude distribution over the span of the cylinder (Re=200) corresponding to a parallel mode, an oblique mode and the difference mode. The spanwise phase distributions are shown in Fig. 3. Vortex dislocations appear in the velocity maps at the difference frequency. The low-frequency difference mode is symmetric, resulting from the interaction of two antisymmetric waves as predicted by the symmetry rules.

Furthermore, the phase variation of the difference mode is determined by the interacting waves. The oblique mode can be expressed as $u'_1 = f(z)\exp(i\omega_1 t) + c.c.$ and the parallel mode as $u'_2 = g(z)\exp(i\omega_2 t) + c.c.$, where f(z) is complex for the oblique wave and g(z) is real for the parallel wave. Then the phase of the difference mode is obtained from the product of $u'_1 u'_2$, so $\theta(z) = -\mathrm{Tan}^{-1}(\mathrm{img}(f)/\mathrm{Re}(f))$, which is the negative of the phase of the parent oblique mode. The phase data in Fig. 3 show exactly this behavior, thus the low-frequency mode is the product of the nonlinear interaction of the oblique and parallel modes. The parent modes were found to decay beyond x/D=2, while the difference mode amplifies until x/D=10 before it begins to decay. The symmetry and the spanwise phase distribution were not changed during the growth of the wave. In this way the wave interactions in the near wake imprint their structure on the far-wake region.

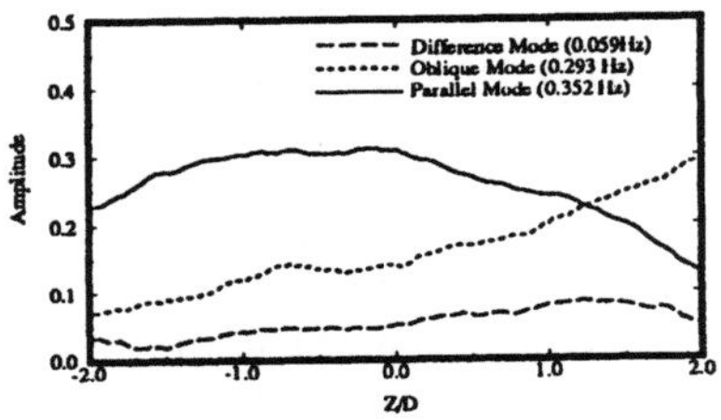

Fig. 2. Spanwise amplitude distribution

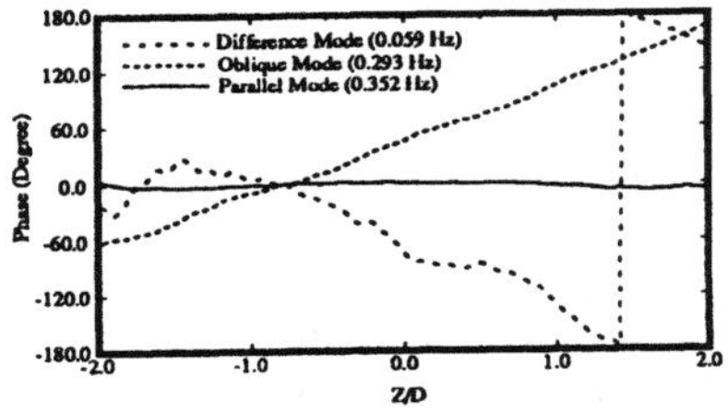

Fig.3. Spanwise phase distribution, x/D=2

Conclusions

The symmetry of interacting three-dimensional wave triads is governed by the same rules as the two-dimensional wave interactions. The low-frequency mode obtained by interaction between a parallel shedding wave and a lower-frequency oblique wave is symmetric about the x-axis. It grows to a larger amplitude than its parent waves, reaches a maximum, then decays at a slower rate to become the dominant disturbance.

References

1. Chow, K. - private communication.

2. Gaster, M.: Vortex shedding from slender cones at low Reynolds numbers. *J. Fluid Mech.* **38**, (1969) 565-578.

3. Gaster, M. : Vortex shedding from circular cylinders at low Reynolds numbers. *J. Fluid Mech.* **46** (1971) 749-756.

4. Gerich, D.;Eckelmann, H.: Influence of End Plates and Free Ends on the Shedding Frequency of Circular Cylinders*J. Fluid Mech.* **122**, (1982) 109-121.

5. Koenig, M; Eisenlohr, H.; Eckelmann, H.: Visualizaiton of the spanwise cellular structure of the laminar wake of wall-bounded circular cylinders. *Phys. Fluids A* **4** (1992) 869-872.

6. Lewis, C.; Gharib, M.: An Exploration of the Wake Three Dimensionalities Caused by a Local Discontinuity in Cylinder Diamter *Phys.Fluids A* (1992)

7. Mansy, H.; D. Williams: Symmetry of Interacting Modes in a Cylinder Wake *Phys.Fluids A* **3**, (1991) 2047-2049.

8. Marasli, B.; Champagne, F.H.; Wygnanski, I.J.: Modal decomposition of velocity signals in a plane, turbulent wake*J.Fluid Mech.* **198**,(1988) 255.

9. Williams, D.; Mansy, H. ; Amato, C. : The response and symmetry properties of a cylinder wake subjected to localized surface excitation *J.Fluid Mech.* **234**, (1992) 71-96.

10. Williamson, C.: Defining a universal and continuous Strouhal-Reynolds number relationship for the laminar vortex shedding of a circular cylinder*Phys.Fluids* **31**, (1988) 2742-2744.

11. Williamson, C.H.K.: Oblique and parallel modes of vortex shedding in the wake of a circular cylinder at low Reynolds numbers. *J. Fluid Mech.* **206** (1989) 579-627.

An Experimental Study of the Three-Dimensional Structure of the Wake of Circular Cylinders in the Laminar and Transitional Reynolds Number Range

M. KÖNIG and H. ECKELMANN

Max-Planck-Institut für Strömungsforschung,
Bunsenstr. 10, 3400 Göttingen, Germany

Summary

Experiments carried out in the laminar range ($50 < Re < 170$) of a Kármán vortex street give evidence for the existance of different modes of oblique vortex shedding in the wake of circular cylinders. The occurance of these discrete modes depends on the end conditions. In the irregular Reynolds number range ($170 < Re < 230$) the Strouhal number data collapse on one single curve which is an extension of one of the modes of the laminar range.

Introduction

The cylinder wake has been subject to intensive research for a long time. Yet, it is surprising that still new phenomena can be found. Gerich and Eckelmann [1] showed that the cylinder boundaries alter the vortex shedding near the cylinder ends. For the laminar Reynolds number range ($50 < Re < 170$) Williamson [2], Eisenlohr and Eckelmann [3], Hammache and Gharib [6] as well as König, Eisenlohr and Eckelmann [4], [5] showed that by manipulating the geometry of the cylinder ends parallel or different oblique modes of vortex shedding can be induced. In this range, between the Strouhal numbers for oblique (index Θ) and parallel shedding (index 0), the following relation [2] holds

$$Sr_\Theta = Sr_0/cos(\Theta) \tag{1}$$

(Θ: angle between cylinder and vortex axes). König *et al.* observed up to four different frequencies in the wake occuring simultaneously along the span of a cylinder and designated the accompanying modes as **P**arallel, **I**ntermediate, e**X**treme and **E**nd cell. The number of cells with different frequencies and hence different shedding angles in the near wake depends on the thickness of the boundary layer that has developed on the confining walls of the cylinder. Fitting the cylinders with end plates leads to only one discontinuity in the Strouhal-Reynolds number relationship with a jump from the P– to the I–mode. If the same cylinder ends on either side in a laminar boundary layer of $\delta \approx 8d$ (d: cylinder diameter), two discontinuities are observed, i.e. an additional jump to the X–mode is obtained. At the transition from the laminar to the irregular range (at $Re \approx 170$) a further discontinuity was found by Williamson [2].

Experimental setup

Hot-wire measurements were carried out in an open-return type wind-tunnel with a closed test section (1400mm high and 280mm wide). The velocity profile is constant over the entire width of the test section up to the boundary layers which are laminar in all cases described here. The cylinders (1, 1.5 and 2.5 mm diameter) were mounted horizontally and are limited either by end plates or the wind-tunnel walls. In all cases, the aspect ratio was greater than 100. DANTEC 55M01 hot-wire-anemometers were used to measure the fluctuation velocities in the wake while a Nicolet 660A fast-Fourier-transform analyzer was employed to obtain spectra of the anemometer output.

Experimental Results

To investigate the behaviour of the different oblique modes introduced by König *et al.* at the transition to the irrregular range cylinders were fitted with differently sized end plates or were simply mounted between the wind tunnel walls. The cylinders fitted with end plates showed, like in [2], at the end of the laminar shedding range ($Re \approx 170$) a decrease in Strouhal number and a collapse on a new mode in the irregular range. Working with a cylinder bounded by the wind tunnel walls leads to extreme mode shedding. However, at $Re \approx 170$ an increase of the Strouhal number and a collapse of the data on the same mode as in the previous case is observed. The described Sr–Re dependencies are depicted in Fig.1a. Note, that when the direct vicinity of the boundaries is excluded (end cells), in the irregular Reynolds number range the Sr–Re relationship is independent of the end conditions. Since an extrapolation of the resulting Sr–Re dependency into the laminar range fits exactly into the gap between the I– and X–mode it was questioned whether it should be possible to select a special end condition which leads to a continious *Sr–Re* relationship all along the regular and irregular range. For this purpose, the diameters of the end plates have been varied systematically. It was found that a $d = 1mm$ cylinder fitted with endplates of $D = 160mm$ leads to a continious variation of the Strouhal number from the regular to the irregular range (Fig.1b). Note that for a whole variety of end conditions only a limited number of discrete *Sr–Re*–modes result. Furthermore, at a given Reynolds number, these modes are equidistant. This behavior can be expressed as

$$Sr_n = Sr_0 - n\Delta Sr(Re), \qquad n = 0, 1, 2, 3, \tag{2}$$

where n denotes the Strouhal number of the nth mode, with $n = 0$ for the P–mode and $n = 1$ for the I–mode. The new mode which extends into the irregular range corresponds to $n = 2$ and the X–mode to $n = 3$ (Fig.2). The end cells introduced by [1] also fit into this mode scheme. Depending on the end conditions of the cylinder the end cell Strouhal numbers collapse on modes $n = 6$ to $n = 8$ (not shown in Fig.2).

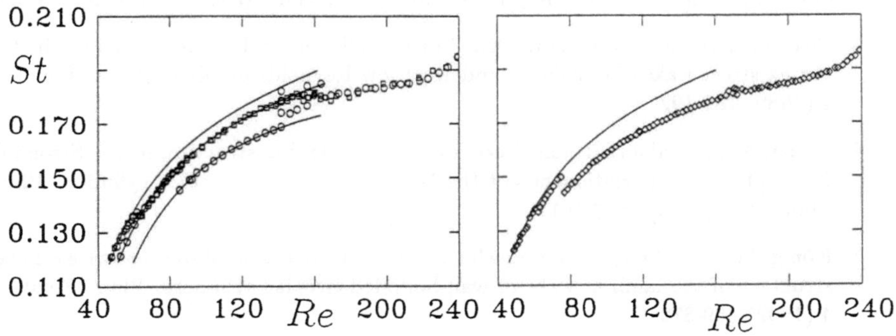

Fig.1a (left): *Sr–Re*-dependencies for a 1.5*mm* cylinder bounded by a 40*mm* end plate (□, **I**–mode) and a 2.5*mm* cylinder bounded by the tunnel walls (o, **X**–mode). For both cases the data collapse in the irregular range. The **P**–mode is included for comparison.

Fig.1b (right): *Sr–Re*–dependencies for a 1.0*mm* cylinder bounded by a 160*mm* end plate (◇). The *Sr–Re* curve extends from the laminar to the irregular range. Note the small kink in the curve that indicates the transition between the two ranges.

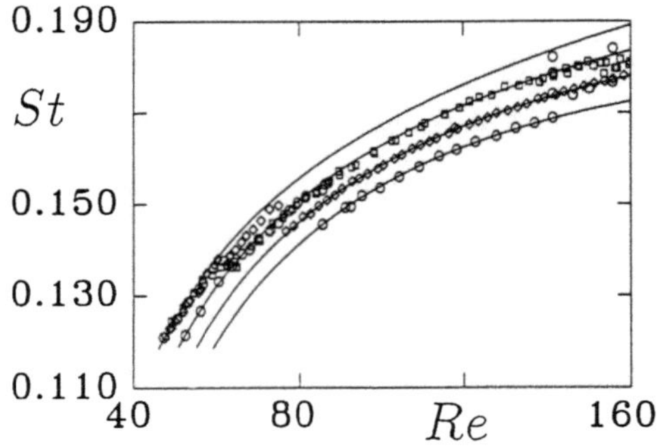

Fig.2. Mode scheme in the laminar range resulting from different end conditions. Uppermost curve: P–mode (n=0), second curve from top: I–mode (n=1), third curve from top: mode that extends into the irregular range (n=2), lowermost curve: X–mode (n=3).

References

1. Gerich, D. and Eckelmann, H.: Influence of end plates and free ends on the shedding frequency of circular cylinders. J. Fluid Mech. 122 (1982) 109-121.
2. Williamson, C.H.K.: Oblique and parallel modes of vortex shedding in the wake of circular cylinders at low Reynolds numbers. J. Fluid Mech. 206 (1989) 579-627.
3. Eisenlohr, H. and Eckelmann, H.: Vortex splitting and its consequences in the vortex street wake of circular cylinders at low Reynolds numbers. Phys. Fluids A 1 (1989) 189-192.
4. König, M., Eisenlohr, H. and Eckelmann, H.: The fine structure in the Strouhal-Reynolds number relationship of the laminar wake of a circular cylinder. Phys. Fluids A 2 (1990) 1607-1614.
5. König, M., Eisenlohr, H. and Eckelmann, H.: Visualization of the spanwise cellular structure of the laminar wake of wall-bounded circular cylinders. Phys. Fluids A 4 (1992) 869-872.
6. Hammache, M. and Gharib, M.: A novel method to promote parallel vortex shedding in the wake of circular cylinders. Phys. Fluids A 1 (1989) 1611-1614.

The Effect of Axial Motion on the Wake of a Cylinder in Steady Uniform Flow

C. G. LEWIS and M. GHARIB

University of California at San Diego
Mail Code 0411
San Diego California 92093

Summary:
A series of experiments was conducted to determine the influence of a steady streamwise vorticity component on the formation and structure of the wake vortices. The streamwise component of vorticity is created by the axial motion of a cylinder in a steady uniform flow. The direction of the motion controls the sign of the streamwise vorticity. These experiments can serve as a simplified model for flows in which a spanwise component of flow is important, such as towed cables, swept wings, and yawed bodies.

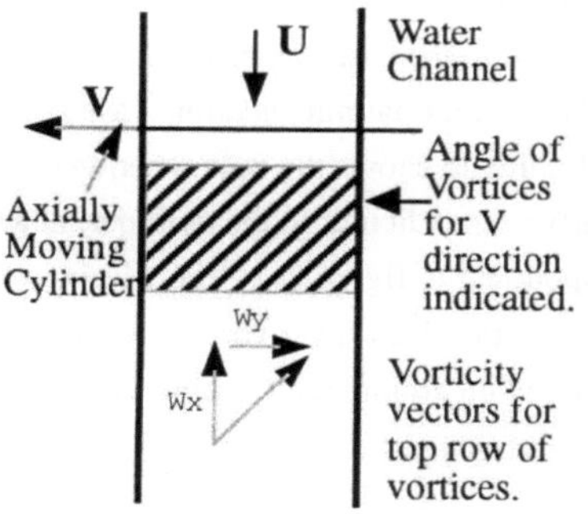

Fig. 1. Experimental Setup.

Background:

The experiments undertaken here use a novel method for actively varying the amount of streamwise vorticity which enters the wake vortices. The experimental approaches (including ours) used prior to this study usually involved uncontrolled amounts of streamwise vorticity which came from geometrical effects such as a step change in the cylinder diameter, yaw, tapering, or imbalanced end conditions. To date, there are no experimental studies which introduce streamwise vorticity into a wake in a controlled dynamic manner. We have designed and tested an apparatus which does this successfully through the axial motion of a circular cylinder. These experiments can serve as a simplified model for flows in which a spanwise component of flow is important, such as towed cables, swept wings, and yawed bodies.

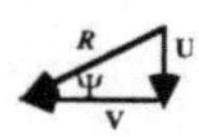

We input a steady level of streamwise vorticity utilizing the concept sketched in figure 1. From the frame of the body, there is a spanwise component of velocity. This spanwise component of flow produces vorticity in the streamwise direction (ω_x), while the uniform flow component produces the spanwise component of vorticity (ω_y). The notation for the velocities is shown to the right. The streamwise uniform velocity is designated **U**, the spanwise velocity of the cylinder is **V**. The magnitude of the sum of these velocity vectors is ***R***, and the direction of ***R*** is Ψ, which is also called the magnitude angle. The shedding angle of the vortices in the wake is θ, which can be different from Ψ.

Figure 2 shows the general behavior of the vortex street as the component **V** is increased, while **U** is held fixed (Reynolds number = 100). The vortices in the wake respond to the increase in **V** by shedding at increasing angles of inclination, while the shedding frequency decreases. Figure 2b shows steep angled shedding, for a magnitude angle of 68 degrees. As the magnitude angle Ψ increases beyond 75, the presence of a short wavelength core modulation becomes noticeable, as can be seen in figure 3c. In addition, the oblique shedding appears to become unstable, and switches back and forth between a very steep angled "trailing mode:", and a less angled "oblique mode". This instability appears for shedding angles steeper than 45 degrees, for which the mean value of the streamwise vorticity is greater than the mean value of the spanwise vorticity. Ultimately, the vortex street breaks down, as can be seen in figure 2d. Spectra for this case is completely chaotic.

Side views of the vortex street are shown in figure 3, which roughly correspond to the cases represented in figures 2a and 2c. The familiar natural shedding case is shown in figure 3a. Figure 3b shows the case when the vortices are shed nearly perpendicular to the cylinder, in a trailing mode (corresponding to figure 2c). The vortex core is labeled, as well as the shear layer vortices which can be seen to have rolled up around the forming vortex. The location of the cylinder is just to the left of the picture border.

Many vortices are being formed simultaneously along the span of the cylinder, in contrast to the single vortex formed in the parallel case. The evolution of the streamwise vorticity in the trailing mode ($\theta \geq 80$ degrees) is shown in figure 4. Figure 4a shows a spanwise cross-section of the vortex formation in the very near wake. In the figure, the tails of vortices of both sign are visible at the surface of the cylinder. Figures 4b and 4c show the roll-up of the initial vortices into complete streamwise spirals.

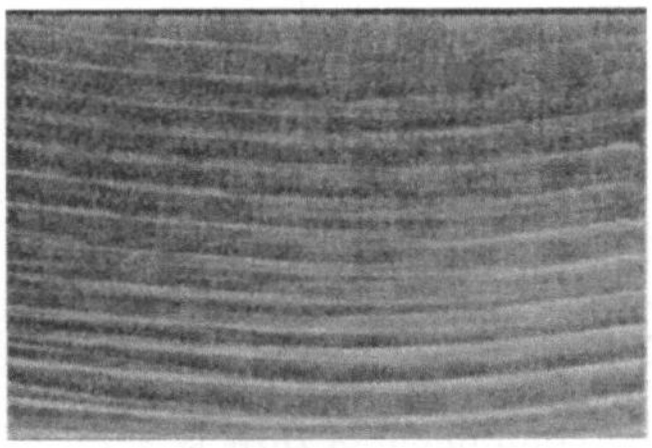

a) Natural Shedding

b) $\psi = 68$ degrees.

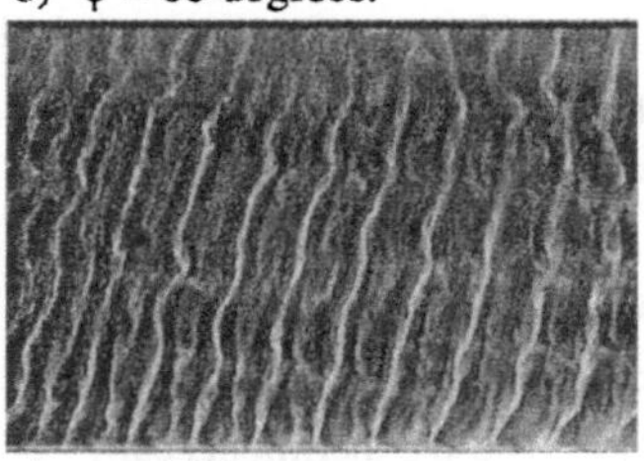

c) $\psi = 80$ degrees.

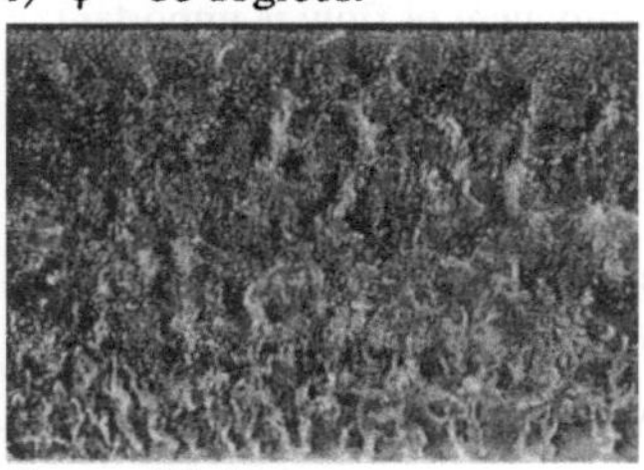

d) $\psi = 84$ degrees.

fig. 2. Progression of wake behavior through natural, oblique, trailing, and chaotic shedding behaviors.

The precise relationship between shedding angle θ and magnitude angle Ψ is shown in figure 5. The sign of θ is always the same as the sign of Ψ. The anti-symmetry of

a) Side view of the natural case.

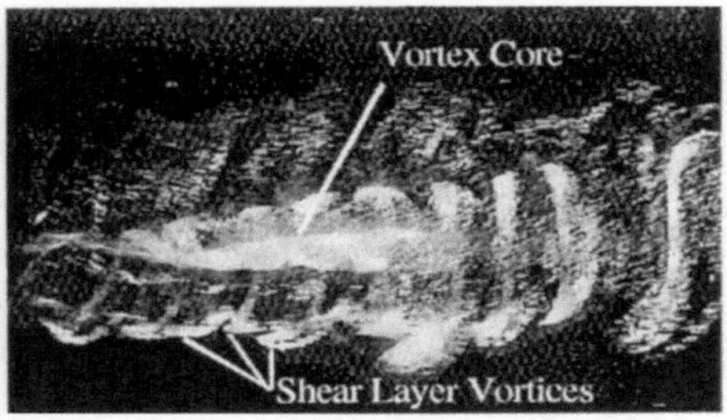

b) Side view of the trailing vortex.

Fig. 3. Side views of the vortex street corresponding to the cases in Fig. 2a and 2c.

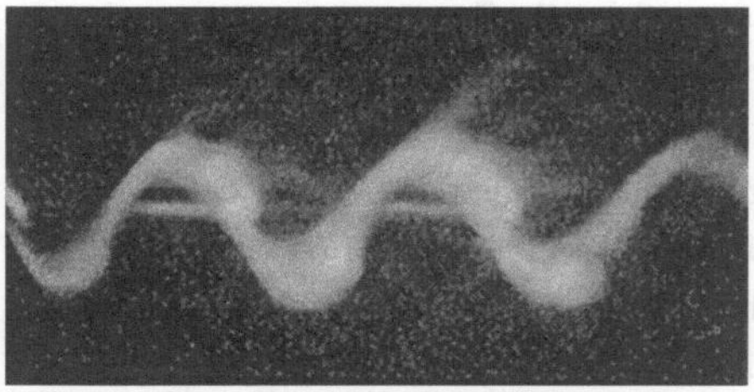

a) Spanwise cross-section at cylinder surface.

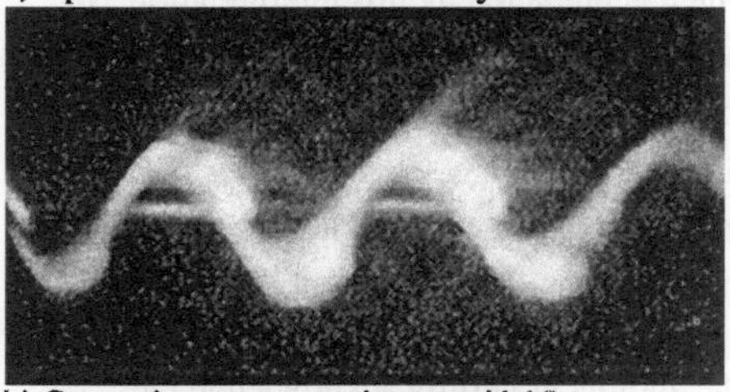

b) Spanwise cross-section at x/d 10.

c) Spanwise cross-section at x/d 25.

Fig. 4. Spanwise cross-sections showing flow visualization of the trailing vortex structures at several downstream locations.

the data shows that the magnitude of Ψ controls the magnitude of θ, for a particular Reynolds number. The scatter in the data is due to a Reynolds number effect. The sign of Ψ determines the sign of the streamwise vorticity which is being introduced to the wake, which in turn determines the direction of the oblique shedding angle θ.

Discussion:

These experiments are similar, via Galilean transformation, to the yawed cylinder case. In our case, if the freestream velocity is held constant, then the ratio of the Stouhal number for flow with an axial motion (equivalent to a yawed cylinder) to the 2-D Strouhal number has been found to vary as the cosine of the magnitude (or yaw) angle, for angle less than 45 degrees, as can be seen in figure 6. This data has very little spread in comparison to previous experimental efforts which were undertaken by physically yawing the cylinder[1,2,3]. Our technique enables us to use control methods to minimize end effects. Also, we are able test very high magnitude angles, which is not really

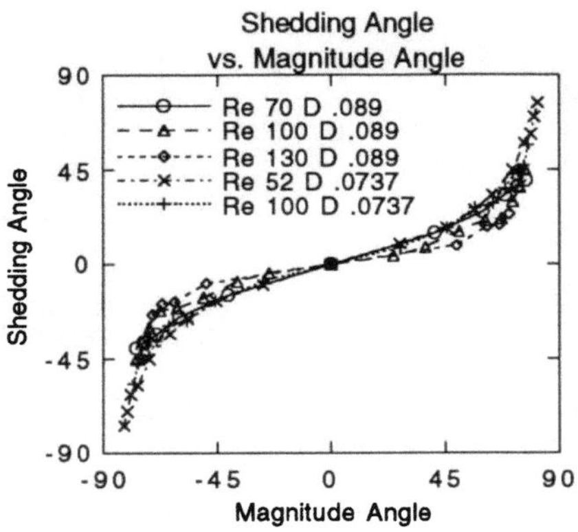

Fig. 5.

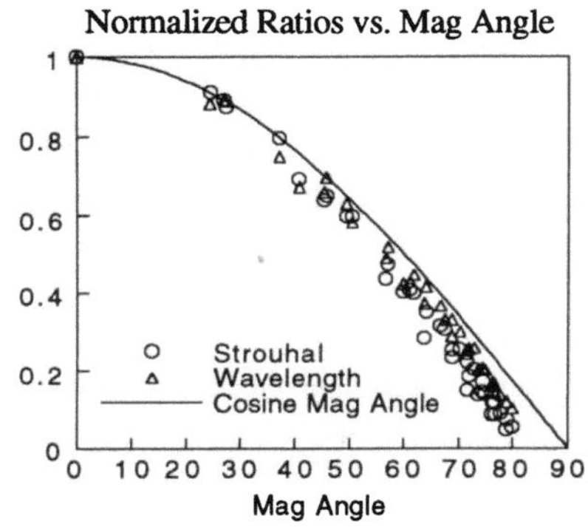

Fig. 6.

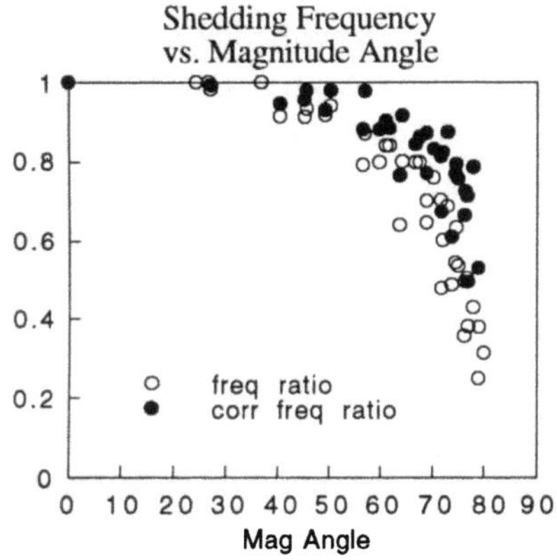

Fig. 7.

feasible for physically yawed bodies. The spread in our data is due to the existence of the short wavelength core modulation (Fig 2c), and, for the magnitude angles ≥ 80 degrees, the instability between the trailing mode and the oblique mode. Even with these sources of variation, this technique considerably reduces the spread in the data, and verifies that the cosine law is a reasonable fit to the data.

However, this cosine law cannot be interpreted as proof that the flow component normal to the cylinder controls the shedding frequency, as has been conjectured based on the independence principal. As can be seen in figure 7, even though the normal component of the flow remains constant, the frequency ratios show large variations from the ratio of one which is predicted by this interpretation. In fact, even if the frequency of the shedding is corrected using the cosine correction[4] (a different cosine law which corrects for the oblique shedding angle θ, not the magnitude angle Ψ), the ratios still do not go to one.

The authors gratefully acknowledge the support of the Office of Naval Research, Division of Ocean Engineering, contract number N0014-90-J-1314.

---

[1]Hanson, A. R.; Vortex Shedding from Yawed Cylinders. AIAA Journal **4** (1966), 738.
[2]Van Atta, C. W.; Experiments on Vortex Shedding from Yawed Circular Cylinders. AIAA Journal. **6** (1968) 931.
[3]Ramberg, S. E.; The effects of yaw and finite length upon the vortex wakes of stationary and vibrating cylinders. J. Fluid Mech. **128** (1983) 81.
[4]Williamson, C.H.K.; Oblique and parallel modes of vortex shedding in the wake of a circular cylinder at low Reynolds numbers. J. Fluid Mech. **206** (1989) 579.

On the Evolution of Three-Dimensionalities in Laminar Bluff-Body Wakes

M. HAMMACHE and M. GHARIB

Department of Applied Mechanics and Engineering Sciences
University of California San Diego

Summary

Wind tunnel experiments of the wake of a circular cylinder in the Reynolds number range of 140-160 are conducted to study the effect of near-wake vortex shedding geometry, namely the parallel and oblique shedding, on the downstream evolution of the wake and on its three-dimensional transition to turbulence. At Reynolds numbers below transition, the initial Karman vortex street has completely decayed by x/d=150 and the wake reorganizes itself into new vortical structures in the far-wake. These structures are found to be parallel in the case of a parallel near-wake, whereas oblique near-wakes lead to secondary structures which are strongly three-dimensional. This three-dimensionality originates from the flow conditions in the near-wake and is responsible for a premature transition to turbulence.

Our past experimental studies of the parallel and oblique vortex shedding from circular cylinders in the Reynolds number range 40-160 [1] have revealed some important differences between these two shedding geometries. A gradual increase of the flow velocity ultimately leads to the three-dimensional transition of the wake to turbulence. However, oblique and parallel vortex streets undergo this transition through distinct mechanisms. Typically, an oblique wake becomes turbulent by Re≈160, whereas a parallel wake remains laminar and parallel up to higher values of Re. Figures 1 and 2, both representing the wake of the same cylinder at Re=160, clearly illustrate the premature transition to turbulence of oblique wakes in comparison to parallel wakes. This finding was the motivation for the present work which is aimed at understanding the nature of the three-dimensional features present in parallel and oblique wakes prior to the transition to turbulence.

Although these three-dimensionalities originate in the near-wake of the cylinder, their

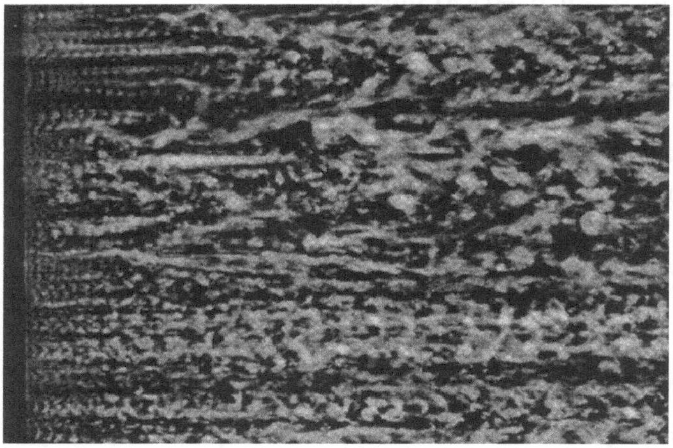

Figure 1: Turbulent flow at Re=160 corresponding to a flow without control of the cylinder end conditions.

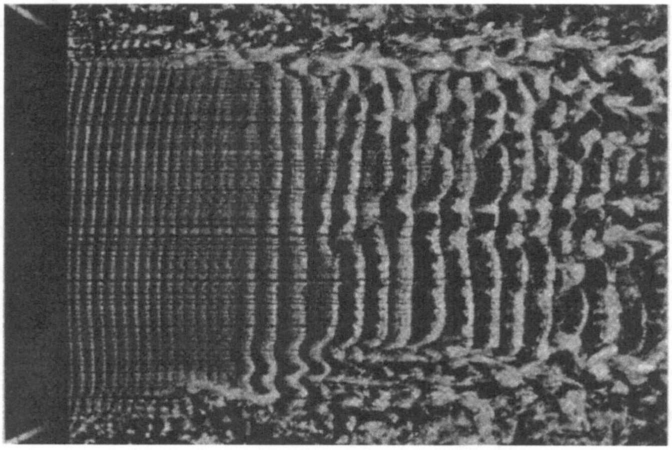

Figure 2: Laminar flow at Re=160 corresponding to a flow with control of the cylinder end conditions.

influence on the downstream evolution of the wake is considerable. At Reynolds numbers above 140 the initial Karman vortex street decays very rapidly, so that past x/d=100-150 the vortex shedding frequency peak (f_1) can no longer be detected. Past this region of decay, the wake reorganizes itself into new vortical structures (the secondary structures), with a passage frequency f_2 which is close to $f_1/2$. This proximity of f_2 to the subharmonic of f_1 lead some investigators to believe that these structures were the result of a pairing process of the Karman vortices. However, Cimbala et al. [2] showed that it is instead hydrodynamic stability that controls the downstream reorganization of the wake. Various frequencies experience cycles of growth and decay with the downstream distance. As the flow evolves downstream, the local mean velocity profile becomes unstable to lower and lower frequencies.

The effect of the near-wake vortex shedding geometry, i.e. parallel versus oblique, on the downstream evolution of the wake is examined here. The technique used to induce parallel vortex shedding as well as an investigation of the origin of oblique vortex shedding are described in detail in [1]. It should be noted that although the data shown here corresponds to specific values of the Reynolds number, the results apply to the range Re=140-160. The cases described here are typical of what is observed in this range and are only used as examples to illustrate the observed phenomena.

The rate of decay of the frequency of the initial Karman vortices (f_1) with the downstream distance is basically the same for the two vortex shedding geometries. At Re=142, f_1 can no longer be detected by x/d=140. Flow visualization photographs show that the secondary structures that appear when the wake reorganizes itself are different for the parallel and oblique near-wake geometries: While the far-wake secondary structures which evolve from a parallel near-wake are also parallel as seen in Figure 3, those which evolve from an oblique near-wake start off generally parallel but quickly develop a strong modulation in the spanwise direction as shown in Figure 4. The effect of this modulation becomes more pronounced with the downstream distance. The secondary structures ultimately break up into a regular pattern of "lumps", with a wavelength λ_2 which can easily be measured from the photographs.

This spanwise wavelength in the far-wake is found to match the spanwise wavelength λ_1 in the near-wake, i.e. the spanwise distance between two consecutive Karman vortices of the same sign. A plot of λ_1 versus λ_2 for various angles and Reynolds numbers is represented in Figure 5. Clearly, the spanwise wavelengths in the near-wake and the far-wake are equal. We conjecture that the spanwise wavelength survives the decay of the initial oblique Karman vortex street and strongly modulates the secondary structures which emerge to form the far-wake.

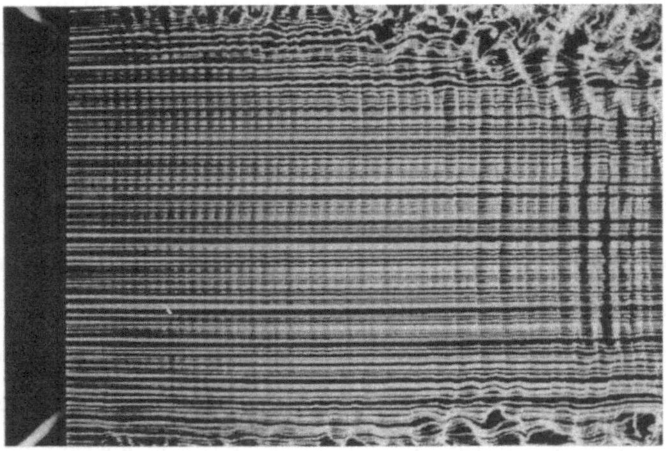

Figure 3: Downstream evolution of parallel shedding at Re=140.

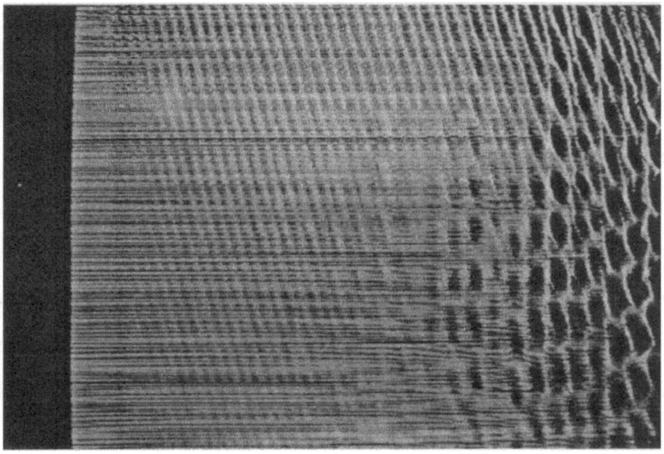

Figure 4: Downstream evolution of oblique shedding at Re=140.

Hot-wire measurements show that the three-dimensionalities described above also affect the spanwise distribution of the mean velocity at various downstream locations in the far-wake. As shown in Figure 6, this distibution is uniform for the parallel case, whereas it is periodic for the oblique case . It was also determined that the spanwise wavelength of the mean velocity distribution matches the spanwise wavelength (λ_2) of the lumps seen in the flow visualization.

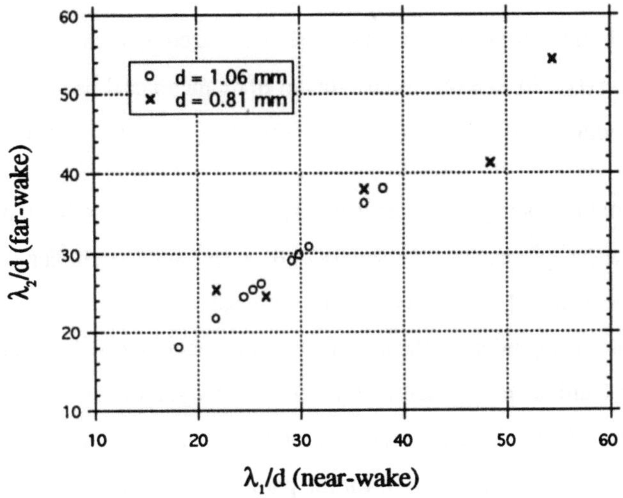

Figure 5: Far-wake spanwise wavelength versus near-wake wavelength, normalized by cylinder diameter, for oblique shedding at Re=140.

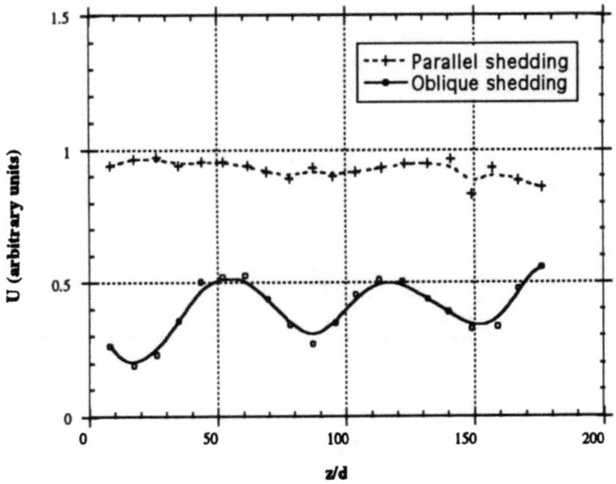

Figure 6: Spanwise distribution of the mean velocity for parallel and oblique shedding at Re=140, x/d=360.

Hot-wire velocity fluctuation spectra were measured at various downstream locations for the parallel and the oblique vortex shedding. The time averaged spectra at three locations, namely x/d=50, 170 and 410 are represented in Figures 7 and 8 to illustrate the results:

-The parallel and oblique near-wakes are characterized by distinct frequencies of vortex shedding (f_1): At Re=142 and a cylinder of diameter d=0.081 cm, f_1= 617 Hz for parallel shedding and 593 Hz for oblique shedding with an angle $\theta = 15^{o}$.

-The dominant frequency in the far-wake (f_2) is the same for both the parallel and oblique near-wake geometries: At Re=142, $\theta = 15^{o}$ and a cylinder of diameter d=0.081 cm, f_2= 295 Hz at x/d=250.The passage frequency of the secondary structures is thus independent of the initial shedding frequency. This further supports the finding of Cimbala et al. that it is indeed hydrodynamic stability rather than pairing that controls the downstream evolution of the wake, despite the fact that the ratio f_2/f_1 is very close to 1/2 (0.478 for the parallel case and 0.497 for the oblique case).

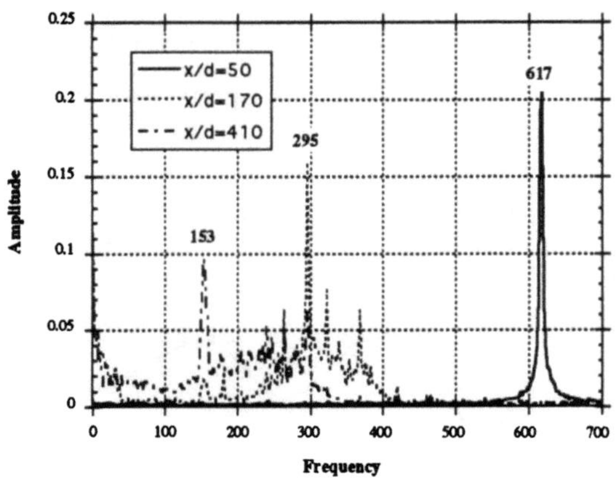

Figure 7: Averaged velocity fluctuation spectra for parallel shedding at x/d=50, 170 and 410. Re=142.

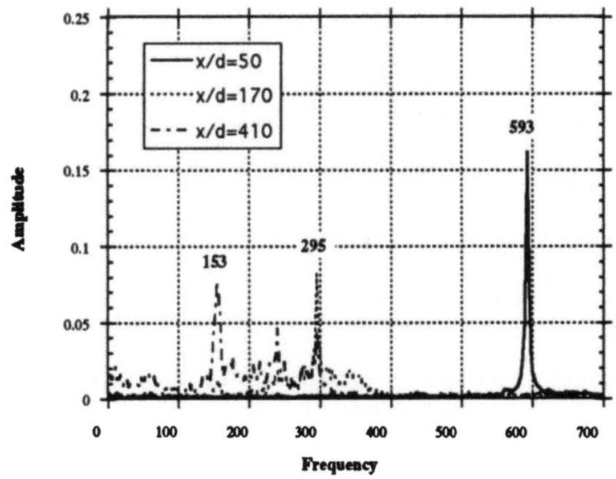

Figure 8: Averaged velocity fluctuation spectra for oblique shedding at x/d=50, 170 and 410. Re=142, θ=15°.

Note that the frequency peak f_2 decays farther downstream and the wake reorganizes itself again into tertiary structures with an even lower frequency f_3 = 153 Hz as measured at x/d=410. It is possible that the wake goes through further reorganizations at lower frequencies farther downstream, but we limited our study to x/d=50-410.

In conclusion: The presence of three-dimensionality in oblique vortex streets, which originates immediately in the base of the cylinder leads to an early transition to turbulence. In contrast, a parallel wake remains laminar until three-dimensionality of the vortex filaments becomes important.

This work was supported by a grant from the Office of Naval Research, Division of Ocean Engineering, contract No. N0014-90-J-1314.

References

1. Hammache, M., Gharib, M.: An experimental study of the parallel and oblique vortex shedding from circular cylinders. J. Fluid Mech. vol. 232, (1991) 567.

2. Cimbala, J.M., Nagib, H.M., Roshko, A.: Large structure in the far-wakes of two-dimensional bluff bodies. J. Fluid Mech. vol. 190, (1988) 265.

Experiments on Vortex Shedding Behind Linearly Tapered Cylinders

Paul S. Piccirillo and C.W. Van Atta *
University of California, San Diego
La Jolla, California USA 92093

September 1, 1992

Abstract

Experiments were performed to study the flow behind a linearly tapered cylinder, in order to gain more information about laminar vortex shedding behind the cylinder, the evolution of the velocity field in the far wake of the cylinder, and also chaotic vortex shedding behind the cylinder. For laminar vortex shedding, scalings were found for the shedding frequency, difference frequency, and shedding cell size. The behavior of the cylinders was found be less orderly as the amount of taper increased. In the far wake, the velocity fluctuations evolve in two phases to chaos, partially by transferring energy from high to low frequency. In the chaotic case, the rigid positioning of the cell boundaries breaks down, so that the number of vortices shed by cylinder becomes a monotonic function of spanwise position.

1 Experiments Performed

The study of flows behind bluff bodies has a long and rich history. In recent years there has been much work done on non-uniform bluff bodies, such as stepped[1] and linearly tapered[2,3,4] cylinders. In this work we examine both the initial vortex shedding by the cylinder, and how the shed vortices evolve in the far wake of the cylinder.

Four cylinders were used in the ware and wind tunnel experiments, with taper ratios varying from 50-to-1 to 100-to-1. The cylinder Reynolds numbers in the experiments were primarily in the laminar vortex shedding regime. Two experiments were done in the chaotic vortex shedding regime, with centerspan Reynolds numbers (Re_{cs}) of 180 and 270. Flow visualization was performed in both regimes, using a fluorescein dye-natural oil mixture which was washed off the cylinder and illuminated with a laser. For quantitative results, a rake of hot wires was placed behind the cylinder in a wind tunnel.

In order to understand the nature of the flow, a varied set of data reduction techniques was employed. The standard velocity time traces and Fourier analysis were used to understand the mean behavior the flow, along with techniques from non-linear analysis, such as Poincare sections and phase space reconstructions. In order to better understand the dynamics of the

*also Scripps Institution of Oceanography

flow, and especially the vortex splits, both the Hilbert transform and the wavelet transform were also used. All the employed quantitative techniques were used in close cooperation with the flow visualization.

2 Near Wake Results

The most important quantities in the mean flow to be investigated are the shedding frequencies in the flow, and the difference between the shedding frequencies in adjacent shedding cells, which we call the modulation frequency. The shedding frequencies in the laminar range for all the cylinders were collapsed onto one curve (shown in figure 1), by converting the shedding frequencies into local Strouhal numbers. The modulation frequency was found to be nearly a constant along the span of the cylinder, and was scaled onto one curve using a global scaling (figure 2). This scaling used a geometric correction to account for the taper ratio difference between the cylinders.

The flow behind a tapered cylinder is composed of discrete shedding cells. The mean size of these cells was found to collapse when Re_{c_s} was greater than 100. The individual cell lengths varied in size, with the cell lengths being roughly self-similar spanwise along the cylinder. The vortex shedding was found to be independent of the aspect ratio of the cylinder. Also, new cells were found to be created at the edges of the vortex shedding region of the flow, either next to an area not shedding vortices, or next to an end cell. Finally, the effect of end conditions on the overall flow were found to be small.

Flow visualization of the vortex splits showed that at the split, there exists a cross-street connection, in which a vortex line shed from one side of the cylinder connected to a vortex line shed from the other side. Preceding the vortex split in time, a kink develops in the vortex lines, perhaps due to a secondary instability in the flow. Also, following the vortex split, a "hole" appears in the flow visualization results, due to the high stress in the region of the vortex split, caused by the discontinuity in the vorticity at the split. Hilbert transform results suggest that the flow transmits information well along the cylinder span, producing an elongated elliptical shape for the signal amplitude around the vortex split, as opposed to the results of Dallard and Browand[5], who find a circular shape for the amplitude around a vortex split in a mixing layer.

Hilbert transform and wavelet results show that the flow behind the 50-to-1 cylinder is different in some respects from the other three studies. The flow is less well ordered in its near wake, as seen in the behavior of its modulation frequency, which is more disorganized and has more amplitude variation than for the other three cylinders. Also, the Hilbert transform results show the 50-to-1 cylinder has a less repeatable structure from vortex split to vortex split. On the basis of this cylinder flow's similarity to Gaster's[6,7] and Van Atta and Piccirillo's[1] results, we postulate that more tapered cylinders have a 'leaky' cell structure, in which non-adjacent shedding cells have important effects on one another. Less tapered cylinders have a 'tight' cell structure, in which only adjacent cells affect one another. For a more complete account of these results, see Piccirillo and Van Atta[4].

3 Far Wake Results

The far wake of a tapered cylinder is very different from the far wake of a straight cylinder, due to the modulation frequency. The interaction of the modulation frequency with the shedding frequencies in the far wake leads to the rapid creation of many peaks in the velocity spectrum of the flow within 25 mean diameters downstream of the cylinder. These peaks are all spaced by the modulation frequency, suggesting that the modulation frequency is the key component in their creation. Since all the peaks are still only sums and differences of the shedding frequency and the modulation frequency, the flow is still quasi-periodic at this point, a result confirmed by examining phase space reconstructions of the flow.

Between 25 and 35 mean diameters downstream, the flow behavior changes, from one in which the energy in transferred into discrete peaks, to one in which the energy is transferred in a more continuous way. As a result, within 50 mean diameters of the cylinder, the flow has becomes nearly completely chaotic, with almost no residual (i.e. individual peaks) remaining from the near wake flow. At this point, wavelet analysis shows that the flow consists of small, discrete regions which still contain vigorous vortex activity, separated in time by regions where there is little vortex activity. The vigorous vortex region corresponds in time to the region away from the vortex split in the near wake, suggesting the vortex split is primarily responsible for the evolution of the flow to chaos.

It must be emphasized at this point that the evolution of chaos in the far wake is an active process. Chaos is not the result of the flow decaying to the point where little signal remains. In order to show this, we plot the evolution of the spectral magnitudes of the modulation and vortex shedding frequencies against donwstream location in figure 3. The magnitude of the modulation frequency rises as the flow evolves downstream, indicating that energy is being transfered to it. Only after the magnitudes of the modulation and shedding frequencies become approximately equal does the modulation frequency begin to decay.

4 Chaotic Results

As stated above, two runs were made in the chaotic vortex shedding regime. Many of the results obtained here were similar to those gotten in the laminar regime. For example, for adjacent measuring stations, the signals still moved phasewise relative to one another in a relatively orderly way. However, the spectrum showed the broad peaks associated with chaos, as shown in figure 4. The reason for this change is that the positions of the vortex splits spanwise along the cylinder was no longer fixed as in the laminar regime. Thus, the cellular behavior of the flow has broken down, leading to a situation in which the number of shed vortices per unit time become a monotonic function of spanwise position. Also, Hilbert transform results showed a loss of spanwise coherency. Thus, the flow is no longer effectively transmitting information spanwise along the cylinder. This suggests that the presence of cells in the laminar case is linked to the flow's ability to transmit information along the cylinder.

References

[1] Lewis, C.G. and Gharib, M. 1992 An exploration of the wake three dimensionalities caused by a local discontinuity in cylinder diameter. *Phys. Fluids A* **4** 104-117.

[2] Van Atta, C.W. and Piccirillo, P. 1990 Topological defects in vortex streets behind tapered circular cylinders at low Reynolds numbers. *New Trends in Non-linear Dynamics and Pattern Forming Phenomena*, ed P. Coullet and P. Huerre, 243-250.

[3] Piccirillo, Paul S. and Van Atta, C.W. 1991 An Experimental Study of Flow Behind a Uniformly Tapered Cylinder. *AIAA Paper 91-0750* AIAA 29th Aerospace Sciences Meeting, Reno.

[4] Piccirillo, Paul S. and Van Atta, C.W. 1992 An experimental study of vortex shedding behind linearly tapered cylinders at low Reynolds numbers. *J. Fluid Mech.* in press.

[5] Gaster, M. 1969 Vortex shedding from slender cones at low Reynolds numbers. *J. Fluid Mech.* **38** 565-576.

[6] Gaster, M. 1971 Vortex shedding from circular cylinders at low Reynolds numbers. *J. Fluid Mech.* **46** 749-756.

[7] Dallard, T. and Browand, F.K. 1992 Scale transitions at defect sites in the mixing layer: Application of the 2-D arc Wavelet transform. submitted to *J. Fluid Mech.*.

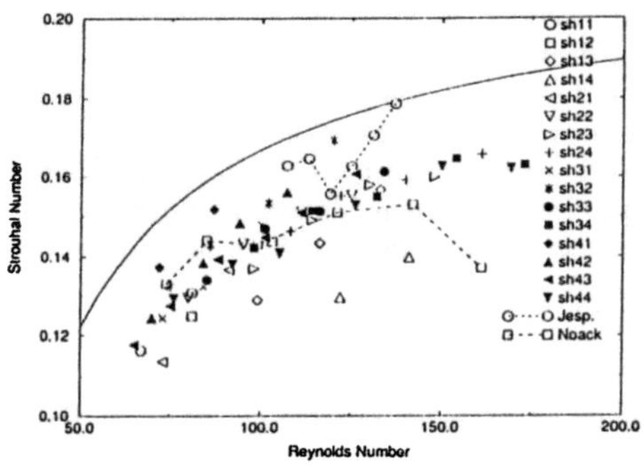

Figure 1: Shedding frequency Strouhal number vs. *Re*

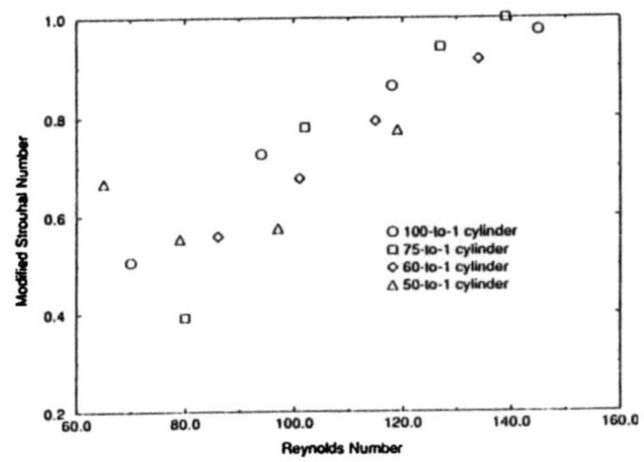

Figure 2: Modified modulation frequency Strouhal number vs. Re_{cs}

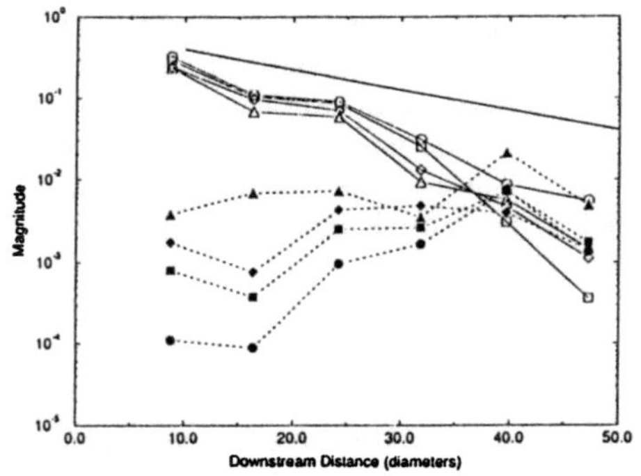

Figure 3: Evolution of spectral magnitudes in the far wake.

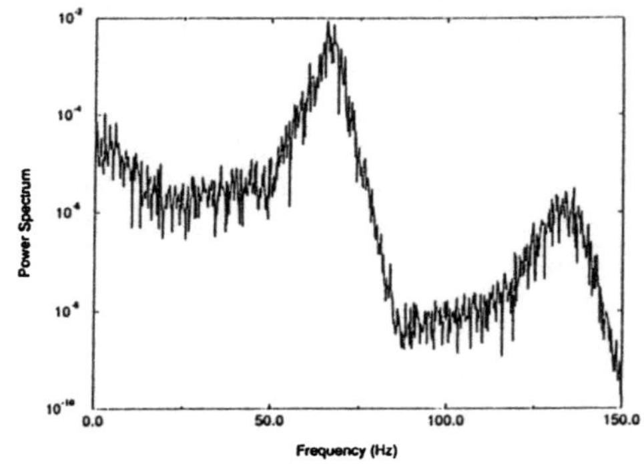

Figure 4: Power spectrum for chaotic vortex shedding.

Vortex Shedding from Slender Cones at Low Reynolds Numbers

A. PAPANGELOU

Department of Engineering
University of Cambridge
Cambridge CB2 1PZ
United Kingdom

1. Introduction

The phenomenon of vortex shedding arising from bluff bodies in a fluid flow has been investigated for over a hundred years. Most previous research has concentrated on circular cylinders normal to the flow, and has shown that the frequency associated with shedding depends on Reynolds number. The present investigation has been concerned with vortex shedding from cones, that is bodies of circular cross-section possessing a slight spanwise taper. The resulting presence of a continuously varying Reynolds number along the span of the cone induces a continuously varying 'natural' shedding frequency. Earlier experiments have shown that, at low Reynolds numbers, spanwise 'cells' are formed along the cone, in each of which the frequency is constant [1].

2. Experiments

Wind-tunnel experiments with six cones of small taper have shown the existence of spanwise cells at local Reynolds numbers of the order of 100 [2]. Figure 1 shows a typical variation of the shedding frequency along a cone's span at these Reynolds numbers. Four main cells and two end cells may be observed, all lying below Roshko's line for the shedding frequency variation of a cylinder [3]. The influence on the flow configuration of

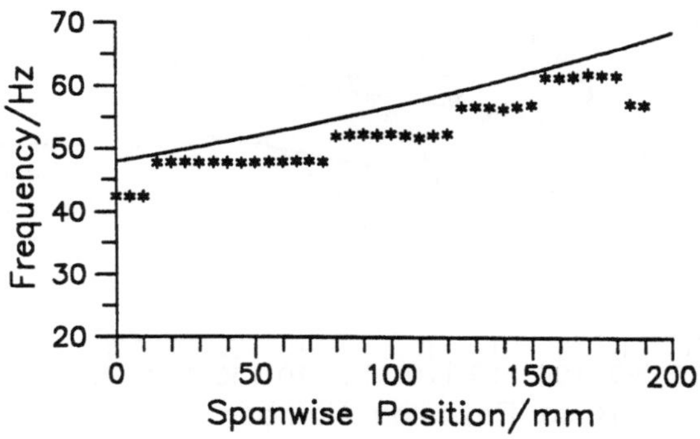

Fig. 1. Experimental spanwise variation of shedding frequency (—— Roshko line)

parameters such as taper angle, speed, and geometry was investigated, and a number of interesting features were noted, including:

a) The frequency jump between adjacent cells was constant along a cone's span, and was a function of air speed and cone taper angle only (being independent of any lengthscale).

b) The flow configuration was dependent on the base Reynolds number and was independent of the tip Reynolds number.

c) A change in either geometry or air speed could result in a cellular realignment. Figure 2 shows the number of cells supported along a cone's span as a function of its length and the base Reynolds number. The jump from a two-cell to a three-cell configuration was investigated, showing that a new cell was a local addition -- no sudden global changes were observed.

d) Flow visualization revealed curved vortices as well as the phenomenon of vortex splitting between adjacent cells [4].

e) Amplitude and phase modulation of the signal were simultaneously present, and are believed to be associated with the coupling mechanism.

f) In order to investigate the possible presence of 'chaos' in this non-linear dynamical system, experimental time series were analysed using recently developed signal processing techniques [5]. Such techniques have been successfully applied to 'closed' fluid systems (such as Taylor-Couette flow). By reconstructing phase portraits from the time series in this way, a deeper insight into the dynamical behaviour of the system could be obtained. However, no evidence of 'chaotic' behaviour was found. This may have been due to excessive levels of background noise, but it is possible that 'open' systems may differ fundamentally from 'closed' systems.

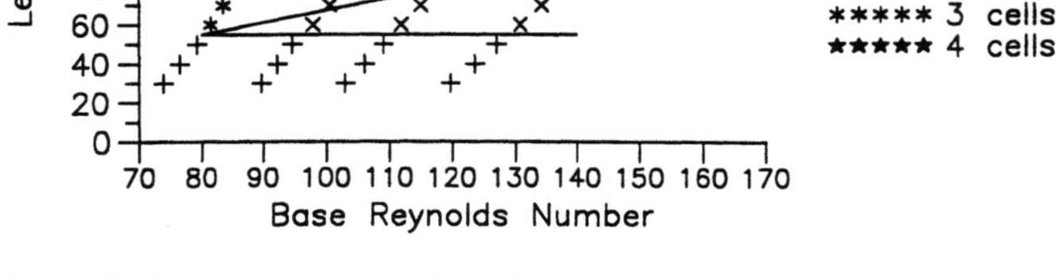

Fig.2. The influence of cone length and base Reynolds number on number of cells

3. Landau-Stuart model

The use of a non-linear oscillator to model vortex shedding was first proposed by Gaster [6]. More recently the Landau-Stuart equation has been used to model vortex shedding [7], [8]. A mathematical model based on the complex Landau-Stuart equation with the addition of a diffusive term to account for the spanwise coupling present was investigated. This equation takes the form:

$$\frac{\partial u}{\partial t} - \lambda \frac{\partial^2 u}{\partial z^2} = \sigma u - \tfrac{1}{2} l \mid u^2 \mid u \qquad u, \sigma, l \text{ complex}, \lambda \text{ real}$$

and was integrated numerically. At sufficiently high coupling coefficients, λ, the numerical simulations showed the presence of cells, whose frequencies were always below that of the 'natural' frequency (observed in the absence of coupling) -- see figure 3. Frequency jumps

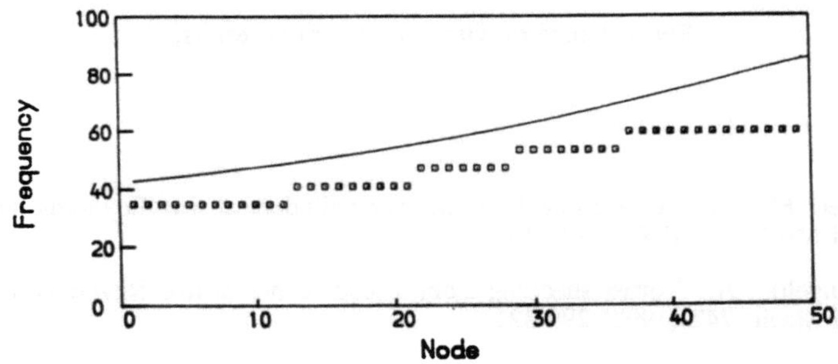

Fig.3. Numerical spanwise variation of shedding frequency (—— 'natural' freq.)

between adjacent cells were observed to be constant, and the response of the cells to changes of geometry and speed yielded behaviour similar to that of the experiments. The emergence of a new cell was a local and not a global event -- at the instant of a cell's birth, the other cell boundaries and cell frequencies remained unaffected, as in the experiments. Figure 4 shows a typical isometric view of the flow obtained from these simulations.

4. Conclusions

The addition of a new parameter (the varying local diameter) to the classical problem of the flow past a circular cylinder has resulted in a flow which exhibits many new characteristics. The presence of spanwise coupling in the flow, a phenomenon not yet fully understood, leads to the formation of cells. The behaviour and characteristics of these cells have been investigated, and some similarities with the flow past a circular cylinder have been noted. Analysis of experimental time series did not reveal the presence of 'chaos', though this has not been ruled out. Finally, a mathematical model has been proposed for the system which has successfully modelled many of the experimentally determined observations.

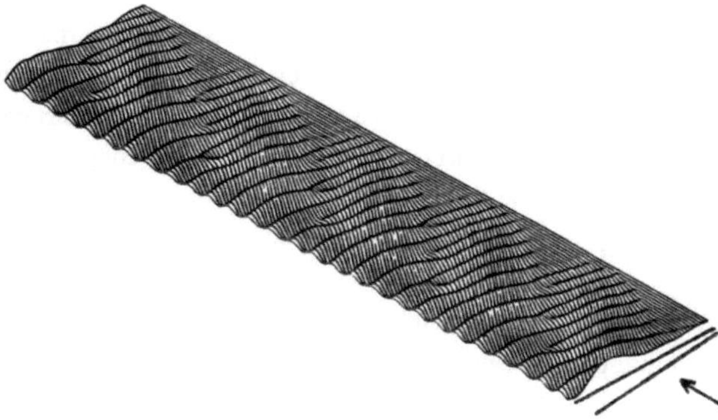

Fig.4. Isometric view of numerical results.

References

1. Gaster, M.: Vortex shedding from circular cylinders at low Reynolds numbers. J. Fluid Mech. 46 (1971) 749-756.

2. Papangelou, A.: Vortex shedding from slender cones at low Reynolds numbers. J. Fluid Mech. 242 (1992) 299-321.

3. Roshko, A.: On the development of turbulent wakes from vortex streets. NACA Rep. 1191 (1954).

4. Eisenlohr, H.; Eckelmann, H.: Vortex splitting and its consequences in the vortex street wake of cylinders at low Reynolds number. Phys. Fluids A 1 (1989) 189-192.

5. Broomhead, D.S.; King, P.: Extracting qualitative dynamics from experimental data. Physica 20D (1986) 217-236.

6. Gaster, M.: Vortex shedding from slender cones at low Reynolds numbers. J. Fluid Mech. 38 (1969) 565-576.

7. Provansal, M.; Mathis, C.; Boyer, L.: Bénard-von Kármán instability: transient and forced regimes. J. Fluid Mech. 182 (1987), 1-22.

8. Sreenivasan, K.R.; Strykowski, P.J.; Olinger, D.J.: On the Hopf bifurcation and Landau-Stuart constants associated with vortex 'shedding' behind circular cylinders. Private communication conveyed by R. Narasimha.

Springer-Verlag and the Environment

We at Springer-Verlag firmly believe that an international science publisher has a special obligation to the environment, and our corporate policies consistently reflect this conviction.

We also expect our business partners – paper mills, printers, packaging manufacturers, etc. – to commit themselves to using environmentally friendly materials and production processes.

The paper in this book is made from low- or no-chlorine pulp and is acid free, in conformance with international standards for paper permanency.

MIX
Papier aus verantwortungsvollen Quellen
Paper from responsible sources
FSC® C105338

If you have any concerns about our products,
you can contact us on
ProductSafety@springernature.com

In case Publisher is established outside the EU,
the EU authorized representative is:
Springer Nature Customer Service Center GmbH
Europaplatz 3, 69115 Heidelberg, Germany

Printed by Libri Plureos GmbH
in Hamburg, Germany